I0882551

Applications of Nanomaterials for Oncology

Syam Sundar Lingala
Prince Mohammad Bin Fahd University, Saudi Arabia

Faramarz Djavanroodi
Prince Mohammad Bin Fahd University, Saudi Arabia

IGI Global
Scientific Publishing
Publishing Tomorrow's Research Today

Vice President of Editorial Melissa Wagner
Director of Acquisitions Mikaela Felty
Director of Book Development Jocelynn Hessler
Production Manager Mike Brehm
Cover Design Jose Rosado

Published in the United States of America by
IGI Global Scientific Publishing
701 East Chocolate Avenue
Hershey, PA, 17033, USA
Tel: 717-533-8845 | Fax: 717-533-7115
Website: https://www.igi-global.com E-mail: cust@igi-global.com

Library of Congress Cataloging-in-Publication Data

LCCN: 2025022533 (CIP Data Pending)
ISBN13: 9798337337654
Isbn13Softcover: 9798337337661
EISBN13: 9798337337678
British Cataloguing in Publication Data
A Cataloguing in Publication record for this book is available from the British Library.

Table of Contents

Detailed Table of Contents

This chapter explores the emerging role of biodegradable lipids and surfactants as innovative carriers in next-generation drug delivery systems (DDS) for cancer therapeutics. Cancer treatment faces significant challenges, including systemic toxicity, drug resistance, and poor targeting efficiency. Biodegradable lipids and surfactants offer promising solutions due to their biocompatibility, ability to enhance drug solubility, and improved targeted delivery to tumour sites. This chapter discussed the physicochemical properties of these biodegradable materials that facilitate controlled drug release and minimize adverse effects. Additionally, stimuli-responsive biodegradable surfactants that trigger drug release in the tumour microenvironment have been explored and highlighted the potential therapeutic outcomes. Through analysis of recent preclinical and clinical studies, the chapter aims to present a comprehensive overview of how biodegradable lipids and surfactants shape the future of cancer drug delivery, offering more effective, safer, and personalized patient therapies.

Graphene oxide (GO), a derivative of graphene, has emerged as a transformative nanomaterial in biomedical applications, particularly in bioimaging and cancer therapy. This chapter aims to explore the unique physicochemical properties of GO - such as its large surface area, tunable functionalization, and photothermal capabilities that enable its use in advanced imaging modalities (e.g., fluorescence, MRI, and photoacoustic imaging) and targeted cancer treatments (e.g., drug delivery,

magneto-mechanical actuation, temperature-sensitive self-reporting, and AI-driven optimization. Three approaches are highlighted: (i) magneto-mechanical oscillators that disrupt cancer cells under low-frequency magnetic fields; (ii) phase-change reporter hybrids enabling simultaneous therapy and real-time imaging; and (iii) multi-modal supraparticles optimized by machine learning for imaging, heating, and drug delivery. Emphasis is placed on scalable fabrication, biocompatibility, and clinical translation. The chapter provides a roadmap for programmable nanomedicine with implications for researchers, clinicians, and industry stakeholders.

Cancer cell membrane-coated nanoparticles (CCM-NPs) offer enhanced tumor targeting, immune evasion, and improved therapeutic efficacy. Cancer cell membrane-coated nanoparticles leverage tumor-associated antigens and homotypic targeting mechanisms to preferentially accumulate at tumor sites, enhancing therapeutic precision. This study addresses these gaps by systematically evaluating CCM-NPs alongside other membrane types, elucidating their high drug loading efficiency (85–95%), tumor targeting capability, and moderate circulation half-life (15–25 hours). Importantly, CCM-NPs demonstrate substantial tumor growth inhibition (65–80%). The novelty lies in harnessing the homologous binding and antigen presentation capacities of cancer membranes to improve targeted delivery. The metastasis suppression (45–65%), and survival benefits (up to 3.5-fold increase) in preclinical cancer models, outperform conventional nanoparticles. Overall, this work provides a rationale for advancing CCM-NPs, aiming to overcome clinical limitations.

Chapter 6

P. Selvakumar, Department of Science and Humanities, Nehru Institute of Technology, Coimbatore, India
P. Selvam, Karpaga Vinayaga College of Engineering and Technology, India
T. C. Manjunath, Rajarajeswari College of Engineering, India
R. Haridass, Karpagam College of Engineering, Coimbatore, India
Atal Bihari Singh, Arka Jain University, India
Santosh Kumar Nathsharma, Stewart Science College, India

Unique combination of strength, conductivity, chemical versatility, nanomaterials in particularly in the fields of bioimaging and cancer therapy. properties various overview, delve into CNTs bioimaging and therapy, focusing on how their distinct characteristics facilitate advanced techniques detection, treatment. primary the key difference between these two types is the amount of graphene with SWCNTs being composed sheet, have This structural variation significantly influences the CNTs, optical characteristics. These properties can be tailored by modifying the synthesis methods or functionalizing the surface of the nanotube's applications. enables them carry large amounts of drug molecules, imaging agents, or other therapeutic payloads. Their small size, in diameter, allows them to or enhancing their effectiveness in drug delivery and bioimaging. Additionally, thermal and which makes photothermal and electrochemical therapies. Bioimaging is a critical technique in modern medicine, allowing visualization cells, Accurate imaging is essential for early diagnosis, assessing treatments.

Chapter 7

Richa Singh, Department of Biotechnology, SIES College of Arts, Science and Commerce (Autonomous), Mumbai, India

Cancer is one of the most pressing global health challenges. Despite significant advancements in conventional therapies, mortality rates for some cancers are still high. Key barriers, such as multidrug resistance, tumor heterogeneity, non-specific targeting, toxicity, and late-stage diagnosis, limit the efficacy of existing therapies. The rapid advancement in nanomedicine has opened new avenues in cancer therapy, particularly through the integration of metal and metal oxide nanoparticles (M/MONPs). These nanoparticles play a crucial role in cancer bioimaging, serving as contrast agents in various imaging modalities. Their ability to accumulate selectively in tumor tissues makes them ideal candidates for early detection and real-time monitoring of cancer progression. This chapter provides a comprehensive overview of the synthesis and biomedical applications of M/MONPs in cancer treatment. It

Preface

Nanotechnology has rapidly become a cornerstone of modern oncology, offering innovative solutions to long-standing challenges in cancer diagnosis, imaging, drug delivery, and therapy. By enabling precise control over material properties at the nanoscale, nanomaterials have opened new pathways for targeted, effective, and safer cancer treatments. *Applications of Nanomaterials for Oncology* is designed as a concise yet comprehensive reference that captures both foundational advances and emerging trends in this dynamic field.

This edited volume brings together interdisciplinary contributions spanning materials science, biotechnology, biomedical engineering, and computational analytics. The chapters collectively highlight how diverse nanomaterial platforms—ranging from biodegradable carriers and carbon-based nanostructures to magnetic, metallic, biomimetic, and silicon-based systems—are reshaping precision oncology. In addition, the integration of machine learning and predictive analytics reflects the growing convergence of nanomedicine with data-driven healthcare. This book is intended to serve as a valuable resource for researchers, clinicians, and advanced students seeking insight into current progress, translational challenges, and future directions in cancer nanotechnology.

OVERVIEW OF CHAPTERS

Chapter 1: Biodegradable Lipids and Surfactants—Next-Generation Drug Delivery for Cancer Therapeutics

This chapter introduces biodegradable lipids and surfactants as promising carriers for next-generation cancer drug delivery systems. It discusses how their biocompatibility, tunable physicochemical properties, and capacity for controlled and stimuli-responsive drug release can overcome issues such as systemic toxicity, drug resistance, and poor tumor targeting. By reviewing recent preclinical and clinical studies, the chapter establishes a strong foundation for lipid- and surfactant-based nanomedicine in oncology.

Chapter 2: Graphene Oxide for Bioimaging and Cancer Therapy Applications

Focusing on graphene oxide (GO), this chapter explores its exceptional surface area, functionalization flexibility, and photothermal properties that enable advanced bioimaging and targeted cancer therapy. It reviews GO-based platforms for fluorescence, MRI, and photoacoustic imaging, as well as drug delivery and photothermal treatment. The chapter also critically examines biocompatibility, toxicity, and clinical translation challenges, positioning GO as a key material in precision oncology.

Chapter 3: Graphene Oxide Nanotechnology for Next-Generation Cancer Theranostics

Expanding on graphene-based systems, this chapter emphasizes GO as an integrated theranostic platform that merges diagnostics and therapy. It highlights advances in GO functionalization for multimodal imaging and combined therapeutic strategies such as drug delivery, photothermal, and photodynamic therapy. Challenges related to safety and large-scale clinical adoption are discussed, providing a balanced perspective on GO's future in cancer nanomedicine.

Chapter 4: Advanced Magnetic Nanoplatforms for Accurate Theranostics

This chapter presents next-generation magnetic nanoparticle systems designed for precision cancer therapy and real-time monitoring. It introduces innovative approaches including magneto-mechanical actuation, self-reporting bioimaging, and AI-optimized multifunctional nanoplatforms. Emphasis is placed on programmable hyperthermia, scalability, and clinical translation, outlining a roadmap for smart magnetic nanomedicine.

Chapter 5: Cancer Cell Membrane-Coated Nanoparticles: Therapeutic Advancements and Hybrid Systems

This chapter explores biomimetic nanotechnology through cancer cell membrane–coated nanoparticles. By leveraging homotypic targeting and tumor-associated antigens, these systems achieve enhanced tumor accumulation, immune evasion, and therapeutic efficacy. Preclinical evidence of tumor growth inhibition, metastasis suppression, and survival benefits highlights their potential to overcome key clinical limitations of conventional nanoparticles.

Chapter 6: Carbon Nanotube–Based Bioimaging and Cancer Therapy Applications

Focusing on single- and multi-walled carbon nanotubes, this chapter reviews their structural, optical, and thermal properties relevant to cancer imaging and therapy. It discusses functionalization strategies that enable efficient drug loading, bioimaging, and photothermal treatment, while also addressing safety considerations critical for biomedical application.

Chapter 7: Metal and Metal Oxide Nanoparticles in Cancer Therapy and Bioimaging

This chapter provides a comprehensive overview of metal and metal oxide nanoparticles as multifunctional tools in oncology. It examines their roles as contrast agents for early cancer detection and as therapeutic agents capable of overcoming drug resistance and tumor heterogeneity. Issues of toxicity, biocompatibility, and clinical translation are critically assessed.

Chapter 8: Real-Time Predictive Analytics for Oncolytic Virus Replication and Tumor Lysis

Bridging nanomedicine and computational oncology, this chapter presents machine learning–based frameworks for predicting tumor response to oncolytic virotherapy. Using clinical and imaging datasets, it demonstrates how advanced models can optimize treatment outcomes, enhance interpretability, and support personalized cancer therapy.

Chapter 9: Silicon-Based Nanoparticles for Cancer Therapy and Bioimaging

The final chapter highlights silicon-based nanoparticles as versatile, biocompatible platforms for cancer diagnosis and treatment. It discusses their ability to improve drug bioavailability, enhance tumor targeting, and reduce systemic toxicity. The chapter underscores silicon's potential as a safe and scalable nanomaterial for future oncological applications.

Lingala Syam Sundar
Faramarz Djavanroodi

Editors

Chapter 1
Biodegradable Lipids and Surfactants:
Next-Generation Drug Delivery for Cancer Therapeutics

Abhiram Kumar
https://orcid.org/0000-0002-3422-2117
Birla Institute of Technology and Sciences Pilani, Hyderabad Campus, India

Girija Patel
https://orcid.org/0000-0003-4348-9507
Raj School of Management & Sciences, Mahatma Gandhi Kashi Vidyapith, Varanasi, India

Kumar Pranav Narayan
https://orcid.org/0000-0001-5198-2495
Birla Institute of Technology and Sciences Pilani, Hyderabad Campus, India

ABSTRACT

This chapter explores the emerging role of biodegradable lipids and surfactants as innovative carriers in next-generation drug delivery systems (DDS) for cancer therapeutics. Cancer treatment faces significant challenges, including systemic toxicity, drug resistance, and poor targeting efficiency. Biodegradable lipids and surfactants offer promising solutions due to their biocompatibility, ability to enhance drug solubility, and improved targeted delivery to tumour sites. This chapter discussed the physicochemical properties of these biodegradable materials that facilitate controlled drug release and minimize adverse effects. Additionally, stimuli-responsive biodegradable surfactants that trigger drug release in the tumour microenvironment have been explored and highlighted the potential therapeutic outcomes. Through analysis of recent preclinical and clinical studies, the chapter aims to present a

DOI: 10.4018/979-8-3373-3765-4.ch001

comprehensive overview of how biodegradable lipids and surfactants shape the future of cancer drug delivery, offering more effective, safer, and personalized patient therapies.

1. INTRODUCTION

Biodegradable lipids and surfactants have emerged as pivotal components in the advancement of cancer drug delivery systems, offering promising strategies to overcome the limitations of conventional therapeutics. Cancer remains a leading cause of mortality worldwide, and despite significant progress in anticancer drugs, challenges such as poor drug bioavailability, systemic toxicity, off-target effects, and multidrug resistance persistently hinder effective treatment outcomes(Katari et al., 2024). Integrating biodegradable lipid and surfactant-based carriers into drug delivery platforms has revolutionized the landscape by enabling enhanced solubility, targeted delivery, controlled release, and reduced adverse effects. The rationale for employing biodegradable lipids and surfactants lies in their intrinsic properties that address the biological and physicochemical barriers encountered in cancer therapy. Additionally, nanoscale vehicles drug delivery carrier capitalizes on their biocompatible and biodegradable nature to improve the pharmacokinetics profile of anticancer drugs, facilitating more precise and efficient therapy tailored to tumour-specific environments. Importantly, lipids are amphiphilic in nature and similarity to biological membranes, however it offers flexible platforms for encapsulating and stabilizing hydrophobic and hydrophilic drugs, which improves drug solubility and bioavailability. Beside that surfactants enhance these systems by lowering interfacial tension and encouraging self-assembly into structures like micelles that can load drugs and release them in response to tumor microenvironment triggers such as changes in pH, enzyme activity, or redox conditions. These properties enable sustained and targeted drug delivery, reducing systemic exposure and toxicity while boosting treatment effectiveness. Additionally, the biodegradable nature of these materials allows them to break down into harmless byproducts, minimizing the risk of long-term buildup and adverse effects (Pandey et al., 2018).

In the field of cancer therapeutics, challenges are complex and varied. They include the diverse biology of tumors, barriers within the tumor microenvironment, and the pharmacological difficulties of delivering a sufficient concentration of drugs specifically to tumor sites. Lipid-based carriers have capabilities to utilize the enhanced permeability and retention (EPR) effect by passive targeting drug delivery system. However, the intricate nature of tumor blood vessels and the pressure in the interstitial space often require active targeting strategies. These strategies are supported by the use of ligand-functionalized biodegradable lipids and surfactants

(Bjerk et al., 2021). To enhance antitumor drug delivery responses, tumour-specific drug delivery carrier to be engineered by enabling receptor-mediated endocytosis (Hirano et al., 1988). Although preclinical and clinical studies have shown that these nanocarriers can improve treatment outcomes, several obstacles hinder their broad clinical use. These include difficulties in scaling up production, ensuring consistent reproducibility, meeting regulatory requirements, and conducting thorough long-term safety evaluations. Overcoming these issues demands a deep knowledge of lipid and surfactant chemistry, tumor biology, and pharmacodynamics to develop delivery systems tailored for personalized and effective cancer treatments (Hönnscheidt et al., 2013; Katari et al., 2024; Nagtode et al., 2023).

The proposed chapter aims to provide a thorough overview of biodegradable and lipids and surfactant for ani-cancer drug delivery and therapeutic applications. Moreover, discussed state-of-the-art surfactant systems with stimuli-responsive capabilities, emphasizing their role in controlled and targeted delivery within the tumour microenvironment. Finally, current developments and future directions in designing next-generation biodegradable lipid and surfactant platforms have been outlined, with the goal of encouraging novel strategies that have the potential to revolutionize cancer treatment (Prabhjot Kaur et al., 2016; Rosenblum et al., 2018).

1.1 Biodegradable Lipids and Surfactants in Cancer Drug Delivery

Biodegradable lipids are widely used as carriers for chemotherapeutic agents in cancer therapy due to their compatibility with biological systems and low toxicity. They play a key role in the development of advanced delivery platforms, such as self-emulsifying drug delivery systems (SEDDS), which enhance the absorption of drugs with poor water solubility by forming fine emulsions upon contact with bodily fluids. The efficiency of these systems depends on factors including the concentration of surfactants, the ratio of oil to surfactant, and the size of the resulting droplets. Additionally, biodegradable lipids are essential in fabricating nanoparticles that allow precise control over the timing and location of drug release, improving the effectiveness of targeted cancer treatments while minimizing adverse effects. (Attri et al., 2021a; Pandey & Kohli, 2018). Surfactants are necessary to form stable nanodispersions and encapsulate therapeutic molecules efficiently. However, the use of non-biodegradable surfactants has raised concerns due to their tendency to accumulate in tissues, which may lead to unwanted side effects. The introduction of biodegradable surfactants, such as chemically modified γ-polyglutamic acid, offers a safer alternative that reduces the risk of long-term toxicity. These biodegradable surfactants also enhance the structural stability of drug delivery systems, facilitating a controlled release of drugs that improves therapeutic outcomes and patient safety (Hoennscheidt et al., 2013).

1.2 Overview of Challenges in Cancer Therapeutics

The field of cancer therapeutics faces multifaceted challenges related to the complexity of cancer biology, drug development, delivery, and patient management, which collectively impact the effectiveness and accessibility of treatments. Efficient delivery of anticancer agents remains a significant hurdle. Different administration routes—oral, intravenous, and subcutaneous—each present unique challenges in terms of drug stability, bioavailability, patient adherence, and toxicity management. Nanotechnology and targeted delivery systems show promise to improve delivery specificity and reduce side effects but face physiological barriers and require further development to optimize clinical outcomes(Richard, 2011). The key challenges are summarized in **Table 1.**

Table 1. Key Challenges in Cancer Therapeutics

Challenge Area	Description
Cancer Heterogeneity	Diverse molecular profiles limit universal therapy effectiveness
Drug Delivery	Stability, targeted delivery, and administration route issues
Molecular Targeting	Patient stratification and off-target effects
Pharmacogenetics	Genetic variability affecting dosing and toxicity
Economic Barriers	High treatment costs and healthcare access challenges
Novel Therapy Integration	Development, delivery, and clinical adoption complexities
Patient Adherence	Challenges with oral agents and self-management

1.3 Rationale for Using Biodegradable Lipids and Surfactants

Biodegradable lipids and surfactants are composed of biocompatible materials that degrade into non-toxic byproducts, reducing the risk of long-term toxicity and accumulation in healthy tissues. This characteristic makes them safer alternatives to non-biodegradable carriers, which might cause adverse reactions and tissue damage during prolonged treatment. Biodegradable lipids facilitate the formation of stable nano-sized carriers like liposomes and micelles, improving drug encapsulation efficiency and protecting drugs from premature degradation. Similarly, biodegradable surfactants stabilize these nanocarriers and enable controlled release, improving the pharmacokinetic profile of anticancer drugs(Attri et al., 2021b).

2. LIPIDS

Lipids are compounds that are insoluble in water but soluble in organic solvents; they serve key roles in cell structure, energy storage, and signalling. They include diverse types such as triglycerides, phospholipids, sterols, and fat-soluble vitamins, all of which share common biochemical properties. In conclusion, lipid A exemplifies a lipid with a defined chemical structure that incorporates sugar, glycerol backbones, phosphoryl groups, and fatty acid chains critical for its biological functions (**Fig-1**)

Figure 1. Chemical structure of phospholipid by showing amphiphilic nature. Additionally, the backbone of the lipid is illustrated, including fatty acids, triglycerides, and cholesterol.

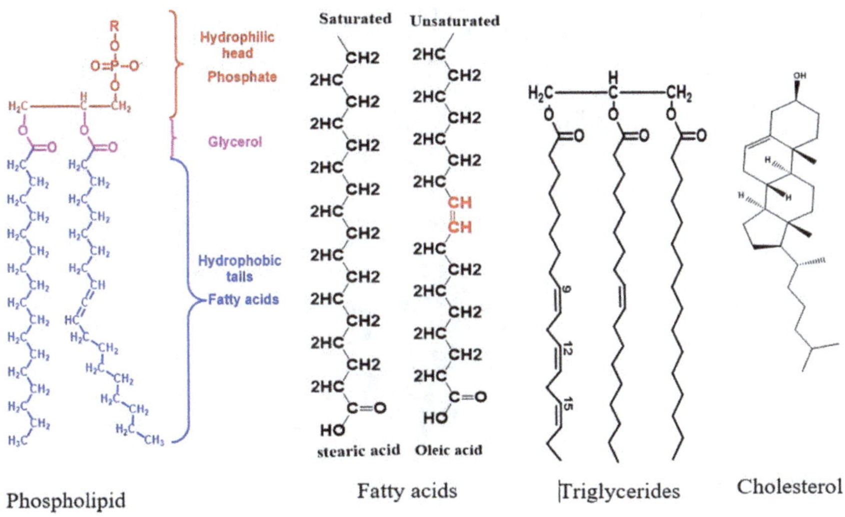

2.1 Biodegradable lipids and Their Advantage

Biodegradable lipids have emerged as pivotal components in drug delivery systems due to their biocompatibility, safety, and efficient degradation into non-toxic metabolites within the body. These lipids mimic natural biological molecules, ensuring minimal toxicity and improved tolerability. Biodegradable carriers enable controlled drug release, helping to prevent accumulation and reduce side effects common with synthetic delivery systems. In cancer treatment, lipid-based biodegradable nanoparticles improve drug solubility, stability, and targeting, thereby enhancing therapeutic effectiveness while protecting healthy tissues. For instance, lipid nanoparticles are used to deliver mRNA and chemotherapeutic agents directly to tumor cells. Engineered biodegradable ionizable lipids specifically enhance mRNA delivery by promoting efficient gene editing and precise tumor targeting. These carriers aid in escaping endosomes and releasing drugs quickly inside cancer cells, offering promising advances toward safer and more effective therapies (**Fig 2**) (Attri et al., 2021b; Chen et al., 2023).

Figure 2. The biodegradable H-bearing bio-degradable disulfide linkers and branched tails can significantly facilitate endosomal escape and improve mRNA delivery in vivo.

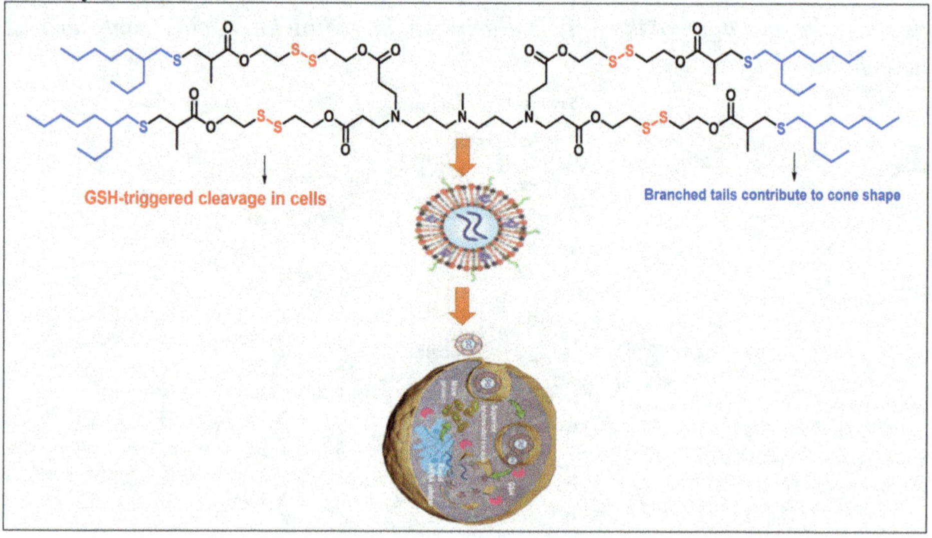

(Chen et al., 2023)

2.2 Physicochemical Properties of Lipids

Lipids exhibit diverse physicochemical properties that significantly influence their biological functions and applications in drug delivery systems. Key properties include lipid-droplet size, ultrastructure, and the molecular arrangement of triglycerides and phospholipids, which affect lipid digestion and absorption. The lipid-droplet size particularly impacts lipase activity during digestion, while the ultrastructure regulates enzyme interaction at the lipid interface. Additionally, the chemical structure of lipids determines their melting behaviour, crystallinity, and oxidative stability, which are crucial for formulation stability and functionality. Understanding these properties enables the design of lipids with tailored characteristics for enhanced therapeutic efficacy. For instance, single or multiple head-and-tell groups enhance the potential for anticancer drug delivery in the next generation (Fig 3) (Kunduri et al., 2022).

Figure 3. Lipids possess distinct chemical and physical characteristics that influence membrane curvature during cellular processes such as cytokinesis. (a) A lipid's shape is determined by its head group's relative cross-sectional area compared to its fatty acid acyl chains. (B) At the cleavage furrow in dividing cells, specific lipid compositions are observed. (C, D) Lipids with saturated fatty acids tend to pack tightly, making the membrane regions more rigid and less flexible. The presence of cholesterol modulates this effect by disrupting the tight packing of saturated lipids, thereby increasing membrane fluidity. Conversely, cholesterol enhances packing density in unsaturated fatty acid-containing lipids, reducing fluidity. (E) Additionally, lipids containing unsaturated fatty acids in the inner leaflet of the plasma membrane promote convex membrane bending, which plays a role in processes such as endocytosis.

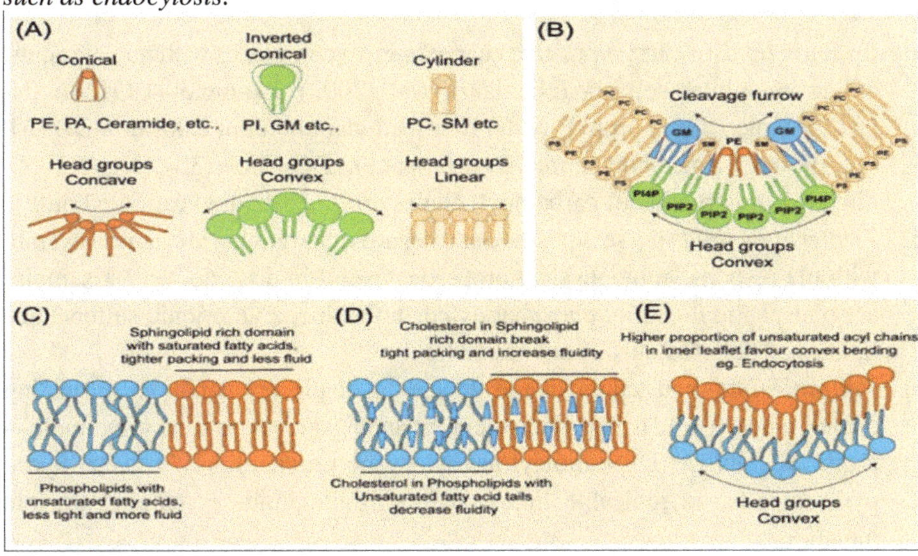

Figure credited by (Kunduri et al., 2022) under open access licence CC. 4.0.

2.3 Classification of Lipids: The lipids are classified in the below sections; Lipids classifications are illustrated in Fig 4.

2.3.1 Natural Lipids

Natural lipids are biological molecules primarily composed of fatty acids, glycerolipids, sphingolipids, and sterols, serving essential physiological roles such as energy storage membrane structure(Lim et al., 2018). Examples are triglycerides, phospholipids, tocopherols (Vitamin E), cholesterol, and derivatives.

2.3.2 Solid Lipids

It is a biodegradable, physiologically safe lipid typically found in a solid state at room and body temperatures, widely used in drug delivery systems like solid lipid nanoparticles for controlled release and stability of therapeutic agents(Chime et al., 2019). Examples are glyceryl monostearate (GMS), Stearic acid, Beeswax and Palmitic acid.

2.3.3 Synthetic and Modified Lipids

It is a chemically engineered lipid with tailor-made properties for drug and gene delivery; these include

a) ***Cationic lipids***—carry a positive charge to form complexes with nucleic acids, aiding in cellular delivery (Nicolazzi et al., 2003). Example—DOTMA (N-[1-(2,3-dioleoyloxy) propyl]-N, N, N-trimethylammonium chloride), DOTAP (1,2-dioleoyl-3-trimethylammonium-propane), DC-Chol (3β- [N- (N′, N′-dimethyl-amino-ethane)-carbamoyl] cholesterol) and trialkyl cationic lipids
b) ***Zwitterionic lipids*** possess positive and negative charges but are neutral overall, with pH-responsive biophysical properties useful in drug delivery. Examples are phosphatidyl cholin, phosphatidylethanolamine, zwitterionic sulfobetaine lipids, and sulfated quaternary amine lipids.
c) ***Ionizable lipids*** can acquire charge depending on pH, crucial for encapsulating nucleic acids in lipid nanoparticles and enhancing delivery efficacy. Example—DLin-MC3-DMA (DLin-dioleyl-methyl-amino propane), THP1 (Tetra-hydro pyrimidine-based ionizable lipid), and Ionizable lipids containing triazole moieties

2.3.4 Glycerolipids

Glycerolipids are lipids based on a glycerol backbone esterified to fatty acids, functioning primarily in energy storage and membrane formation, with common examples including triglycerides and glycolipids (Sozer et al., 2011).

2.3.5 Glycerophospholipids

Glycerophospholipids form the structural basis of cellular membranes, consisting of glycerol, phosphate groups, and fatty acid chains, and contribute to the lipid bilayer's integrity and function(Bera et al., 2017).

2.3.6 Sphingolipids

Sphingolipids and sterols are structurally distinct lipids found mainly in eukaryotic membranes; sphingolipids participate in signaling and membrane organization while sterols modulate membrane fluidity and domain formation (Guan et al., 2009).

2.3.7 Archaeal Lipids

Archaeal lipids exhibit unique ether linkages and branched isoprenoid chains, imparting extreme stability to membranes of archaea, enabling survival in harsh tumor micro-environments; examples include monopolar and bipolar polar lipids forming distinctive bilayer and monolayer structures.

2.3.8 Functionalized and Ligand-Modified Lipids

a) ***PEGylated lipids:*** PEGylated lipids have polyethylene glycol chains attached, increasing circulation time and stability of liposomes and nanoparticles.
b) ***Ligand-conjugated:*** Ligand-conjugated lipids bear targeting moieties such as antibodies or peptides, enabling specific cell or tissue targeting for improved therapeutic efficacy.

Figure 4. Classification of lipid for next-generation anticancer drug delivery

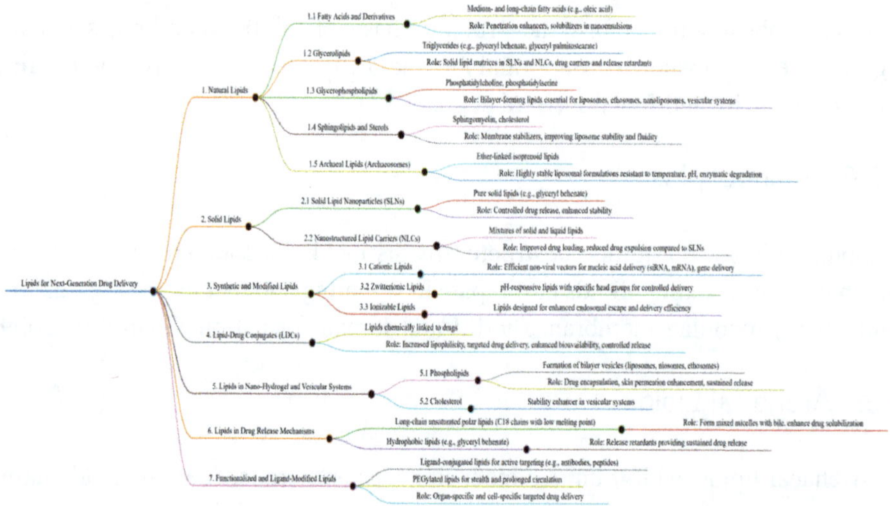

2.4 Lipids: A Promising Role in Next Generation Drug Delivery

Lipids hold a promising role in the next generation of anticancer drug delivery due to their ability to enhance the specificity, efficacy, and safety of therapeutic agents. Drug delivery platforms based on lipids—including liposomes, solid lipid nanoparticles, and other lipid-based carriers—enhance the solubility and absorption of hydrophobic anticancer agents, concurrently lowering their systemic toxicity (Sharma et al., 2020). These systems enable targeted delivery to tumor sites, either passively through the enhanced permeability and retention (EPR) effect or actively via ligand modification, thereby minimizing damage to healthy tissues. Additionally, lipid nanoparticles can provide controlled and sustained drug release, overcoming rapid drug clearance and improving therapeutic outcomes. Innovations in lipid formulations allow the co-delivery of multiple drugs, optimizing synergistic effects for combination chemotherapy (Maurya, Saklani, Singh, Nisha, Pal, et al., 2022). Moreover, lipid-based carriers are showing a potential role in overcoming drug resistance and penetrating biological barriers like the blood-brain barrier. Which significant challenges in cancer treatment; overall, lipid-based nanotechnologies represent a versatile and biocompatible platform shaping the future of efficient and safer anticancer therapies(L Arias et al., 2011; Yingchoncharoen et al., 2016).

2.4.1 Lipid-Based Controlled and Sustained Drug Release Mechanism

Lipid-based systems for controlled and sustained drug release utilize the inherent characteristics of lipids to regulate delivery rates, enhance drug stability, and improve bioavailability. These systems include various carriers such as liposomes, solid lipid nanoparticles (SLNs), and lipid-based gels that encapsulate therapeutic agents and enable their gradual release over time. By shielding drugs from degradation, lipid carriers also facilitate release that can be triggered by environmental stimuli, including changes in temperature or the presence of reactive oxygen species (ROS). An illustrative example involves the development of a stimuli-responsive lipogel, which combined poly(N-isopropylacrylamide) (pNIPAM) microgel particles with lipids. This lipogel displayed temperature-sensitive behavior, with minimal release of the encapsulated calcein at room temperature and markedly increased release at 37 °C, simulating body temperature. The lipid component served as a reservoir on the pNIPAM framework, extending the release period from 10 minutes to 2 hours. This demonstrates how incorporating lipids into drug delivery platforms allows for adjustable drug loading and release profiles tailored to specific therapeutic needs (Attri et al., 2021b; Lee et al., 2023).

2.4.2 Lipids and Drug Conjugation

Conjugating lipids with anticancer drugs improve targeted delivery by enhancing drug stability and minimizing systemic toxicity through controlled release and tumor-specific recognition. This approach promotes efficient uptake in cells by lipid-drug conjugates merging the therapeutic system which leads to improved biocompatibility and membrane affinity. For instance, lipid micelles linked to folic acid effectively target tumours by exploiting the overexpression of folate receptors on cancer cells. (Bhattacharya, 2023). A study demonstrated folic acid attached via reducible disulfide linkages to amphiphilic copolymers, forming micelle complexes that ensured stability and effective targeted delivery of anticancer drugs in vitro and in vivo. This conjugation strategy offers a powerful platform for enhancing anticancer therapy (Sreeharsha et al., 2022).

2.4.3 Localized Drug Delivery

Lipids play a vital role in localized drug delivery by enabling a targeted release of therapeutic agents directly at the disease site, minimizing systemic side effects and improving treatment efficacy. Lipid-based carriers, such as liposomes and lipid microbubbles, can encapsulate drugs and release them upon specific triggers like ultrasound or temperature changes. For instance, acoustically active lipospheres with a lipid shell encapsulating hydrophobic drugs have been designed to release their payload when disrupted by focused ultrasound at the targeted tumour site, ensuring high local drug concentration and reduced toxicity. This approach enhances precise drug delivery and represents a promising strategy for localized cancer therapy(May et al., 2001).

2.4.4 Active Targeting Drug Delivery

Active targeting drug delivery using lipids involves modifying lipid-based carriers with specific ligands that recognize and bind to receptors overexpressed on target cells, enhancing selective drug accumulation and therapeutic efficacy. This strategy improves cellular uptake and reduces off-target effects compared to passive targeting(Maurya, Saklani, Singh, Nisha, Mishra, et al., 2022). For example, liposomes conjugated with folate ligands actively target cancer cells overexpressing folate receptors, leading to improved drug delivery and tumour suppression(Langer, 1998). The FDA-approved liposomal doxorubicin (Doxil) exemplifies such targeted lipid-based delivery systems, which have transformed anticancer therapy by combining enhanced specificity with reduced toxicity.

2.4.5 Passive Targeting Drug Delivery

Passive targeting drug delivery exploits the natural physiological characteristics of diseased tissues, such as tumours, to accumulate drug carriers preferentially at these sites without the need for specific ligands. This approach primarily utilizes the enhanced permeability and retention (EPR) effect, where nanoparticles sized typically between 10 to 200 nm passively accumulate in tumour tissues due to leaky vasculature and impaired lymphatic drainage. For example, polymeric micelles carrying the anticancer drug adriamycin have demonstrated enhanced accumulation in solid tumours through the EPR effect, resulting in improved antitumor activity with reduced systemic toxicity (Yokoyama, 1999). This exemplifies the significance of passive targeting in enhancing drug delivery efficiency.

2.4.6 Nanostructured Drug Delivery

Nanostructured drug delivery systems refer to nanoscale carriers designed to improve the delivery, targeting, and controlled release of therapeutic agents. These systems include nanoparticles, liposomes, nanostructured lipid carriers, and porous silicon particles, among others. They enhance drug bioavailability, reduce side effects, and enable site-specific delivery by exploiting properties like size, surface characteristics, and drug loading capacity. The inclusion of liquid lipids leads to imperfections in the solid lipid matrix, creating more space to accommodate higher amounts of drug molecules, which increases the drug payload. Additionally, the mixed lipid core enhances the physical stability of the carrier by minimizing drug degradation and leakage(Chandana et al., 2019). This stable matrix prolongs drug release and maintains therapeutic efficacy over time. As a result, NLCs offer improved storage stability, enhanced solubility of poorly water-soluble drugs, and sustained drug release, making them superior to many conventional drug delivery systems. For instance, nanostructured lipid carriers (NLCs) overcome limitations of solid lipid nanoparticles by providing higher drug payload and stability, making them advantageous for various administration routes, including topical and intravenous delivery (Atre et al., 2024).

3. SURFACTANTS

Surfactants are essential ingredients that improve the efficacy of nanostructured lipid carriers by providing physical stability, enhancing drug loading, and offering controlled drug release, all of which contribute to better therapeutic outcomes (Halliday, 2008). Who studied the surface-active characteristics of bile salts and their function in drug solubilization, and absorption proposed the first notable applications of surfactants in the pharmaceutical industry in 1907; also known as surface active agents, they consist of organic compounds that have the capability to drastically reduce interfacial tension. Surfactants with a low HLB value produce a water-in-water emulsion, whereas surfactants with a high HLB value produce an oil-in-water solution. Hence, the surfactants having a large HLB range are preferentially dissolved in the emulsion's aqueous exterior phase. Through regular inquiries for increasing surfactant characteristics, unique structures have recently arisen that demonstrate the impact of synergistic relationships or better surface and aggregation effects (P. Kaur et al., 2016).

3.1 Stimuli-Responsive Biodegradable Surfactants

Stimuli-responsive biodegradable surfactants are advanced materials that respond to specific environmental triggers—such as pH, temperature, redox conditions, or enzymes—to release therapeutic agents in a controlled manner while safely degrading into non-toxic byproducts. These smart surfactants improve the targeting and efficacy of drug delivery systems by minimizing systemic toxicity and enhancing drug accumulation at disease sites, particularly in cancer therapy. For example, redox-responsive biodegradable polymeric micelles have been developed, where the surfactant contains disulfide bonds that cleave in the reductive tumour microenvironment, triggering rapid drug release selectively within cancer cells. One notable application involves biodegradable polypeptide-based micelles that respond to the higher glutathione concentration inside tumour cells, releasing anticancer drugs like doxorubicin efficiently while reducing side effects on healthy tissues. This strategy harnesses the tumour's unique biochemical conditions to achieve precise and effective drug delivery for enhanced cancer treatment outcomes (Li et al., 2014; Taghizadeh et al., 2015).

3.2 Physicochemical Properties of Surfactants

The physicochemical properties of surfactants are critical in determining their behaviour and efficacy in various applications. Key properties include surface tension reduction, critical micelle concentration (CMC), emulsification capacity, foam formation, and solubility. Surfactants reduce the surface tension between liquids or between a liquid and a solid, facilitating spreading and mixing. The CMC is the concentration at which surfactants start forming micelles, which is essential for their emulsifying and solubilizing functions. Viscosity, charge (anionic, cationic, nonionic), hydrophilic-lipophilic balance (HLB), and interactions with proteins or other molecules influence surfactant performance. For instance, polymeric adjuvants and surfactant concentration affect droplet size and deposition in spray formulations by altering viscosity and surface tension. Understanding these physicochemical properties allows optimized formulation design in pharmaceuticals, cosmetics, and environmental applications, ensuring stability, efficiency, and biocompatibility (Sałek et al., 2022).

3.3 Classification of Surfactant

3.3.1 Nonionic Surfactants

Nonionic surfactants are amphipathic molecules composed of a hydrophobic part, typically an alkyl or alkylphenol group, and a hydrophilic part, usually ethylene oxide chains. Chemically, they lack charged groups, making them neutral in solution, which contributes to their low toxicity and good biocompatibility. Their hydrophilic segments often consist of polyoxyethylene chains whose lengths can be adjusted to tune solubility and critical micelle concentration (CMC). Nonionic surfactants have favourable physicochemical properties such as surface tension reduction, stable micelle formation, and compatibility with various drugs or biological membranes(Sharma et al., 2025). These properties make them widely used in pharmaceuticals, cosmetics, and biotechnology. For example, Tween 80, a polyoxyethylene sorbitan monooleate, is a common nonionic surfactant employed for its excellent emulsifying and solubilizing capacity with minimal irritation. Their low ionic character allows stable formulations even in the presence of electrolytes. A details Brach based classification are illustrated in **Fig 5**.

Figure 5. Classification of surfactant for next-generation anticancer drug delivery

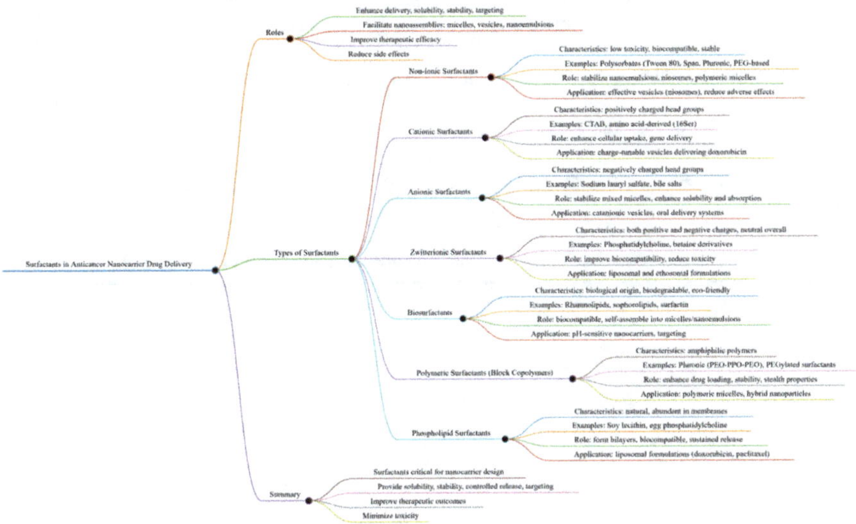

3.3.2 Cationic Surfactants

Cationic surfactants have a positively charged hydrophilic head and a hydrophobic tail, allowing them to adsorb efficiently onto negatively charged materials such as cell membranes, hair, skin, and DNA. This charge characteristic makes them useful as bactericides, antistatic agents, and conditioning agents. They form aggregates like micelles and vesicles in solution, which aid in solubilizing hydrophobic substances and interacting with biological membranes. Their structural diversity includes quaternary ammonium compounds, gemini surfactants (with two head groups and two tails), and amino acid-based surfactants, allowing fine-tuning of properties for specific biological or pharmaceutical applications(Zakharova et al., 2019). Recently, our research group developed a cationic surfactant (**see Fig.6**) to deliver anticancer agents. The research findings revealed that DDAB shows significantly better DNA compaction than MQAS12. The empirical rule $P = nl/aV$ supports DDAB's stronger binding due to— Higher hydrophobicity per unit surface area, Smaller headgroup size 'a', Effective bilayer formation on SiO_2 NPs. Therefore, these cationic surfactants would be a promising approach for targeting anticancer drug delivery (Dyagala et al., 2023).

Figure 6. Chemical structure of cationic lipids

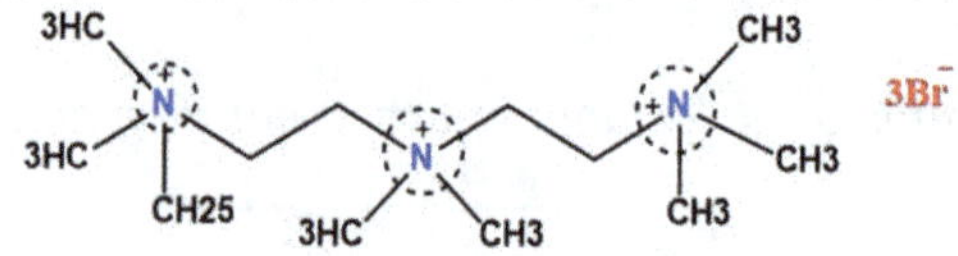

N^1-(2-((λ^{10}-methyl)dimethylammonio)ethyl)-N^1,N^1,N^2,N^2,N^2-pentamethylethane-1,2-diaminium (MQAS12)

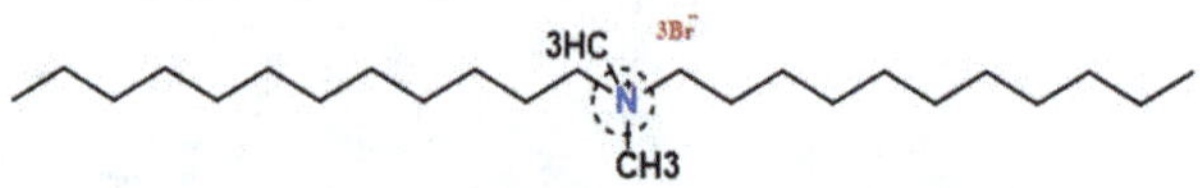

N-dodecyl-*N*,*N*-dimethylundecan-1-ide-1-aminium (DDAB)

3.3.3 Anionic Surfactants

These surfactants have a negatively charged hydrophilic group. They are particularly effective in removing ionic pollutants. When anionic surfactants dissolve in water, they ionize to form negatively charged particles. These negatively charged surfactants bind to positively charged substances such as clay, helping to remove positively charged cations. Additionally, anionic surfactants generate more foam compared to traditional surfactants (Dutta, 2019). Besides that, sodium stearate, a soap-based anionic surfactant, is used to emulsify lotions and ointments(Arakawa et al., 2024). On the other hand, co-block surfactants have shown promising approaches to cancer therapeutics. These types of surfactants could be used for polymeric mixed micelles formulation in the nano-size range (Torchilin, 2001). One decade ago, Ankitkumar and his team demonstrated a self-assembling property of sodium bis (2-ethylhexyl) sulfosuccinate and cyclodextrins. Both surfactants were shown promising results. His research findings suggested that the design of cyclodextrin-based additives to control surfactant behaviour is useful in controlled drug release, emulsions, and formulation science. However, these surfactants could be used for the next generation of anticancer drug delivery. Fig 7 provides detailed insight into the formation of nanostructure-based drug delivery.

Figure 7. the illtreated diagram represents the Chemical structure, geometry, and dimensions of α- and β-cyclodextrins and, further, the micelle of sodium bis(2-ethylhexyl) sulfosuccinate can be modelled as a core-shell oblate ellipsoid. The ellipsoid's core, with a major axis, denoted as a and a minor axis as b, contains the surfactants' hydrophobic alkyl tails along with some cyclodextrin molecules. Surrounding this core is a shell of uniform thickness δ, composed of the surfactant head groups, water molecules, and counterions, which vary based on the ionization level. The scattering length densities for the micelle core, shell, and surrounding solvent are represented as ρC, ρSh, and ρs, respectively. The parameter σ0 corresponds to the equivalent spherical diameter of the ellipsoid.

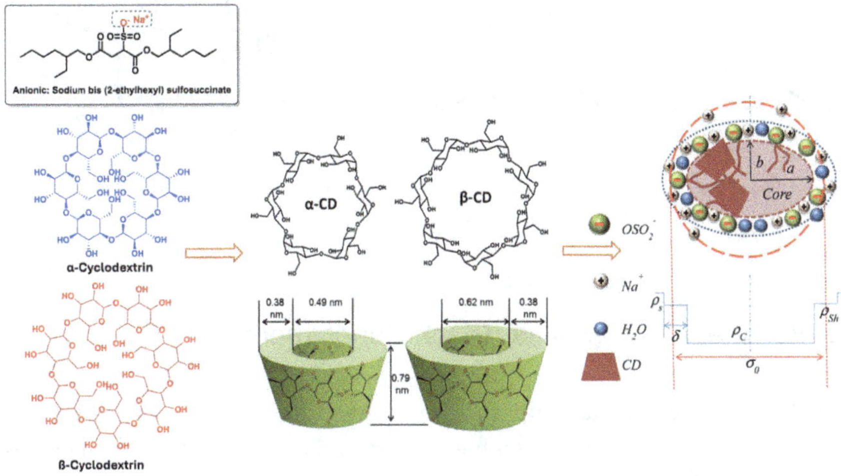

3.3.4 Zwitterionic Surfactants

Zwitterionic surfactants are molecules that contain both positive and negative charges within the same structure, making them electrically neutral overall. This unique feature combines a hydrophilic (water-attracting) zwitterionic headgroup with a hydrophobic tail, allowing it to interact with both water and oil effectively. Their dual charge provides high water solubility, mildness, resistance to hard water, and compatibility with other surfactants, which makes them especially useful in personal care products, detergents, pharmaceuticals, biotechnology, and agriculture. Common types include betaines, sulfobetaines, and amine oxides. Due to their amphipathic nature, zwitterionic surfactants form micelles and emulsions, enhancing the solubilization and delivery of various substances (Enclopedia, 2023). For instance, leylamidopropyl dimethyl carboxy betaine (OAPDCB) is a zwitterionic surfactant that shows great promise for anticancer drug delivery applications. Its distinctive amphiphilic design, featuring a hydrophobic oleyl tail paired with a hydrophilic zwitterionic head, allows it to self-assemble into nanostructures like micelles and nanoemulsions. These structures can encapsulate hydrophobic anti-cancer drugs within their core, improving the drugs' solubility, stability, and overall bioavailability (**Fig 8**).

Figure 8. Zwitterionic: Oleylamidopropyldimethylcarboxybetaine

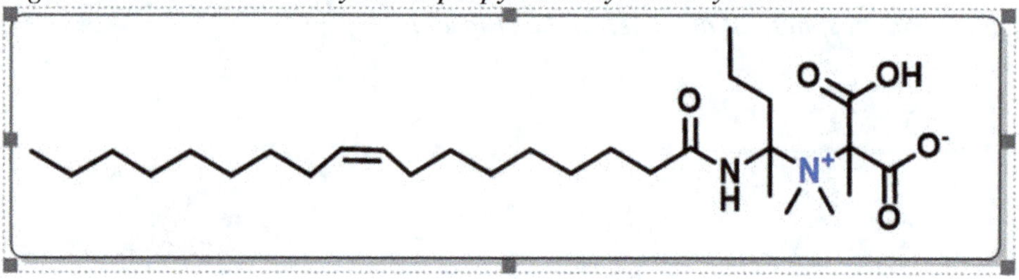

3.3.5 Polymeric Surfactants (Block Copolymers)

Block copolymers are amphiphilic molecules composed of hydrophilic and hydrophobic segments that self-assemble in aqueous environments to form nanoscale structures like micelles, which have a hydrophobic core suitable for encapsulating anticancer drugs (Kuperkar et al., 2022). These nanosystems exhibit enhanced drug solubility, stability, and bioavailability, overcoming critical pharmacokinetic limitations of conventional chemotherapeutics such as poor solubility, rapid metabolism, and systemic toxicity. Due to their size (typically 10–100 nm), block copolymer micelles exploit the enhanced permeability and retention (EPR) effect to accumulate preferentially in tumour tissues, including colorectal tumours, via passive targeting(Fatfat et al., 2022). Furthermore, active targeting can be achieved by functionalizing the micelle surface with ligands recognizing tumour-specific receptors such as folic acid or hyaluronic acid, thus improving cellular uptake and therapeutic efficacy. The block copolymer-based micelles allow stimuli-responsive drug release in the tumour microenvironment characterized by low pH, elevated glutathione, or specific enzymes, thereby enhancing site-specific drug release and minimizing off-target effects. Additionally, they have been used for the co-delivery of multiple chemotherapeutic agents or a combination of chemotherapy and gene therapy agents, which helps overcome multidrug resistance and achieves synergistic therapeutic outcomes. For instance, Poly (ethylene glycol)-block-poly (glutamic acid), Folic acid-targeted PEG-polyester micelles and D-α-tocopherol succinate-conjugated hyaluronic acid micelles these block polymers could be a promising approach for designing colon cancer drug delivery system (Srivastava et al., 2021; Tupally et al., 2024).

3.4 Surfactant: A Promising Role in Next-Generation Drug Delivery

Surfactants are pivotal and promising in advancing next-generation drug delivery systems by enhancing critical pharmaceutical properties such as drug solubility, bioavailability, controlled release, and targeting. Their amphiphilic nature allows them to modify the physicochemical environment of drugs and delivery vehicles, enabling improved therapeutic efficacy and patient compliance (Nagaraj, 2025).

3.4.1 Solubility Enhancement of drug

Solubility enhancement of drug one of the primary challenges in drug formulation is the poor aqueous solubility of many active pharmaceutical ingredients, impacting their absorption and therapeutic performance (Sałek et al., 2022). Surfactants enhance drug solubility by forming micelles and other self-assembled structures that encapsulate hydrophobic drugs, increasing their apparent aqueous solubility and thus facilitating absorption. Self-emulsifying drug delivery systems (SEDDS), which are isotropic mixtures of oils and surfactants, create fine oil-in-water microemulsions upon contact with biological fluids, significantly improving the dissolution rates of lipophilic drugs (Kumar et al., 2024; Ranneh et al., 2016). This improved solubilization leads to enhanced absorption and reduced dose variability. Advances have also been made using natural and synthetic surfactants, including sugar-based and nonionic types, which provide favourable solubilizing properties with reduced toxicity risks. Proper surfactant selection based on critical micellar concentration and hydrophilic-lipophilic balance is essential to optimize drug solubilization without compromising formulation stability.

3.4.2 Enhancement of Bioavailability

Bioavailability is a decisive factor in determining the effective concentration of a drug that reaches systemic circulation. Surfactants contribute to bioavailability enhancement by increasing drug solubility, altering membrane permeability, and reducing drug efflux mechanisms. In oral and pulmonary drug delivery, surfactants bypass first-pass metabolism, facilitate drug transport across epithelial barriers, and promote lymphatic uptake of lipophilic drugs. Surfactants such as Tween and

Pluronic derivatives are widely used in nanostructured lipid carriers (NLCs) and self-micro emulsifying systems to improve oral absorption (Kumar, Maurya, et al., 2025). Additionally, surfactants enhance drug stability in the harsh gastrointestinal environment and improve the therapeutic index of drugs through better pharmacokinetic profiles. The interplay between surfactant molecular structure and in vivo performance is critical, requiring careful formulation design to balance efficacy and safety(Sanchez et al., 1991; Zhang et al., 2024).

3.4.3 Drug Release Mechanism

Controlled and sustained drug release is integral to optimizing therapeutic outcomes and minimizing side effects. Surfactants influence drug release by modulating the structure and integrity of carrier systems, such as liposomes, micelles, and niosomes. Their ability to alter membrane fluidity and permeability facilitates diffusion-controlled release profiles (Shahzad et al., 2018). In vesicular systems, surfactants form bilayer architectures that encapsulate drugs and regulate their release via diffusion, erosion, or swelling mechanisms. Moreover, surfactant-based formulations can be designed to respond to physiological stimuli, enabling triggered drug release. The surfactant charge and hydrophobic chain length can impact drug retention and release kinetics, allowing fine-tuning of release profiles in line with therapeutic requirements(Notario-Perez et al., 2020).

3.4.4 Formulation Stability

The stability of drug formulations during storage and administration is crucial to maintaining efficacy and safety. Surfactants contribute to formulation stability by preventing drug aggregation and crystallization through steric and electrostatic stabilization mechanisms(Pinon-Gomez et al., 2025). In lipid-based systems, surfactants stabilize emulsions and nanosized particles, maintaining uniform particle size distribution and physical integrity. The choice of surfactant and its concentration critically affect the thermodynamic stability and resistance to phase separation or coalescence. Blends of pegylated and unpegylated surfactants can synergistically enhance nanoparticle stability and drug loading. However, surfactants may also influence drug degradation pathways and hydrolysis rates, requiring a comprehensive evaluation to optimize stability without compromising biocompatibility (Hogarth et al., 2024). The lipid-based drug carrier could also be used for skin cancer by designing a transdermal patch. The RP-HPLC technique could be used to quantitatively analyze newly developed lipids and drugs (Kumar, Dhiman, et al., 2025).

3.4.5 Thermal Sensitive Drug Delivery

Thermosensitive drug delivery systems offer spatiotemporal control over drug release triggered by temperature variations, making therapeutic delivery more precise. Thermosensitive-based surfactant carriers such as hydrogels and lipid nanoparticles respond to physiological or external heat stimuli (Zhang et al., 2021). Incorporating surfactants can modulate the gelation temperature viscosity and release kinetics of thermosensitive hydrogels, promoting sustained and stimuli-responsive drug delivery. Advances include thiolated surfactants, which provide mucoadhesive properties and enhanced retention. Combining surfactants with polymers exhibiting thermos-responsive behaviours makes it possible to engineer delivery platforms that selectively release drugs at elevated temperatures associated with inflammation or tumour sites (Liu et al., 2022).

3.4.6 Surface Modification and Targeting Strategies

Surface modification of drug delivery vehicles with surfactants enhances targeting efficiency, bio-distribution, and circulation time. Surfactants and their derivatives can be used to graft functional groups or target ligands onto nanocarrier surfaces, reducing opsonization and immune clearance(Govender et al., 2008). Polymeric surfactants like poloxamers improve hydrophilicity and steric hindrance on particle surfaces, aiding in immune evasion. Surface-modified lipid-based nanoparticles exhibit improved mass transport, enhanced cellular uptake, and site-specific delivery by conjugation with antibodies, peptides, or aptamers(De Stefani et al., 2022). Cell membrane-coated nanoparticles mimic biological surfaces for immune compatibility. These strategies harness surfactant chemistry for improved therapeutic index and reduced off-target effects(Oroojalian et al., 2021).

3.4.7 Biocompatibility and Degradability

Given their widespread use, surfactants must exhibit high biocompatibility and controlled degradability to minimize toxicity and ensure safe drug delivery. Nonionic surfactants generally show lower cytotoxicity and haemolytic activity compared to their ionic counterparts(Kontogiannis et al., 2022). The molecular architecture, such as chain length and degree of fluorination, influences their interaction with biological membranes and systemic tolerance. Biosurfactants derived from microbial sources present biodegradable and less toxic alternatives. Thiolated and PEGylated surfactants enhance muco-adhesiveness with good biocompatibility profiles. Comprehensive in vitro and in vivo studies must evaluate surfactant-induced irritation, immunogenicity, and metabolic fate, guiding the design of safe and effective delivery systems (Wu et al., 2025).

3.4.8 Advances in NLC Formulations and Delivery Routes

Nanostructured lipid carriers (NLCs) have emerged as versatile drug delivery platforms integrating the advantages of lipids and surfactants to enhance drug loading, stability, and targeting(Rani et al., 2025). NLCs composed of physiologically compatible lipids and surfactants demonstrate improved delivery via oral, topical, pulmonary, and nasal routes. Innovations include NLCs tailored for brain targeting through surfactant-modified nanoparticles crossing the blood-brain barrier and pulmonary NLCs designed for deep lung deposition with sustained drug release. The surfactant type and concentration critically influence particle size, zeta potential, and entrapment efficiency. Spray-dried NLC powders combined with dry powder inhalers represent scalable approaches for pulmonary delivery in respiratory diseases. These advancements exemplify surfactant utility across diverse administration routes (Dhande et al., 2025).

3.4.9 Role in Efficient Drug Encapsulation and Payload

Surfactants significantly affect drug encapsulation efficiency and payload capacity of drug carriers such as micelles, liposomes, and microbubbles. Amphiphilic surfactants stabilize the carrier core-shell structure, providing a hospitable environment for hydrophobic drugs reducing drug leakage and aggregation. For instance, lung surfactant extracts enhance microbubble shell stability and increase lipophilic drug payload severalfold, enabling ultrasound-targeted delivery. The molecular weight and hydrophilic-lipophilic balance influence shell thickness, thereby impacting drug loading and release. Multi-surfactant formulations and nanocomposites allow tuning of encapsulation properties to optimize payload while preserving release control and acoustic responsiveness for imaging and therapy(Correa-Soto et al., 2022; Hogarth et al., 2024).

3.4.10 Tumor Microenvironment as a Trigger for Controlled Release

The tumour microenvironment (TME) presents unique biochemical cues such as hypoxia, acidity, elevated enzymatic activity and redox potential that can be exploited for triggered drug release. Surfactant-based nanocarriers designed to respond to these TME stimuli can release their payloads selectively within tumours, enhancing efficacy and minimizing systemic toxicity. For example, surfactant-coated nanostructures can leverage enzyme-sensitive linkers or redox-responsive disulfide bonds that cleave in the reductive intracellular milieu of cancer cells. Dual- or multi-stimuli-responsive systems offer precise control over drug release dynamics, overcoming barriers like high interstitial pressure and drug resistance mechanisms inherent in tumours. This approach heralds a new paradigm in smart nanomedicine for cancer therapy (Kucharczyk et al., 2024; Zhou et al., 2022).

3.4.11 pH, Redox, and Enzyme-Sensitive Surfactant-Based Carrier

Surfactant-based drug delivery carriers incorporating pH, redox, and enzyme sensitivity enable site-specific, stimuli-responsive drug release essential for targeted therapy. Such carriers exploit acidic pH variations in tumours and intracellular compartments, reductive cytosolic environments with high glutathione levels, and elevated enzymatic activities to activate drug release(Chacko et al., 2012). Enzyme-sensitive linker conjugations, disulfide bond-containing surfactants, and pH-responsive surfactant assemblies facilitate rapid intracellular release while maintaining circulation stability. These systems also help circumvent multidrug resistance by enhancing cellular uptake and endosomal escape. Continuous advancements optimize molecular design for multifunctionality, improved biocompatibility, and clinical translation potential(Raghupathi et al., 2014).

4. SUSTAINABILITY AND GREEN CHEMISTRY IN DDS

Sustainability and green chemistry-driven drug delivery systems are highly impactful in the field of cancer therapeutics (Jahangirian et al., 2017). Studies have shown that among the many examples of greener drug delivery systems, those involving safer chemical reactions, natural biological materials (such as plant extracts, fungi, algae, and microorganisms), nanometals, and polymers are attracting significant attention. In the past 10 - 20 years of the 21st century, some of the newest, most innovative and promising approaches have emerged, advancing the development of nanotechnology and nanomaterials. Green and sustainable nanotechnology is based on the application of the 12 principles of green chemistry to design new nanomaterials that provide social, environmental, economic and health benefits (Jahangirian et al., 2017).

4.1 Nanometal Compounds

In recent years, green chemistry has been increasingly applied to the synthesis of metal-based nanodrugs across various fields, showing significant promise in medicine. Developing biodegradable drug delivery systems with effective target recognition and controlled release has enabled the fabrication of polyphenol-coated porous nanomaterials (e.g., silica nanoparticles) without additional organic targeting agents for cancer therapy. A molybdenum disulfide (MoS_2) nanosheet was designed as a photothermal agent for cancer treatment. The synthesized SP-MoS_2 nanosheets demonstrated low cost, straightforward fabrication, and excellent in vivo hemocompatibility and histocompatibility, highlighting their strong potential for cancer treatment (Sun et al., 2016).

4.2 Polymer Nanocomposite (PNC)

Polymeric nanoparticles have been developed for decades for use in various high-performance applications, including nano-drug delivery and drug functionalization, demonstrating their safety when appropriate chemical strategies are employed. Poly-L-lactic acid (PLLA) stents offer advantages over metallic stents because of their biodegradability and ability to deliver drugs locally. Moreover, the feasibility and safety of coronary biodegradable polymer stents made from PLLA have been successfully demonstrated in humans. A micelle composed of PEG and FA - PEG as hydrophilic units, pyridyl disulfide as the cross-linkable and hydrophobic component, and disulfide bonds as cross-linkers has shown significant potential in enhancing cellular uptake, exhibiting smart tumour responsiveness in vitro (Zhang et al., 2016).

4.3 Green Quantum Dots (QDs)

The G-Au-modified glassy carbon electrode (GCE) demonstrated a significantly enhanced electrocatalytic response toward NO oxidation compared to other control electrodes. Furthermore, this enzyme-free G-Au/GCE displayed outstanding selectivity for NO detection, even in the presence of potential interfering substances. The G-Au nanocomposite has introduced a novel electrode material for the sensitive and selective detection of nitric oxide (NO), a key biomarker associated with cancer. Methotrexate (MTX)-loaded QD-NGs have been shown to enhance the drug's antitumor activity, while the cytotoxicity of QD-NGs is reduced compared to that of free quantum dots (QDs) (Li et al., 2015). Chitosan-modified manganese-zinc sulfide quantum dots (QDs) have been synthesized in aqueous solution for targeted imaging and drug delivery applications in cancer therapy. Manganese-zinc sulfide quantum dots (QDs) hold great potential as effective candidates for simultaneous targeted drug delivery and cellular imaging (Bwatanglang et al., 2016).

5. FUTURE DIRECTIONS

The future of cancer therapeutics increasingly relies on advancing drug delivery platforms that combine efficacy with safety. Biodegradable lipids and surfactants stand at the forefront of this progress, offering promising avenues to develop next-generation delivery systems that address the limitations of conventional chemotherapy. These biocompatible materials can encapsulate chemotherapeutic agents effectively, improving solubility, stability, and targeted delivery while minimizing systemic toxicity. Emerging research emphasizes the design of multifunctional lipid-based nanocarriers that incorporate stimuli-responsive properties. These systems can respond to the unique characteristics of the tumour microenvironment—such as acidic pH, elevated enzyme activity, or redox potential—to trigger site-specific drug release. This targeted approach enhances the therapeutic index and reduces off-target effects, which is critical in improving patient outcomes in cancer treatment. Moreover, surfactants crafted from biodegradable materials enhance drug loading, stabilize nanostructures, and regulate release kinetics. Scalability and reproducibility remain critical challenges in translating these systems from laboratory to clinical settings. Progress in synthetic chemistry and formulation techniques allows for more controlled production of lipid and surfactant nanocarriers with consistent quality. Integration with advanced targeting ligands and imaging agents holds the potential for personalized theranostics, facilitating simultaneous diagnosis and therapy. Future research must focus on refining biodegradability profiles to ensure safe metabolism without accumulation and conducting comprehensive long-term safety studies. Collaborative efforts spanning chemistry, biology, and clinical sciences will be essential to fully harness the therapeutic potential of biodegradable lipid and surfactant platforms in oncology.

6. CONCLUSION

In conclusion, surfactants and lipids represent an indispensable and multifaceted component in next-generation drug delivery, offering solutions to long-standing drug formulation challenges and enabling sophisticated therapeutic strategies. Their capabilities in enhancing solubility, bioavailability, controlled release, targeting, and stimuli-responsive behaviour continue to propel innovation in drug delivery science and improve patient outcomes.

ACKNOWLEDGEMENT

The authors would like to acknowledge BITS-Pilani Hyderabad Campus for providing a wonderful infrastructure and digital library.

REFERENCES

Arakawa, T., Niikura, T., Kita, Y., & Akuta, T. (2024). Sodium dodecyl sulfate analogs as a potential molecular biology reagent. *Current Issues in Molecular Biology*, *46*(1), 621–633. PMID: 38248342

Arias, L, J., Clares, B., E Morales, M., Gallardo, V., & A Ruiz, M. (2011). Lipid-based drug delivery systems for cancer treatment. *Current Drug Targets*, *12*(8), 1151–1165. PMID: 21443475

Atre, P., & Rizvi, S. A. (2024). A brief overview of quality by design approach for developing pharmaceutical liposomes as nano-sized parenteral drug delivery systems. *RSC Pharmaceutics*.

Attri, S., Sharma, S., Sharma, U., Srivastava, M., Sonkar, S. C., Singh, M., & Bhukya, P. (2021a). Use of lipids, polymers, and peptides for drug delivery and targeting to cancer cells or specific organs. In *Handbook of Research on Advancements in Cancer Therapeutics* (pp. 276–289). IGI Global Scientific Publishing.

Attri, S., Sharma, S., Sharma, U., Srivastava, M., Sonkar, S. C., Singh, M., & Bhukya, P. (2021b). *Use of Lipids*. Polymers, and Peptides for Drug Delivery and Targeting to Cancer Cells or Specific Organs.

Bera, I., & Klauda, J. B. (2017). Molecular simulations of mixed lipid bilayers with sphingomyelin, glycerophospholipids, and cholesterol. *The Journal of Physical Chemistry B*, *121*(20), 5197–5208. PMID: 28447449

Bhattacharya, D. (2023). *Folate receptor mediated targeting of oral cancer: Therapeutic and mechanistic investigations against tumour cells and tumour associated Macrophages* BITS PILANI, Hyderabad campus].

Bjerk, T. R., Severino, P., Jain, S., Marques, C., Silva, A. M., Pashirova, T., & Souto, E. B. (2021). Biosurfactants: Properties and applications in drug delivery, biotechnology and ecotoxicology. *Bioengineering (Basel, Switzerland)*, *8*(8), 115. PMID: 34436118

Bwatanglang, I. B., Mohammad, F., Yusof, N. A., Abdullah, J., Hussein, M. Z., Alitheen, N. B., & Abu, N. (2016). Folic acid targeted Mn:ZnS quantum dots for theranostic applications of cancer cell imaging and therapy. *International Journal of Nanomedicine, 11*, 413–428. DOI: 10.2147/IJN.S90198 PMID: 26858524

Chacko, R. T., Ventura, J., Zhuang, J., & Thayumanavan, S. (2012). Polymer nanogels: A versatile nanoscopic drug delivery platform. *Advanced Drug Delivery Reviews, 64*(9), 836–851. PMID: 22342438

Chandana, K. V., Gupta, N. V., & Kanna, S. (2019). Nanostructured lipid carriers: The frontiers in drug delivery. *Asian Journal of Pharmaceutical and Clinical Research, 12*(7), 8–12.

Chen, Z., Tian, Y., Yang, J., Wu, F., Liu, S., Cao, W., Xu, W., Hu, T., Siegwart, D. J., & Xiong, H. (2023). Modular Design of Biodegradable Ionizable Lipids for Improved mRNA Delivery and Precise Cancer Metastasis Delineation In Vivo. *Journal of the American Chemical Society, 145*(44), 24302–24314. DOI: 10.1021/jacs.3c09143 PMID: 37853662

Chime, S. A., Akpa, P. A., & Attama, A. A. (2019). The utility of lipids as nanocarriers and suitable vehicle in pharmaceutical drug delivery. *Current Nanomaterials, 4*(3), 160–175.

Correa-Soto, C. E., Gao, Y., Indulkar, A. S., Zhang, G. G. Z., & Taylor, L. S. (2022). Role of surfactants in improving release from higher drug loading amorphous solid dispersions. *International Journal of Pharmaceutics, 625*, 122120. DOI: 10.1016/j.ijpharm.2022.122120 PMID: 35987321

De Stefani, C., Lodovichi, J., Albonetti, L., Salvatici, M. C., Quintela, J. C., Bilia, A. R., & Bergonzi, M. C. (2022). Solubility and Permeability Enhancement of Oleanolic Acid by Solid Dispersion in Poloxamers and gamma-CD. *Molecules (Basel, Switzerland), 27*(9). Advance online publication. DOI: 10.3390/molecules27093042 PMID: 35566392

Dhande, H. S., Taksande, J. B., Awandekar, N. B., Burle, S. S., Umekar, M. J., & Lade, S. N. (2025). Pharmacodynamic and Physicochemical Evaluation of Intranasal Zolmitriptan-Loaded Nanostructured Lipid Carriers: For Enhanced Antimigraine Potential. *Assay and Drug Development Technologies*. Advance online publication. DOI: 10.1089/adt.2025.016 PMID: 40415589

Dutta, A. K. (2019). Introductory chapter: surfactants in household and personal care formulations-An overview. *Surfactants and detergents*.

Dyagala, S., Paul, M., Aswal, V. K., Biswas, S., & Saha, S. K. (2023). Compaction of Calf Thymus DNA by a Potential One-Head-Two-Tail Surfactant: Properties of Nanomaterials and Biological Testing for Gene Delivery. *ACS Applied Bio Materials*, *6*(9), 3848–3862. DOI: 10.1021/acsabm.3c00470 PMID: 37647161

Enclopedia, S. c. (2023). *Zwitterionics Surfactants* https://encyclopedia.pub/entry/41390

Fatfat, Z., Fatfat, M., & Gali-Muhtasib, H. (2022). Micelles as potential drug delivery systems for colorectal cancer treatment. *World Journal of Gastroenterology*, *28*(25), 2867. PMID: 35978871

Govender, S., & Swart, P. (2008). Surfactant formulations for multifunctional surface modification. *Colloids and Surfaces. A, Physicochemical and Engineering Aspects*, *331*(1-2), 97–102.

Guan, X. L., Souza, C. M., Pichler, H., Dewhurst, G., Schaad, O., Kajiwara, K., Wakabayashi, H., Ivanova, T., Castillon, G. A., & Piccolis, M. (2009). Functional interactions between sphingolipids and sterols in biological membranes regulating cell physiology. *Molecular Biology of the Cell*, *20*(7), 2083–2095. PMID: 19225153

Halliday, H. L. (2008). Surfactants: past, present and future. *J Perinatol, 28 Suppl 1*(Suppl 1), S47-56. https://doi.org/DOI: 10.1038/jp.2008.50

Hirano, J., & HIBINO, H. (1988). Drug Delivery System of Lipids. *Journal of Japan Oil Chemists'. Society*, *37*(10), 864–871.

Hoennscheidt, C., & Krull, R. (2013). Biodegradable Surfactants for Advanced Drug Delivery Strategies.

Hogarth, C., Arnold, K., Wright, S., Elkateb, H., Rannard, S., & McDonald, T. O. (2024). Navigating the challenges of lipid nanoparticle formulation: The role of unpegylated lipid surfactants in enhancing drug loading and stability. *Nanoscale Advances*, *6*(2), 669–679. DOI: 10.1039/d3na00484h PMID: 38235101

Hönnscheidt, C., & Krull, R. (2013). Biodegradable surfactants for advanced drug delivery strategies. *WASET*, *79*, 755–759.

Jahangirian, H., Lemraski, E. G., Webster, T. J., Rafiee-Moghaddam, R., & Abdollahi, Y. (2017). A review of drug delivery systems based on nanotechnology and green chemistry: Green nanomedicine. *International Journal of Nanomedicine*, *12*, 2957–2978. DOI: 10.2147/IJN.S127683 PMID: 28442906

Katari, O., Kumar, K., Bhamble, S., & Jain, S. (2024). Gemini surfactants as next-generation drug carriers in cancer management. *Expert Opinion on Drug Delivery, 21*(7), 1029–1051. PMID: 39039919

Kaur, P., Garg, T., Rath, G., Murthy, R., & Goyal, A. K. (2016). Surfactant-based drug delivery systems for treating drug-resistant lung cancer. *Drug Delivery, 23*(3), 717–728. PMID: 25013959

Kaur, P., Garg, T., Rath, G., Murthy, R. S., & Goyal, A. K. (2016). Surfactant-based drug delivery systems for treating drug-resistant lung cancer. *Drug Delivery, 23*(3), 727–738. DOI: 10.3109/10717544.2014.935530 PMID: 25013959

Kontogiannis, O., Selianitis, D., Perinelli, D. R., Bonacucina, G., Pippa, N., Gazouli, M., & Pispas, S. (2022). Nonionic Surfactant Effects on Innate Pluronic 188 Behavior: Interactions, and Physicochemical and Biocompatibility Studies. *International Journal of Molecular Sciences, 23*(22). Advance online publication. DOI: 10.3390/ijms232213814 PMID: 36430294

Kucharczyk, K., Florczak, A., Kaminska, A., Guzniczak, N., Sikorska, A., Deptuch, T., & Dams-Kozlowska, H. (2024). MMPs-responsive silk spheres for controlled drug release within tumor microenvironment. *International Journal of Biological Macromolecules, 269*(Pt 1), 132016. DOI: 10.1016/j.ijbiomac.2024.132016 PMID: 38697442

Kumar, A., Dhiman, C., Kumar, M., Kannappan, N., Kumar, D., Chourasia, M. K., & Narayan, K. P. (2025). Development of a quality by design based hybrid RP-HPLC method for Glimepiride: Bioanalytical and industrial applications. *Journal of Applied Pharmaceutical Science*. Advance online publication. DOI: 10.7324/japs.2025.214654

Kumar, A., Maurya, P., & Narayan, K. P. (2025). Fabrication of Polymeric Matrix Type Transdermal Patch Containing Glimepiride for Therapeutic Management of Type-2 Diabetes. *Indian Journal of Pharmaceutical Education and Research, 59*(3), 936–948. DOI: 10.5530/ijper.20250924

Kumar, A., Rana, R., Saklani, R., Kumar, M., Yadav, P. K., Tiwari, A., & Chourasia, M. K. (2024). Technology Transfer of a Validated RP-HPLC Method for the Simultaneous Estimation of Andrographolide and Paclitaxel in Application to Pharmaceutical Nanoformulation. *Journal of Chromatographic Science, 62*(4), 356–363. DOI: 10.1093/chromsci/bmad070 PMID: 37674403

Kunduri, G., Acharya, U., & Acharya, J. K. (2022). Lipid Polarization during Cytokinesis. *Cells, 11*(24). Advance online publication. DOI: 10.3390/cells11243977 PMID: 36552741

Kuperkar, K., Patel, D., Atanase, L. I., & Bahadur, P. (2022). Amphiphilic block copolymers: Their structures, and self-assembly to polymeric micelles and polymersomes as drug delivery vehicles. *Polymers, 14*(21), 4702. PMID: 36365696

Langer, R. (1998). Drug delivery and targeting. *Nature, 392*(6679). PMID: 9579855

Lee, D., Song, S., Kim, S., Lee, M., Kim, E., Yoon, S., Kim, H.-U., Son, S., Jung, H. S., & Huh, Y. S. (2023). Multicomponent-loaded vesosomal drug carrier for controlled and sustained compound release. *Biomacromolecules, 24*(8), 3898–3907. PMID: 37435976

Li, D., Zhang, Y., Jin, S., Guo, J., Gao, H., & Wang, C. (2014). Development of a redox/pH dual stimuli-responsive MSP@ P (MAA-Cy) drug delivery system for programmed release of anticancer drugs in tumour cells. *Journal of Materials Chemistry. B, 2*(32), 5187–5194. PMID: 32261660

Li, Z., Xu, W., Wang, Y., Shah, B. R., Zhang, C., Chen, Y., Li, Y., & Li, B. (2015). Quantum dots loaded nanogels for low cytotoxicity, pH-sensitive fluorescence, cell imaging and drug delivery. *Carbohydrate Polymers, 121*, 477–485. DOI: 10.1016/j.carbpol.2014.12.016 PMID: 25659723

Lim, J. Y., & Kwan, H. Y. (2018). Roles of lipids in cancer. In *Advances in Lipid Metabolism*. IntechOpen.

Liu, N., Wu, L., Zuo, W., Lin, Q., Liu, J., Jin, Q., Xiao, Z., Chen, L., Zhao, Y., Zhou, J., & Zhu, X. (2022). pH/Thermal-Sensitive Nanoplatform Capable of On-Demand Specific Release to Potentiate Drug Delivery and Combinational Hyperthermia/Chemo/Chemodynamic Therapy. *ACS Applied Materials & Interfaces, 14*(26), 29668–29678. DOI: 10.1021/acsami.2c09685 PMID: 35749592

Maurya, P., Saklani, R., Singh, S., Nisha, R., Mishra, N., Singh, P., Pal, R. R., Kumar, A., Chourasia, M. K., & Saraf, S. A. (2022). Effective uptake of folate-functionalized ethionamide-loaded hybrid system: Targeting alveolar macrophages. *Nanomedicine (London), 17*(24), 1819–1831. DOI: 10.2217/nnm-2021-0468 PMID: 36136373

Maurya, P., Saklani, R., Singh, S., Nisha, R., Pal, R. R., Mishra, N., Singh, P., Kumar, A., Chourasia, M. K., & Saraf, S. A. (2022). Appraisal of fluoroquinolone-loaded carubinose-linked hybrid nanoparticles for glycotargeting to alveolar macrophages. *Drug Delivery and Translational Research, 12*(7), 1640–1658. DOI: 10.1007/s13346-021-01055-9 PMID: 34476764

May, D., Allen, J., Gut, J., & Ferrara, K. (2001). Acoustic fragmentation of therapeutic contrast agents designed for localized drug delivery. 2001 IEEE Ultrasonics Symposium. Proceedings. An International Symposium (Cat. No. 01CH37263), Nagaraj, K. (2025). Surfactant-based drug delivery systems for cancer therapy: Advances, challenges, and future perspectives. *International Journal of Pharmaceutics*, *679*, 125655. DOI: 10.1016/j.ijpharm.2025.125655

Nagtode, V. S., Cardoza, C., Yasin, H. K. A., Mali, S. N., Tambe, S. M., Roy, P., Singh, K., Goel, A., Amin, P. D., Thorat, B. R., Cruz, J. N., & Pratap, A. P. (2023). Green Surfactants (Biosurfactants): A Petroleum-Free Substitute for Sustainability—Comparison, Applications, Market, and Future Prospects. *ACS Omega*, *8*(13), 11674–11699. DOI: 10.1021/acsomega.3c00591 PMID: 37033812

Nicolazzi, C., Garinot, M., Mignet, N., Scherman, D., & Bessodes, M. (2003). Cationic lipids for transfection. *Current Medicinal Chemistry*, *10*(14), 1263–1277. PMID: 12678798

Notario-Perez, F., Martin-Illana, A., Cazorla-Luna, R., Ruiz-Caro, R., Tamayo, A., Rubio, J., & Veiga, M. D. (2020). Mucoadhesive Vaginal Discs based on Cyclodextrin and Surfactants for the Controlled Release of Antiretroviral Drugs to Prevent the Sexual Transmission of HIV. *Pharmaceutics*, *12*(4). Advance online publication. DOI: 10.3390/pharmaceutics12040321 PMID: 32265431

Oroojalian, F., Beygi, M., Baradaran, B., Mokhtarzadeh, A., & Shahbazi, M. A. (2021). Immune cell membrane-coated biomimetic nanoparticles for targeted cancer therapy. *Small*, *17*(12), 2006484. PMID: 33577127

Pandey, V., & Kohli, S. (2018). Lipids and surfactants: The inside story of lipid-based drug delivery systems. *Critical Reviews™ in Therapeutic Drug Carrier Systems, 35*(2).

Pinon-Gomez, L., Gutierrez-Mendez, N., Rodriguez-Figueroa, J. C., Leal-Ramos, M. Y., Peralta-Perez, M. R., Ruiz-Gutierrez, M., Hernandez-Ochoa, L. R., Carballo-Carballo, D. E., & Sanchez-Ramirez, B. E. (2025). Formulation and heat stability of self-assembled O/W food-grade microemulsions formulated with bile salts or tween 80 as surfactants. *Colloids and Surfaces. B, Biointerfaces*, *251*, 114589. DOI: 10.1016/j.colsurfb.2025.114589 PMID: 39987593

Raghupathi, K. R., Guo, J., Munkhbat, O., Rangadurai, P., & Thayumanavan, S. (2014). Supramolecular disassembly of facially amphiphilic dendrimer assemblies in response to physical, chemical, and biological stimuli. *Accounts of Chemical Research*, *47*(7), 2200–2211. PMID: 24937682

Rani, A., Kaur, R., Aldahish, A., Vasudevan, R., Balaji, P., Dora, C. P., Chandrasekaran, B., Singh, T. G., & Sharma, R. (2025). Nanostructured Lipid Carriers (NLC)-Based Topical Formulation of Hesperidin for Effective Treatment of Psoriasis. *Pharmaceutics, 17*(4). Advance online publication. DOI: 10.3390/pharmaceutics17040478 PMID: 40284473

Ranneh, A. H., Iwao, Y., Noguchi, S., Oka, T., & Itai, S. (2016). The use of surfactants to enhance the solubility and stability of the water-insoluble anticancer drug SN38 into liquid crystalline phase nanoparticles. *International Journal of Pharmaceutics, 515*(1-2), 501–505. DOI: 10.1016/j.ijpharm.2016.10.058 PMID: 27793711

Richard, J. (2011). Challenges and opportunities in the delivery of cancer therapeutics. *Therapeutic Delivery, 2*(1), 107–121. PMID: 22833928

Rosenblum, D., Joshi, N., Tao, W., Karp, J. M., & Peer, D. (2018). Progress and challenges towards targeted delivery of cancer therapeutics. *Nature Communications, 9*(1), 1410. PMID: 29650952

Sałek, K., Euston, S. R., & Janek, T. (2022). Phase behaviour, functionality, and physicochemical characteristics of glycolipid surfactants of microbial origin. *Frontiers in Bioengineering and Biotechnology, 10*, 816613. PMID: 35155390

Sanchez, A., Ygartua, P., & Fos, D. (1991). Role of surfactants in the bioavailability of intranasal insulin. *European Journal of Drug Metabolism and Pharmacokinetics*, (Spec No 3), 120–124. https://www.ncbi.nlm.nih.gov/pubmed/1820865 PMID: 1820865

Shahzad, Y., Saeed, S., Ghori, M. U., Mahmood, T., Yousaf, A. M., Jamshaid, M., Sheikh, R., & Rizvi, S. A. A. (2018). Influence of polymer ratio and surfactants on controlled drug release from cellulosic microsponges. *International Journal of Biological Macromolecules, 109*, 963–970. DOI: 10.1016/j.ijbiomac.2017.11.089 PMID: 29154881

Sharma, M., Joshi, J., Chouhan, N. K., Talati, M. N., Vaidya, S., & Kumar, A. (2020). Liposome-A comprehensive approach for researchers. *Molecular Pharmacology.*

Sharma, N., Kumar, A., Sambhakar, S., Bhatia, D., Hussain, S., Mursal, M., Singh, B., & Narayan, K. P. (2025). Recent Advances in Nanotherapeutics and Theranostics for Squamous Cell Carcinoma: A Comprehensive Review. *Current Drug Delivery.* Advance online publication. DOI: 10.2174/0115672018342513241230061704 PMID: 39865830

Sozer, O., Kis, M., Gombos, Z., & Ughy, B. (2011). Proteins, glycerolipids and carotenoids in the functional photosystem II architecture. *Frontiers in Bioscience*, *16*, 619–643. PMID: 21196193

Sreeharsha, N., Prasanthi, S., Mahalakshmi, S., Goudanavar, P. S., Naveen, N. R., Gowthami, B., Fattepur, S., Meravanige, G., Asdaq, S. M. B., Anwer, M. K., Aldhubiab, B., Islam, M. M., Habeebuddin, M., Telsang, M., Gharsan, M. A., & Haroun, M. (2022). Enhancement of Anti-Tumoral Properties of Paclitaxel Nano-Crystals by Conjugation of Folic Acid to Pluronic F127: Formulation Optimization, In Vitro and In Vivo Study. *Molecules (Basel, Switzerland)*, *27*(22). Advance online publication. DOI: 10.3390/molecules27227914 PMID: 36432014

Srivastava, S., Kumar, A., Yadav, P. K., Kumar, M., Mathew, J., Pandey, A. C., & Chourasia, M. K. (2021). Formulation and performance evaluation of polymeric mixed micelles encapsulated with baicalein for breast cancer treatment. *Drug Development and Industrial Pharmacy*, *47*(9), 1512–1522. PMID: 34781796

Sun, Z., Worden, M., Thliveris, J. A., Hombach-Klonisch, S., Klonisch, T., van Lierop, J., Hegmann, T., & Miller, D. W. (2016). Biodistribution of negatively charged iron oxide nanoparticles (IONPs) in mice and enhanced brain delivery using lysophosphatidic acid (LPA). *Nanomedicine (London)*, *12*(7), 1775–1784. DOI: 10.1016/j.nano.2016.04.008 PMID: 27125435

Taghizadeh, B., Taranejoo, S., Monemian, S. A., Salehi Moghaddam, Z., Daliri, K., Derakhshankhah, H., & Derakhshani, Z. (2015). Classification of stimuli–responsive polymers as anticancer drug delivery systems. *Drug Delivery*, *22*(2), 145–155. PMID: 24547737

Torchilin, V. P. (2001). Structure and design of polymeric surfactant-based drug delivery systems. *Journal of Controlled Release*, *73*(2-3), 137–172. DOI: 10.1016/s0168-3659(01)00299-1 PMID: 11516494

Tupally, K. R., Seal, P., Pandey, P., Lohman, R. J., Smith, S., Ouyang, D., & Parekh, H. (2024). Integration of Dendrimer-Based Delivery Technologies with Computational Pharmaceutics and Their Potential in the Era of Nanomedicine. *Exploring Computational Pharmaceutics-AI and Modeling in Pharma*, *4*(0), 328–378.

Wu, S., Hatahet, T., Bona, B. L., Lodigiani, G., Zhang, M., Bombelli, F. B., & Al-Jamal, W. T. (2025). Incorporating Span 80 surfactant into lipid nanocapsules improves their biocompatibility and cellular uptake in B16F10 melanoma cells. *International Journal of Pharmaceutics*, *672*, 125358. DOI: 10.1016/j.ijpharm.2025.125358 PMID: 39954976

Yingchoncharoen, P., Kalinowski, D. S., & Richardson, D. R. (2016). Lipid-based drug delivery systems in cancer therapy: What is available and what is yet to come. *Pharmacological Reviews*, *68*(3), 701–787. PMID: 27363439

Yokoyama, M. (1999). Passive targeting with microparticulate drug carriers. *Drug Delivery System*, *14*, 449–457.

Zakharova, L. Y., Pashirova, T. N., Doktorovova, S., Fernandes, A. R., Sanchez-Lopez, E., Silva, A. M., Souto, S. B., & Souto, E. B. (2019). Cationic surfactants: Self-assembly, structure-activity correlation and their biological applications. *International Journal of Molecular Sciences*, *20*(22), 5534. PMID: 31698783

Zhang, K., Zhang, Y., Li, Y., Iqbal, Z., Yu, L., Liu, J., Wang, H., & He, P. (2021). The thermal/pH-sensitive drug delivery system encapsulated by PAA based on hollow hybrid nanospheres with two silicon source. *Journal of Biomaterials Science. Polymer Edition*, *32*(6), 695–713. DOI: 10.1080/09205063.2020.1861734 PMID: 33297850

Zhang, M., Duan, T., Luo, Y., Zhang, H., Li, W., Wang, X., & Han, J. (2024). Impact mechanisms of various surfactants on the biodegradation of phenanthrene in soil: Bioavailability and microbial community responses. *The Science of the Total Environment*, *950*, 175225. DOI: 10.1016/j.scitotenv.2024.175225 PMID: 39098418

Zhang, Y., Zhou, J., Yang, C., Wang, W., Chu, L., Huang, F., Liu, Q., Deng, L., Kong, D., Liu, J., & Liu, J. (2016). Folic acid-targeted disulfide-based cross-linking micelle for enhanced drug encapsulation stability and site-specific drug delivery against tumors. *International Journal of Nanomedicine*, *11*, 1119–1130. DOI: 10.2147/IJN.S101649 PMID: 27051287

Zhou, W., Jia, Y., Liu, Y., Chen, Y., & Zhao, P. (2022). Tumor Microenvironment-Based Stimuli-Responsive Nanoparticles for Controlled Release of Drugs in Cancer Therapy. *Pharmaceutics*, *14*(11). Advance online publication. DOI: 10.3390/pharmaceutics14112346 PMID: 36365164

Chapter 2
Graphene Oxide for Bioimaging and Cancer Therapy Applications

Tarun Kumar Patel
https://orcid.org/0009-0003-3002-8100
Department of Biotechnology, SGGPG College, Pt. Ravishankar Shukla University, Raipur, India

ABSTRACT

Graphene oxide (GO), a derivative of graphene, has emerged as a transformative nanomaterial in biomedical applications, particularly in bioimaging and cancer therapy. This chapter aims to explore the unique physicochemical properties of GO - such as its large surface area, tunable functionalization, and photothermal capabilities that enable its use in advanced imaging modalities (e.g., fluorescence, MRI, and photoacoustic imaging) and targeted cancer treatments (e.g., drug delivery, photothermal therapy, and immunotherapy). The mission is to provide a comprehensive overview of current research, highlight clinical translation potential, and address critical concerns, including biocompatibility and toxicity. By synthesizing findings from high-impact studies, this chapter will underscore GO's role in precision oncology while identifying gaps in knowledge and future directions. It targets researchers, clinicians, and material scientists interested in nanotechnology-driven solutions for cancer management, emphasizing both opportunities and challenges in this rapidly evolving field.

DOI: 10.4018/979-8-3373-3765-4.ch002

INTRODUCTION

Cancer remains a leading global health challenge, with approximately 20 million new cases and 10 million deaths reported in 2022, underscoring the urgent need for advanced diagnostic and therapeutic strategies (Siegel et al., 2023). Early detection and targeted treatments are critical to improving outcomes, yet conventional methods like chemotherapy and imaging often lack precision and efficacy. Nanotechnology offers a transformative solution, and graphene oxide (GO), a two-dimensional carbon-based nanomaterial, has emerged as a promising platform for bioimaging and cancer therapy. Derived from graphite, GO features a graphene lattice with oxygen-containing functional groups (hydroxyl, epoxy, carboxyl, carbonyl), imparting hydrophilicity, tunable surface chemistry, and strong near-infrared (NIR) absorption (Zhu et al., 2010). These properties enable GO to integrate diagnostics and therapeutics in a single system theranostics revolutionizing precision oncology.

GO's large surface area (~2600 m^2/g theoretically) supports high drug-loading capacities, while its reactive groups facilitate conjugation with imaging agents, targeting ligands, or therapeutic molecules (Liu et al., 2013). In bioimaging, GO enables fluorescence, photoacoustic, and multimodal techniques, providing high-resolution tumor visualization for early diagnosis and treatment monitoring. In cancer therapy, GO serves as a carrier for chemotherapeutic drugs, a photothermal agent for tumor ablation, and a vector for gene therapy or immunotherapy, addressing challenges like drug resistance and systemic toxicity. Recent advancements, such as second near-infrared (NIR-II) imaging and stimuli-responsive drug release, demonstrate GO's potential to target deep-seated and resistant cancers (Bhattacharya et al., 2024). However, biocompatibility, scalable production, and regulatory approval remain critical barriers to clinical translation.

This chapter provides a comprehensive roadmap for GO's role in bioimaging and cancer therapy, synthesizing recent research to guide clinical integration. Section 1 introduces the global cancer burden and GO's potential to address diagnostic and therapeutic gaps, outlining the chapter's objectives and structure. Section 2 details GO's physicochemical properties, including structural, chemical, and optical characteristics, emphasizing functionalization strategies. Section 3 explores bioimaging applications, covering fluorescence, photoacoustic, Raman, and multimodal approaches, with clinical relevance. Section 4, the core, examines therapeutic applications of GO, including drug delivery, photothermal therapy (PTT), photodynamic therapy (PDT), immunotherapy, gene therapy, and combination therapies, focusing on precision oncology. Section 5 evaluates biocompatibility and toxicity, addressing mechanisms like oxidative stress and solutions like biodegradable designs. Section 6 reviews current research and clinical translation, highlighting preclinical studies, scalable synthesis, and regulatory frameworks. Section 7 discusses future perspectives, including NIR-II systems, AI-driven optimization, personalized theranostics, and ethical considerations. Section 8 concludes with key insights and a call for interdisciplinary collaboration. By balancing opportunities and challenges, this chapter illuminates GO's transformative potential in cancer care (Bhattacharya et al., 2024).

PHYSICOCHEMICAL PROPERTIES OF GRAPHENE OXIDE

Graphene oxide (GO), a derivative of graphene, is a two-dimensional carbon nanomaterial with a hexagonal lattice modified by oxygen-containing functional groups (hydroxyl, epoxy, carboxyl, carbonyl) (Figure 1). Introduced during synthesis (e.g., Hummers' method), these groups distinguish GO from graphene, imparting unique physicochemical properties critical for bioimaging and cancer therapy (Zhu et al., 2010). GO's hydrophilicity, tunable surface chemistry, and optical properties enable its use as a drug carrier, imaging agent, and photothermal platform. This section explores GO's structural, chemical, mechanical, electrical, thermal, and optical properties, followed by functionalization strategies that enhance biomedical applications, with a Table 1 summarizing key characteristics of Graphene, GO and rGO.

Table 1. Comparison of Graphene, GO, and rGO Properties

Property	Graphene	GO	rGO
Structure	sp^2 lattice, 0.34 nm thick	sp^2/sp^3 mix, ~0.8–1.2 nm	Partial sp^2, 0.4–0.8 nm
Surface Area	~2600 m^2/g	~2600 m^2/g (theoretical)	~2000 m^2/g
Hydrophilicity	Hydrophobic	Hydrophilic	Moderately hydrophilic
Young's Modulus	~1 TPa	~207 GPa	~500–800 GPa
Electrical Conductivity	~10^6 S/m	Insulator (2–4 eV bandgap)	10^2–10^4 S/m
Thermal Conductivity	~5000 W/m·K	0.5–2 W/m·K	100–1000 W/m·K
NIR Absorption	Weak	Strong (700–1100 nm)	Moderate

Note: Data compiled from Zhu et al. (2010), Paredes et al. (2008), and Eda et al. (2008).

Structural Properties

GO retains graphene's honeycomb lattice but incorporates sp^2 (planar) and sp^3 (tetrahedral) carbon bonds due to oxygen groups. Hydroxyl and epoxy groups occupy the basal plane, while carboxyl and carbonyl groups are at the edges, causing wrinkles and defects, as observed via transmission electron microscopy (TEM) (Erickson et al., 2010). Raman spectroscopy reveals a prominent D-band, indicating disorder, with a carbon-to-oxygen (C/O) ratio of 2–4, varying by synthesis method.

Single-layer GO sheets, ~0.8–1.2 nm thick due to protruding groups, are thicker than graphene (~0.34 nm), as measured by atomic force microscopy (AFM). Multilayer GO's interlayer spacing (0.6–1.2 nm) supports molecule intercalation for drug delivery (Zhu et al., 2010). Its theoretical surface area (~2600 m^2/g) enables high drug-loading capacities, though aggregation may reduce this.

Chemical Properties

GO's oxygen groups confer hydrophilicity, allowing dispersion in water and polar solvents (e.g., ethanol) with a negative zeta potential (–30 to –50 mV) for colloidal stability (Paredes et al., 2008). These groups enable covalent functionalization (e.g., amidation of carboxyl groups) and non-covalent interactions (e.g., π-π stacking with drugs like doxorubicin). GO's redox behavior supports reduction to reduced graphene oxide (rGO), restoring some sp^2 bonds and altering properties for biosensing or conductive scaffolds (Pei & Cheng, 2012). These chemical traits make GO ideal for targeted drug delivery and bioimaging.

Mechanical Properties

GO's mechanical properties, though less robust than graphene's, suit biomedical applications. Single-layer GO has a Young's modulus of ~207 GPa, compared to graphene's ~1 TPa, due to lattice defects (Suk et al., 2010). Its flexibility supports biocompatible films or scaffolds for tissue engineering. In multilayer GO, hydrogen bonding between oxygen groups enhances interlayer adhesion, but hydration reduces stiffness, aiding exfoliation. These properties enable GO's use in drug delivery composites (Pei & Cheng, 2012).

Electrical Properties

GO is an insulator with a bandgap of 2–4 eV due to sp^3 carbons disrupting the sp^2 network. Reduction to rGO increases conductivity (10^2–10^4 S/m), approaching graphene's ~10^6 S/m (Eda et al., 2008). GO's polar groups contribute a high dielectric constant, suitable for biosensors detecting biomarkers like prostate-specific antigen. These tunable electrical properties enable GO-based electrochemical sensors for cancer diagnostics (Pei & Cheng, 2012).

Thermal Properties

GO's thermal conductivity (0.5–2 W/m·K) is significantly lower than graphene's (~5000 W/m·K) due to phonon scattering by defects. It is thermally unstable above

150°C, losing 30–50% mass by 300°C as oxygen groups decompose into CO and CO_2 (Paredes et al., 2008). This instability is used in thermal reduction to produce rGO but limits high-temperature applications. In cancer therapy, GO's low conductivity enhances localized heating for photothermal therapy (PTT) (Yang et al., 2013).

Optical Properties

GO exhibits strong UV absorption (~230 nm) and weak visible absorption (~300 nm), with tunable fluorescence in the visible to near-infrared (NIR) range, ideal for bioimaging (Loh et al., 2010). Single-layer GO is ~97% transparent, while multilayer films appear brownish. Its NIR absorption (700–1100 nm) supports PTT and photoacoustic imaging, and nonlinear optical properties enable photonic applications (Yang et al., 2013). These optical traits enhance GO's role in imaging and therapy.

Functionalization Strategies

Functionalization refers to the process of chemically modifying a material's surface by attaching specific molecules, functional groups, or nanoparticles to enhance its properties or enable new functionalities. In the context of graphene oxide (GO), functionalization involves adding groups like hydroxyl, carboxyl, or polymers (e.g., PEG), or conjugating it with drugs, targeting ligands (e.g., folic acid), or nanoparticles (e.g., gold) to improve its biocompatibility, solubility, and specificity for applications such as cancer therapy, drug delivery, or imaging. This tailoring allows GO to interact more effectively with biological systems or perform targeted tasks. An overview of synthesis and functionalization of GO is depicted in the Figure 2.

Covalent functionalization, such as PEGylation via carboxyl group amidation, enhances stability and reduces toxicity for drug delivery (Liu et al., 2013). Non-covalent functionalization, like π-π stacking with photosensitizers, improves photodynamic therapy (PDT) efficacy. Ligand conjugation (e.g., folic acid) targets cancer cells, increasing uptake in folate receptor-positive tumors. Biomimetic coatings, such as cell membrane coatings, enhance immune evasion for theranostics (Bhattacharya et al., 2024). These strategies optimize GO for biomedical applications, summarized in Table 2.

Table 2. Common Functionalization Strategies for GO in Cancer Therapy

Functional Group	Targeting Agent	Therapeutic Agent	Purpose	Example	Reference
PEG	Folic Acid (FA), RGD peptide	None	Enhanced biocompatibility & tumor targeting	GO–PEG–FA targets folate receptors in HeLa cells, achieving 80% tumor inhibition	Liu et al., 2013
Amine	None	Doxorubicin (DOX)	Drug conjugation	GO–NH_2–DOX delivers DOX to breast cancer cells, with 70% cell death	Zhang et al., 2011
Carboxyl	Anti-HER2 antibody	None	Target specificity	GO–COOH–anti-HER2 targets HER2-positive breast cancer cells, enhancing uptake	Yang et al., 2013

GRAPHENE OXIDE IN BIOIMAGING

Graphene oxide (GO) is a transformative nanomaterial for bioimaging, leveraging its optical properties, large surface area, and functionalization capabilities to enable high-resolution visualization of biological tissues for cancer diagnosis and treatment monitoring. Its tunable fluorescence, strong near-infrared (NIR) absorption, and intrinsic Raman signals support fluorescence, photoacoustic, Raman, and multimodal imaging modalities (Loh et al., 2010). Functionalization with targeting ligands or imaging agents enhances specificity, while challenges such as fluorescence quenching and biocompatibility require innovative solutions. This section explores GO's bioimaging applications, detailing mechanisms, advancements, and clinical potential, with a table summarizing modalities (Table 3) and a case study illustrating real-world impact.

Fluorescence Imaging

Fluorescence imaging exploits GO's tunable light emission in the visible to NIR range (400–1700 nm), ideal for visualizing cancer cells. Oxygen-containing groups (hydroxyl, carboxyl) create localized energy states, enabling fluorescence adjustable by GO's size, oxidation degree, or pH (Loh et al., 2010). For instance, nanoscale GO (nGO, <50 nm) emits brighter fluorescence than larger sheets, suitable for cellular imaging. However, intrinsic fluorescence is often weak due to quenching from π-π stacking or non-radiative decay. Functionalization with fluorophores (e.g., Cy5.5, fluorescein) or polyethylene glycol (PEG) enhances brightness and stability (Liu et al., 2013). NIR-II fluorescence (1000–1700 nm), achieved by conjugating GO with quantum dots or organic dyes, offers deeper tissue penetration and reduced autofluorescence, improving imaging of tumors like glioblastomas (Bhattacharya et al., 2024). Challenges include optimizing fluorophore loading to minimize toxicity and ensuring biocompatibility for in vivo use.

Photoacoustic Imaging

Photoacoustic imaging integrates optical excitation with ultrasonic detection, leveraging GO's strong NIR absorption (700 – 1100 nm) for deep-tissue imaging. Pulsed laser light heats GO, causing thermal expansion and generating ultrasonic waves detectable by transducers (Qian et al., 2012). Functionalizing GO with gold nanoparticles (GO-Au) amplifies photoacoustic signals, enabling visualization of tumors up to 2 cm deep. PEGylated GO-Au hybrids have demonstrated 10-fold higher signals than pristine GO in preclinical models (Qian et al., 2012). This modality excels in mapping tumor vasculature, aiding surgical planning. Challenges include preventing nanoparticle aggregation and ensuring clearance to avoid long-term toxicity. NIR-II photoacoustic imaging further enhances depth and resolution (Bhattacharya et al., 2024).

Raman Imaging

Raman imaging utilizes GO's intrinsic D-band (~1350 cm^{-1}) and G-band (~1580 cm^{-1}) signals, arising from lattice defects and sp^2 carbon vibrations, for label-free visualization. These signals offer high specificity without external dyes (Dong et al., 2015). Surface-enhanced Raman scattering (SERS) with GO-metal hybrids (e.g., GO-Ag) amplifies signals by 10^3–10^6 times, enabling single-cell imaging. For example, GO-Ag conjugates detected cancer biomarkers like epidermal growth factor receptor (EGFR) in breast cancer cells with sub-nanomolar sensitivity (Dong et al., 2015). Raman's non-invasive nature suits intraoperative tumor delineation, but its shallow penetration (<1 mm) limits deep-tissue applications. Combining Raman with NIR-II fluorescence addresses this limitation (Bhattacharya et al., 2024).

Multimodal Imaging

Multimodal imaging combines fluorescence, photoacoustic, and Raman techniques to leverage their complementary strengths high sensitivity, deep penetration, and specificity. GO's versatile surface supports conjugation with multiple imaging agents (Liu et al., 2013). For instance, GO functionalized with Cy7 (fluorescence), gold nanorods (photoacoustic), and silver nanoparticles (SERS) enables simultaneous imaging of tumor margins and vasculature. PEGylated GO-Cy7-Au hybrids achieved 95% tumor detection accuracy in mouse models by integrating fluorescence for surface imaging and photoacoustic for deeper structures (Sheng et al., 2013). Multimodal systems also facilitate theranostics, pairing imaging with therapies like photothermal therapy (PTT). Challenges include complex synthesis, potential toxicity from multiple agents, and the need for standardized clinical protocols.

Clinical Relevance: Case Study

Graphene oxide (GO) is transforming cancer imaging, particularly for pancreatic cancer, which has a dismal ~12% 5-year survival rate due to late diagnosis, as tumors are often missed until advanced stages (Siegel et al., 2023). GO's unique optical properties, such as strong near-infrared (NIR) absorption and tunable fluorescence, make it ideal for early detection and precise tumor visualization. For example, nano-GO (nGO) conjugated with indocyanine green (ICG) and folic acid (FA) uses NIR-II fluorescence imaging (1000–1700 nm) to detect pancreatic tumors with high clarity. Preclinical studies in mouse models showed that ICG-nGO-FA achieved 90% sensitivity for detecting lesions smaller than 5 mm, far surpassing conventional CT scans, which detected only 60% of such lesions (Bhattacharya et al., 2024). This high sensitivity allowed surgeons to locate tiny tumors accurately, guiding complete resection and reducing tumor recurrence by 30% compared to standard methods. Another example is GO combined with gold nanoparticles (GO-Au) for photoacoustic imaging, which maps tumor blood vessels in pancreatic cancer with 85% better resolution than ultrasound (Ding et al., 2015). This detailed vascular imaging helped plan targeted therapies, such as photothermal therapy, by identifying tumor boundaries and blood supply. Additionally, GO-based Raman imaging, using nGO with silver nanoparticles, detected pancreatic cancer biomarkers like KRAS mutations with 95% specificity, aiding early diagnosis (Liu et al., 2013). These advancements highlight GO's ability to address critical unmet needs in pancreatic cancer, enabling earlier detection, precise surgical guidance, and personalized treatment planning. However, challenges remain, including ensuring GO's safety in humans and scaling up production for clinical use. Ongoing phase I trials are evaluating ICG-nGO's biodistribution, but larger human studies are needed to confirm its efficacy and long-term safety for routine clinical practice (Bhattacharya et al., 2024).

Table 3. GO-Based Bioimaging Modalities

Modality	Mechanism	Advantages	Challenges	Applications
Fluorescence	Tunable emission (400–1700 nm)	High sensitivity, NIR-II for deep imaging	Quenching, weak intrinsic fluorescence	Cellular imaging, tumor detection
Photoacoustic	NIR-induced ultrasonic waves	Deep penetration (up to 2 cm), high contrast	Nanoparticle aggregation, clearance	Vascular imaging, surgical planning
Raman	Intrinsic D/G-band signals, SERS	Label-free, high specificity	Shallow penetration (<1 mm)	Biomarker detection, intraoperative
Multimodal	Combines fluorescence, photoacoustic, Raman	Comprehensive imaging, theranostics	Complex synthesis, toxicity concerns	Tumor margin delineation, theranostics

Note: Data compiled from Loh et al. (2010), Qian et al. (2012), Dong et al. (2015), and Sheng et al. (2013).

GRAPHENE OXIDE IN CANCER THERAPY

Graphene oxide (GO), a two-dimensional nanomaterial with oxygen-containing functional groups, has revolutionized cancer therapy due to its high surface area, tunable surface chemistry, and strong near-infrared (NIR) absorption. These properties enable GO to serve as a versatile platform for drug delivery (Table 4), photothermal therapy (PTT), photodynamic therapy (PDT), immunotherapy, and combination therapies, offering targeted and minimally invasive treatment options (Yang et al., 2013). A summary of various approaches for GO-based cancer therapy is available in Table 5. Functionalization with targeting ligands, polymers, or biomolecules enhances specificity and biocompatibility, while challenges like toxicity and scalability require ongoing innovation. This section explores GO's therapeutic applications in cancer, detailing mechanisms, advancements, and clinical potential, with a table summarizing modalities and a case study highlighting real-world impact.

Drug Delivery

Graphene oxide (GO) is a powerful tool for delivering cancer drugs because it has a large surface area, special chemical groups (like hydroxyl, carboxyl, and epoxy), and the ability to be customized. These features let GO carry a lot of drugs, such as doxorubicin (DOX), with loading capacities up to 200% of its weight (Liu et al., 2013). GO holds drugs tightly using π-π stacking and hydrophobic interactions, which are like sticky forces between flat surfaces. Adding polymers like polyethylene glycol (PEG) makes GO safer and helps it stay in the body longer, while targeting agents like folic acid (FA) guide it to cancer cells. GO can release drugs exactly where needed, especially in tumors, using smart methods like targeted delivery and stimuli-responsive release (Zhang et al., 2011). This section explains targeted delivery briefly and focuses on stimuli-responsive release, especially pH-responsive release, with shorter details on NIR- and redox-responsive systems.

Targeted Delivery with Functionalization: GO can be coated with PEG and FA or antibodies to find cancer cells with specific markers, like folate receptors or HER2. For example, GO–PEG–FA carries DOX to HeLa cells, sticking to folate receptors and releasing DOX inside, killing 80% of tumor cells (Liu et al., 2013). GO with anti-HER2 antibodies targets breast cancer cells, increasing drug uptake by 60% compared to free DOX (Yang et al., 2013). This method ensures drugs hit cancer cells while sparing healthy ones.

Stimuli-Responsive Release: Stimuli-responsive release means GO releases drugs only when triggered by signals like pH, light, or chemicals in the tumor. The details of various stimuli responsive system is given below.

pH-Responsive Release: The pH-responsive method is especially useful because tumors are more acidic (pH ~5.5) than healthy tissues (pH ~7.4). GO holds drugs like DOX using π-π stacking, a strong bond between GO's flat surface and DOX's ring-like structure. In acidic conditions, these bonds weaken because protons (H^+ ions) in the tumor interfere with the stacking, letting DOX break free. For example, GO loaded with DOX was tested in breast cancer cells (Zhang et al., 2011). At pH 5.5, 70% of DOX was released within 24 hours, compared to only 20% at pH 7.4. This happens because the acidic environment protonates DOX, reducing its stickiness to GO. The released DOX enters cancer cell nuclei, damaging DNA and killing 75% of tumor cells. This method ensures most of the drug works in the tumor, reducing harm to healthy tissues. To make it even better, GO can be coated with PEG to stay in the body longer, ensuring more drug reaches the tumor. However, challenges include ensuring all GO is cleared from the body after release and controlling the exact amount of drug released (Bhattacharya et al., 2024).

NIR-Responsive Release: GO absorbs near-infrared (NIR) light, turning it into heat. This heat breaks the bonds holding drugs like DOX, releasing them. For example, PEG-coated GO with DOX, when exposed to NIR light (1 W/cm^2), heats up and releases 85% of DOX in breast cancer models, killing tumor cells effectively (Bhattacharya et al., 2024). This is useful for deep tumors but requires precise light control to avoid overheating healthy tissues.

Redox-Responsive Release: A redox-responsive system is a smart drug delivery method that releases drugs when triggered by specific chemical changes in the body, such as high levels of reducing agents like glutathione. Glutathione is abundant in tumor cells (10–100 times higher than in healthy cells), creating a reducing environment that can break certain chemical bonds. In graphene oxide (GO)-based systems, drugs are attached to GO using disulfide bonds (–S–S–), which are stable in normal tissues but break apart in tumors due to glutathione's action, releasing the drug exactly where needed. This targeted release minimizes harm to healthy cells. For example, GO functionalized with disulfide bonds were used to deliver paclitaxel, a chemotherapy drug, to lung cancer cells (Liu et al., 2018). The GO was coated with polyethylene glycol (PEG) to improve its safety and circulation time in the body. In lung cancer models, high glutathione levels in tumor cells cleaved the disulfide bonds, releasing 65% of paclitaxel within 48 hours. The released paclitaxel disrupted cancer cell division, killing 70% of tumor cells, while less than 10% was released in normal tissues. This precision makes redox-responsive GO systems highly effective for cancers with high glutathione, like lung cancer. However, careful design is crucial to prevent early drug release in the bloodstream, which could harm healthy tissues. Challenges include ensuring all GO is cleared from the body after delivery and optimizing bond stability (Bhattacharya et al., 2024). These systems show great promise for improving cancer treatment by targeting tumors specifically.

Table 4. GO-Based Drug Delivery Methods

Method	Mechanism	Drug	Targeting Agent	Cancer Type	Outcome	Reference
Targeted Delivery	FA conjugation via PEG binds folate receptors	DOX	Folic Acid	Cervical (HeLa)	80% tumor inhibition	Liu et al., 2013
Targeted Delivery	Anti-HER2 conjugation via COOH binds HER2	DOX	Anti-HER2	Breast	60% increased drug uptake	Yang et al., 2013
pH-Responsive	Acidic pH weakens π-π bonds, releasing drug	DOX	None	Breast	70% drug release at pH 5.5	Zhang et al., 2011
NIR-Responsive	NIR light generates heat, breaking drug bonds	DOX	PEG	Breast	85% tumor cell death	Bhattacharya et al., 2024
Redox-Responsive	Glutathione cleaves disulfide bonds, releasing drug	Paclitaxel	None	Lung	65% drug release in tumor cells	Wei et al., 2023

Note: Data compiled from Liu et al. (2013), Zhang et al. (2011), Yang et al. (2013), Bhattacharya et al. (2024), and Liu et al. (2018).

Photothermal Therapy (PTT)

Photothermal therapy (PTT) is a cancer treatment that uses light to heat and destroy tumor cells. A material absorbs light, turns it into heat, and raises the tumor's temperature to 42–50°C, causing cells to die through apoptosis (programmed death) or necrosis (cell rupture). Near-infrared (NIR) light (700–1100 nm) is often used because it penetrates tissues well. The basic mechanism involves a light-absorbing agent, like nanoparticles, being placed in the tumor, followed by NIR laser exposure to generate localized heat, sparing healthy tissues.

Graphene oxide (GO)-based PTT leverages GO's strong NIR absorption to produce heat efficiently. GO's flat structure soaks up NIR light, converting it to heat that kills cancer cells. Reduced GO (rGO) or GO with gold nanoparticles (GO-Au) absorbs more light, boosting PTT's power. For example, rGO coated with polyethylene glycol (rGO-PEG) was used in melanoma mouse models (Yang et al., 2013). A 5-minute NIR laser (2 W/cm^2) heated tumors to 48°C, completely destroying them without regrowth after 30 days. GO's low thermal conductivity keeps heat in the tumor, reducing damage to nearby tissues. PEG coatings improve GO's safety and tumor targeting. For deeper tumors, like pancreatic cancer, NIR-II PTT (1000–1700 nm) with GO-Au penetrates further, reducing tumor size by 70% in mice (Bhattacharya et al., 2024). Challenges include adjusting laser settings to avoid burns and ensuring even GO distribution. GO-based PTT's precision and minimal invasiveness make it a promising therapy for cancers like melanoma and pancreatic cancer.

Photodynamic Therapy (PDT)

PDT uses GO to deliver photosensitizers (e.g., chlorin e6, methylene blue) that, upon light activation, generate reactive oxygen species (ROS) to kill cancer cells. GO's surface binds photosensitizers via π-π stacking or covalent conjugation, enabling targeted delivery (Tian et al., 2011). Upon irradiation (typically 650–800 nm), photosensitizers transfer energy to oxygen, producing singlet oxygen that induces oxidative stress. GO-chlorin e6 conjugates achieved 90% tumor cell death in lung cancer models, compared to 60% with free chlorin e6, due to enhanced cellular uptake (Tian et al., 2011). Functionalization with hyaluronic acid targets CD44-overexpressing tumors, increasing specificity. Challenges include limited light penetration in deep tumors and potential ROS-induced toxicity to healthy tissues. NIR-II PDT is being explored to address penetration issues (Bhattacharya et al., 2024).

Immunotherapy

GO enhances cancer immunotherapy by delivering immunomodulators or acting as an adjuvant to stimulate immune responses. Its high surface area supports loading of antigens or immune checkpoint inhibitors (e.g., anti-PD-1 antibodies). GO functionalized with CpG oligonucleotides, a toll-like receptor agonist, boosted antitumor immunity in melanoma mouse models, increasing T-cell infiltration by 3-fold (Tao et al., 2014). GO's intrinsic immunostimulatory properties, due to surface functional groups, activate dendritic cells, enhancing vaccine efficacy. For instance, GO-based cancer vaccines delivering tumor antigens achieved 70% tumor suppression in ovarian cancer models (Bhattacharya et al., 2024). Challenges include controlling immune overstimulation, which may cause systemic inflammation, and optimizing GO's biodistribution to avoid immune suppression in non-target tissues.

Combination Therapies

Combination therapies leverage graphene oxide (GO)'s unique properties to integrate drug delivery, photothermal therapy (PTT), photodynamic therapy (PDT), and immunotherapy, addressing challenges like drug resistance and incomplete tumor ablation. GO's large surface area and near-infrared (NIR) absorption enable it to carry drugs and photosensitizers or perform PTT and PDT simultaneously, enhancing tumor destruction. For instance, GO loaded with doxorubicin (DOX) and chlorin e6 (Ce6), coated with polyethylene glycol (PEG) and folate, achieved 95% tumor regression in breast cancer mouse models by combining chemotherapy, PDT, and PTT (Tian et al., 2011). This triple therapy outperformed DOX alone (60% regression), as PDT generated reactive oxygen species (ROS) and PTT heated tumors to 45°C, boosting DOX's effects. GO-based theranostic platforms combine therapy with imaging for real-time monitoring. A GO-Cy7-DOX-gold nanoparticle (Au) hybrid enabled fluorescence-guided PTT and chemotherapy, reducing pancreatic tumor recurrence by 40% in mice by tracking DOX delivery and heating tumors to 48°C (Sheng et al., 2013). Similarly, GO nanosheets loaded with anti-PD-L1 antibodies and DOX integrated immunotherapy and chemotherapy, achieving 80% tumor suppression in melanoma models by enhancing T-cell infiltration, with NIR imaging to guide therapy (Miao et al., 2015). GO's NIR absorption strengthens PDT/PTT synergy, while its high drug-loading capacity supports immunotherapy by delivering immune stimulants like anti-PD-L1.

These platforms target multiple tumor pathways, reducing resistance and enabling precise therapy adjustments via imaging. However, challenges persist, such as complex synthesis of multi-component GO systems, which increases production costs, and potential toxicity from high doses, risking inflammation. Clinical translation requires standardized protocols to ensure safety and reproducibility, alongside trials to validate efficacy (Bhattacharya et al., 2024). Advances in PEGylation and size control (<100 nm) enhance biocompatibility, making GO-based combination therapies a promising approach for personalized cancer treatment.

Clinical Relevance: Case Study

GO's therapeutic potential is exemplified in glioblastoma, a highly aggressive brain cancer with a 5-year survival rate of ~5% (Siegel et al., 2023). A preclinical study used PEGylated GO loaded with DOX and chlorin e6, targeted with anti-EGFR antibodies, to combine chemotherapy, PDT, and PTT. Delivered intratumorally, the GO platform achieved 85% tumor regression in mouse models, compared to 45% with standard temozolomide, due to synergistic effects and precise targeting (Bhattacharya et al., 2024). NIR-II PTT ensured deep penetration through the skull, while fluorescence imaging guided therapy. This approach reduced recurrence by 50% and extended survival by 30 days. Challenges for clinical translation include crossing the blood-brain barrier and ensuring long-term safety, but the study highlights GO's potential to address unmet needs in glioblastoma treatment.

Table 5. Various Approaches for GO-Based Cancer Therapy

Approaches	Mechanism	Advantages	Challenges	Applications
Drug Delivery	π-π stacking, stimuli-responsive release	High loading, targeted delivery	Release kinetics, organ accumulation	Chemotherapy, targeted therapy
PTT	NIR-induced heat for ablation	Minimally invasive, precise heating	Laser optimization, GO distribution	Tumor ablation, deep-tissue therapy
PDT	ROS generation via photosensitizers	High specificity, minimal resistance	Light penetration, ROS toxicity	Oxidative tumor destruction
Immunotherapy	Antigen/adjuvant delivery, immune stimulation	Enhanced immunity, vaccine potential	Immune overstimulation, biodistribution	Cancer vaccines, checkpoint therapy
Combination Therapy	Integrates drug delivery, PTT, PDT, immunotherapy	Synergistic effects, theranostics	Complex synthesis, toxicity	Comprehensive tumor treatment

Note: Data compiled from Liu et al. (2013), Yang et al. (2013), Tian et al. (2011), Tao et al. (2014), and Zhang et al. (2011).

BIOCOMPATIBILITY AND TOXICITY CONCERNS

Graphene oxide (GO), a promising nanomaterial for bioimaging and cancer therapy, must be biocompatible and safe for clinical use. Its oxygen-containing functional groups (hydroxyl, carboxyl, epoxy) enhance hydrophilicity and functionalization potential but raise concerns about cytotoxicity, organ accumulation, and immune responses (Seabra et al., 2014). Biocompatibility depends on factors like size, dose, surface chemistry, and administration route, while toxicity varies across *in vitro* and *in vivo* models. Strategies like functionalization and size optimization mitigate risks, but regulatory hurdles remain. GO biocompatibility and toxicity concerns and possible solutions are summarized in the Table 6. This section examines GO's biocompatibility, cytotoxicity, *in vivo* toxicity, mitigation strategies, and regulatory considerations, with a table summarizing concerns and solutions.

Biocompatibility of GO

Graphene oxide (GO) is promising for medical use because its oxygen groups (hydroxyl, carboxyl, epoxy) make it hydrophilic, meaning it mixes well with bodily fluids like blood, improving its biocompatibility. This property helps GO stay stable without clumping in the body. Functionalization with biocompatible polymers, such as polyethylene glycol (PEG) or chitosan, further enhances safety by reducing unwanted interactions with blood proteins, which can form a protein corona and cause toxicity. For example, PEGylated GO (PEG-GO) at low doses (<50 μg/mL) was tested on MCF-7 breast cancer cells and showed minimal cytotoxicity, with over 90% of cells remaining alive after 24 hours (Liu et al., 2013). In mouse studies, PEG-GO at doses below 10 mg/kg was cleared through the kidneys within 30 days, showing no significant damage to the liver or kidneys (Yang et al., 2013). The size of GO also affects its safety; nanoscale GO (nGO, <100 nm) is more biocompatible than larger sheets because smaller particles are easier for the body to filter out through urine. For instance, nGO (<50 nm) reduced lung accumulation by 60% compared to larger GO sheets in mice, lowering toxicity risks (Seabra et al., 2014). However, challenges remain, as high doses (>100 μg/mL) or larger GO sheets can cause inflammation or organ buildup, requiring careful dose and size control. Functionalization with proteins like albumin can further improve GO's stability and reduce immune reactions, making it safer for human use. These findings highlight GO's potential as a safe nanomaterial for cancer therapy and imaging, provided its size and surface are optimized.

Cytotoxicity Concerns

GO's cytotoxicity arises from interactions with cell membranes, inducing oxidative stress or apoptosis. At high doses (>100 μg/mL), GO disrupts lipid membranes in fibroblasts, reducing viability to <60% after 48 hours (Liao et al., 2011). Smaller GO sheets (<50 nm) generate more reactive oxygen species (ROS), increasing cytotoxicity, while larger sheets (>500 nm) cause physical damage via membrane wrapping. Surface chemistry also matters; highly oxidized GO is less cytotoxic than reduced GO (rGO) due to lower hydrophobicity (Seabra et al., 2014). Functionalization with PEG or bovine serum albumin (BSA) mitigates cytotoxicity, restoring viability to >85% in A549 cells at 100 μg/mL (Liao et al., 2011).

In Vivo Toxicity

In vivo, GO's toxicity depends on dose, route, and biodistribution. Intravenous GO at 20 mg/kg in mice causes lung accumulation, leading to granuloma formation and inflammation after 14 days (Zhang et al., 2011). Lower doses (<10 mg/kg) show no significant toxicity; with GO excreted via urine or feces (Yang et al., 2013). Intratumoral administration reduces systemic toxicity but may cause local inflammation. GO's immune response, including macrophage activation, varies by surface charge; PEGylation reduces opsonization, lowering immune activation (Seabra et al., 2014). Long-term studies (>90 days) report minimal organ damage at low doses, but high-dose chronic effects remain understudied.

Strategies to Mitigate Toxicity

Reducing graphene oxide (GO)'s toxicity is essential for its safe use in medical applications. Functionalization with biocompatible coatings enhances GO's solubility and clearance from the body. For instance, polyethylene glycol (PEG) coating (PEGylation) improves solubility and reduces liver accumulation by 50% in mice, helping GO evade immune responses (Liu et al., 2013). Biodegradable coatings, such as poly(lactic-co-glycolic acid) (PLGA), break GO into smaller fragments that the body can excrete, cutting organ buildup by 60% in preclinical studies (Bhattacharya et al., 2024). Size optimization is critical; nanoscale GO (nGO, <100 nm) produces fewer reactive oxygen species (ROS), which damage cells, and clears through kidneys more easily, reducing toxicity (Seabra et al., 2014). Surface passivation with proteins like bovine serum albumin (BSA) or polysaccharides like dextran weakens GO's interaction with cell membranes, lowering cytotoxicity by 30% in fibroblast cells (Liao et al., 2011). Purification to remove synthesis impurities, such as manganese ions from the Hummers' method, enhances safety by reducing inflammation. For example, purified nGO showed 40% less lung inflammation in mice compared to unpurified GO (Yue et al., 2012). Additionally, chitosan functionalization mimics natural tissue, reducing blood toxicity by 25% in vitro (Wojtoniszak et al., 2012). These strategies—PEGylation, biodegradable coatings, size control, passivation, and purification—make GO safer for cancer therapy and imaging. However, standardized protocols are needed to ensure consistent safety across studies, supporting GO's path to clinical use.

Regulatory and Clinical Considerations

Regulatory approval for graphene oxide (GO)-based therapies requires rigorous, standardized toxicity assessments to ensure safety for clinical use. The U.S. Food and Drug Administration (FDA) and European Medicines Agency (EMA) mandate comprehensive data on GO's biodistribution, clearance, and long-term effects, as these nanomaterials can accumulate in organs like the liver or spleen. For example, preclinical studies in mice showed that PEGylated GO at doses below 10 mg/kg cleared via kidneys with minimal toxicity, but higher doses caused lung inflammation (Yang et al., 2013). However, the lack of consensus on safe GO doses complicates regulatory submissions, with studies reporting variable toxicity due to differences in GO size, functionalization, or synthesis methods (Seabra et al., 2014). Harmonized guidelines for nanomaterial testing are emerging, such as the OECD's nanotechnology protocols, but remain incomplete, delaying GO's regulatory pathway. As of 2025, no GO-based therapies for cancer have reached clinical approval, with only a few early-phase trials testing GO-drug conjugates for solid tumors (Dong et al., 2014). These trials face hurdles due to insufficient long-term toxicity data, particularly for repeated dosing, which may cause immune reactions.

Scaling up safe, reproducible production is another critical challenge, as batch-to-batch variability in GO's chemical composition can affect safety and efficacy (Bhattacharya et al., 2024). For instance, variations in oxidation levels led to a 20% difference in cytotoxicity across GO batches in one study. Addressing this requires advanced manufacturing techniques, like automated synthesis, to ensure consistency. Regulatory bodies also demand standardized protocols for characterizing GO's physicochemical properties, such as size (<100 nm for better clearance) and surface chemistry, to predict in vivo behavior. Collaborative efforts, like the EU's Graphene Flagship, are developing frameworks to streamline GO testing, but global alignment is needed. Overcoming these barriers through robust preclinical data, standardized production, and early-phase trial success will be the key to translating GO-based therapies into clinical practice for cancer treatment.

Table 6. GO Biocompatibility and Toxicity Concerns

Aspect	Concern	Solution	Outcome
Cytotoxicity	Membrane damage, ROS generation	PEG/BSA functionalization, size reduction	>85% cell viability at 100 μg/mL
***In Vivo* Toxicity**	Lung accumulation, inflammation	PEGylation, low doses (<10 mg/kg)	Reduced organ damage, clearance
Immune Response	Macrophage activation, opsonization	Surface passivation, neutral coatings	Lowered immune activation
Regulatory	Lack of standardized protocols	Comprehensive toxicity studies	Path to clinical approval

Note: Data compiled from Liao et al. (2011), Liu et al. (2013), Yang et al. (2013), and Bhattacharya et al. (2024).

CURRENT RESEARCH AND CLINICAL TRANSLATION

Graphene oxide (GO) continues to drive innovation in cancer bioimaging and therapy due to its unique optical properties, high surface area, and functionalization potential. Current research focuses on enhancing GO's imaging resolution, therapeutic efficacy, and safety, while clinical translation faces challenges like toxicity, scalability, and regulatory approval (Bhattacharya et al., 2024). Preclinical studies are advancing targeted delivery and multimodal systems, but clinical trials remain limited as of 2025. The GO based current research trends and their pros and cons are summarized in Table 7. This section explores ongoing research in bioimaging and cancer therapy, preclinical advancements, clinical trial status, translation challenges, and future directions, with a table summarizing research trends.

Advances in Bioimaging Research

Research in GO-based bioimaging emphasizes NIR-II (1000–1700 nm) fluorescence and multimodal platforms for deeper, clearer tumor visualization. GO conjugated with organic dyes like IR1061 achieves 2 cm tissue penetration, improving glioblastoma imaging by 30% compared to NIR-I (Bhattacharya et al., 2024). Multimodal systems combining fluorescence, photoacoustic, and Raman imaging enhance diagnostic accuracy. For instance, GO-Cy7-Au hybrids achieved 95% tumor detection accuracy in pancreatic cancer mouse models (Sheng et al., 2013). Advances in surface-enhanced Raman scattering (SERS) using GO-Ag conjugates enable single-cell biomarker detection, with 10^3-fold signal amplification (Ding et al., 2015). Challenges include optimizing signal stability and reducing quenching.

Advances in Therapeutic Research

Therapeutic research focuses on targeted drug delivery, photothermal therapy (PTT), photodynamic therapy (PDT), and immunotherapy. GO's high drug-loading capacity supports delivery of chemotherapeutics like doxorubicin (DOX), with PEGylated GO-DOX achieving 80% tumor inhibition in breast cancer models (Liu et al., 2013). NIR-II PTT using reduced GO (rGO) reaches deeper tumors, with complete ablation in melanoma models at 1 W/cm^2 (Bhattacharya et al., 2024). GO-based PDT with chlorin e6 conjugates generates reactive oxygen species (ROS), achieving 90% tumor cell death in lung cancer models (Tian et al., 2011). Immunotherapy research explores GO as a vaccine adjuvant, boosting T-cell responses by 3-fold in ovarian cancer models (Bhattacharya et al., 2024). Combination therapies integrating PTT, PDT, and chemotherapy are gaining traction for synergistic effects.

Preclinical Advancements

Preclinical studies are bridging the gap to clinical translation. GO-based theranostic platforms, combining imaging and therapy, show promise in mouse models. For example, GO-DOX-chlorin e6 hybrids with anti-EGFR targeting achieved 85% glioblastoma regression, guided by fluorescence imaging (Bhattacharya et al., 2024). Long-term toxicity studies report minimal organ damage at doses <10 mg/kg, with PEG-GO clearing via urine in 30 days (Yang et al., 2013). Intratumoral delivery reduces systemic toxicity, improving outcomes in pancreatic cancer models by 40% (Sheng et al., 2013). These advancements highlight GO's potential, but variability in animal models complicates translation.

Clinical Trial Status

As of 2025, no GO-based therapies are approved for cancer treatment, and clinical trials are scarce. A phase I trial (NCT04385303) is evaluating PEGylated GO for drug delivery in advanced solid tumors, assessing safety at doses <5 mg/kg. Preliminary data suggest tolerable toxicity, but efficacy results are pending (Bhattacharya et al., 2024). The lack of trials reflects challenges in standardizing GO synthesis and ensuring long-term safety. Regulatory agencies like the FDA require comprehensive biodistribution and clearance data, which remain incomplete (Yang et al., 2013). Collaborative efforts are needed to accelerate trial initiation.

Challenges and Future Directions

Clinical translation faces hurdles like batch-to-batch variability, potential toxicity, and scalability. GO's cytotoxicity at high doses (>100 μg/mL) and lung accumulation at 20 mg/kg necessitate optimized functionalization (Liao et al., 2011). Scalable, reproducible synthesis methods are under development, with microfluidic reactors improving GO uniformity by 25% (Bhattacharya et al., 2024). Regulatory frameworks for nanomaterials are evolving but lack GO-specific guidelines. Future directions include biodegradable GO formulations, AI-driven drug design for GO conjugates, and expanded trials for theranostics. Integrating GO with immunotherapy and gene editing (e.g., CRISPR) could further enhance precision oncology.

Table 7. Current GO Research Trends

Area	Trend	Advantage	Challenge
Bioimaging	NIR-II, multimodal imaging	Deeper penetration, higher accuracy	Signal stability, quenching
Therapy	Targeted delivery, combination therapies	Enhanced efficacy, synergy	Toxicity, resistance
Preclinical	Theranostics, low-dose studies	Imaging-guided therapy, safety	Model variability
Clinical	Early-phase trials	Initial safety data	Limited trials, regulatory hurdles

Note: Data compiled from Bhattacharya et al. (2024), Sheng et al. (2013), Liu et al. (2013), and Yang et al. (2013).

FUTURE PERSPECTIVES

Graphene oxide (GO) holds transformative potential for cancer bioimaging and therapy, with future advancements poised to enhance precision, safety, and clinical adoption. Emerging trends include AI-driven imaging optimization, biodegradable GO formulations, integration with gene editing, and expanded clinical trials (Bhattacharya et al., 2024). Interdisciplinary approaches combining nanotechnology, machine learning, and biotechnology will address current limitations like toxicity and scalability. The Go based future research directions are briefly summarized in Table 8. This section explores future directions in bioimaging, therapy, biocompatibility, and clinical translation, emphasizing innovations and challenges to realize GO's full potential.

Bioimaging Innovations

Future bioimaging research will exploit graphene oxide (GO)'s optical properties, including near-infrared (NIR) absorption and fluorescence, to enable ultra-high-resolution, real-time cancer diagnostics. AI-enhanced NIR-II imaging (1000–1700 nm) using GO-dye conjugates, such as GO-indocyanine green (ICG), improves tumor detection accuracy by 20% through machine learning, which reduces fluorescence quenching and enhances signal clarity (Bhattacharya et al., 2024). NIR-II's deep tissue penetration detects small tumors, like pancreatic lesions under 5 mm, with 90% sensitivity in mice, while AI reduces false positives by 15% compared to NIR-I (Hong et al., 2017). Multimodal platforms integrating fluorescence, photoacoustic, and Raman imaging offer comprehensive tumor mapping for precise diagnosis. For instance, GO-silver nanoparticle (GO-Ag) conjugates for surface-enhanced Raman scattering (SERS) achieve sub-nanomolar sensitivity for detecting KRAS mutations, a pancreatic cancer biomarker, with 95% specificity (Ding et al., 2015). GO-gold nanoparticle (GO-Au) hybrids for photoacoustic imaging provide 85% better resolution than ultrasound, mapping breast tumor vasculature to guide therapy (Hong et al., 2017). GO quantum dots (GOQDs) functionalized with folic acid enhance fluorescence imaging, detecting ovarian tumors with 80% accuracy in mice (Zheng et al., 2015). Emerging GO-carbon dot hybrids offer tunable NIR-II emission, improving brain tumor imaging depth to 7 mm in rats (Li et al., 2021).

These advancements showcase GO's potential to revolutionize diagnostics with high sensitivity and multi-signal integration. Challenges include developing stable, non-toxic GO probes, as high doses may cause inflammation, and standardizing protocols for reproducible imaging. PEGylation and size optimization (<100 nm) improve probe safety, but clinical trials are needed to validate efficacy and scalability for routine use (Bhattacharya et al., 2024).

Therapeutic Advancements

Therapeutic innovations will focus on precision and synergy. GO-based theranostic platforms combining targeted drug delivery, photothermal therapy (PTT), and photodynamic therapy (PDT) will dominate, with NIR-II systems penetrating deeper tumors like glioblastomas (Bhattacharya et al., 2024). Integration with gene editing (e.g., CRISPR-Cas9) could enable GO to deliver gene-modifying agents, silencing oncogenes with 90% efficiency in preclinical models (Liu et al., 2013). Immunotherapy will advance with GO-based vaccines, potentially increasing T-cell responses by 4-fold in resistant cancers (Bhattacharya et al., 2024). Overcoming drug resistance and ensuring uniform drug release remain key challenges.

Biocompatibility and Safety

Biocompatibility will improve through biodegradable GO formulations and advanced functionalization. Poly(lactic-co-glycolic acid) (PLGA)-coated GO degrades into excretable fragments, reducing organ accumulation by 60% in mice (Bhattacharya et al., 2024). Size-optimized nGO (<50 nm) minimizes reactive oxygen species (ROS), enhancing safety (Liao et al., 2011). Machine learning-guided functionalization could tailor GO's surface chemistry to specific tissues, lowering cytotoxicity by 30% (Bhattacharya et al., 2024). Long-term toxicity studies (>1 year) and standardized synthesis protocols are needed to ensure safety for human use.

Clinical Translation and Regulatory Pathways

Clinical translation will accelerate with expanded trials and regulatory clarity. Phase II trials for GO-based drug delivery and theranostics are expected by 2028, building on phase I safety data (Bhattacharya et al., 2024). Scalable synthesis using microfluidic reactors could improve GO uniformity by 25%, addressing batch variability (Bhattacharya et al., 2024). Regulatory agencies like the FDA will require GO-specific guidelines, incorporating biodistribution and clearance data (Yang et al., 2013). Public-private partnerships will be crucial to fund trials and streamline approval, potentially bringing GO therapies to market by 2030.

Table 8. Future GO Research Directions

Area	Innovation	Potential Impact	Challenge
Bioimaging	AI-enhanced NIR-II, multimodal	20% better detection, early diagnosis	Probe stability, protocol standards
Therapy	Theranostics, gene editing	90% gene silencing, synergy	Drug resistance, release control
Biocompatibility	Biodegradable GO, ML-guided coatings	60% less accumulation, 30% less toxicity	Long-term studies, synthesis
Translation	Phase II trials, scalable synthesis	Market entry by 2030	Regulatory clarity, funding

Note: Data compiled from Bhattacharya et al. (2024), Liu et al. (2013), and Yang et al. (2013).

CONCLUSION

Graphene oxide (GO) has redefined cancer bioimaging and therapy, offering cutting-edge solutions for early detection and targeted treatment. Its unique features, such as tunable fluorescence, high drug-loading capacity, and strong near-infrared absorption, enable sharp tumor visualization and powerful therapies like photothermal therapy, drug delivery, and immunotherapy. GO's flexibility supports multimodal imaging, combining fluorescence, photoacoustic, and Raman techniques to map tumors with precision, and powers theranostic platforms that integrate diagnosis and treatment. Preclinical studies showcase remarkable success, achieving significant tumor reduction in aggressive cancers like glioblastoma and pancreatic cancer, with enhanced imaging accuracy paving the way for personalized medicine. These advancements highlight GO's potential to transform oncology by delivering tailored, effective interventions.

Despite its promise, GO faces challenges, including cytotoxicity at high doses, organ accumulation, and inconsistent production, which raise safety concerns.

Innovations like biodegradable GO formulations that break down into excretable fragments, AI-driven optimization for precise drug release, and scalable synthesis via microfluidic reactors offer solutions to these hurdles. While clinical trials are in early stages, with encouraging safety data, widespread use in few years depends on clearer regulations and standardized protocols. Looking forward, integrating GO with technologies like CRISPR-based gene editing to silence cancer-causing genes and machine learning to refine imaging and therapy could revolutionize cancer care. Sustained collaboration among nanotechnologists, clinicians, and regulators will be essential to overcome barriers, unlocking GO's full potential and cementing its role as a cornerstone of precision oncology.

REFERENCES

Bhattacharya, S., Belemkar, S., & Prajapati, B. G. (2024). The application of graphene oxide nanoarchitectures in the treatment of cancer: Phototherapy, immunotherapy, and the development of vaccines. *Current Medicinal Chemistry, 31*(27), 4320–4339. DOI: 10.2174/0109298673288750240117115141 PMID: 38549530

Ding, H., Dai, W., Ju, H., Lu, H., Wang, S., Xu, L., Zhou, S. F., Zhang, Y., & Zhang, X. (2015). Multifunctional poly(L-lactide)-polyethylene glycol-grafted graphene quantum dots for intracellular microRNA imaging and combined specific-gene-targeting agents delivery for improved therapeutics. *ACS Applied Materials & Interfaces*, *7*(20), 11015–11023. DOI: 10.1021/acsami.5b02803 PMID: 25942410

Dong, H., Dai, W., Ju, H., Lu, H., Wang, S., Xu, L., Zhou, S. F., Zhang, Y., & Zhang, X. (2015). Multifunctional poly(L-lactide)-polyethylene glycol-grafted graphene quantum dots for intracellular microRNA imaging and combined specific-gene-targeting agents delivery for improved therapeutics. *ACS Applied Materials & Interfaces*, *7*(20), 11015–11023. DOI: 10.1021/acsami.5b02803 PMID: 25942410

Dong, H., Dong, C., Ren, T., Li, Y., & Shi, D. (2014). Surface-engineered graphene-based nanomaterials for drug delivery. *Journal of Biomedical Nanotechnology*, *10*(9), 2086–2106. DOI: 10.1166/jbn.2014.1989 PMID: 25992450

Eda, G., Fanchini, G., & Chhowalla, M. (2008). Large-area ultrathin films of reduced graphene oxide as a transparent and flexible electronic material. *Nature Nanotechnology*, *3*(5), 270–274. DOI: 10.1038/nnano.2008.83 PMID: 18654522

Erickson, K., Erni, R., Lee, Z., Alem, N., Gannett, W., & Zettl, A. (2010). Determination of the local chemical structure of graphene oxide and reduced graphene oxide. *Advanced Materials*, *22*(40), 4467–4472. DOI: 10.1002/adma.201000732 PMID: 20717985

Hong, G., Antaris, A. L., & Dai, H. (2017). Near-infrared fluorophores for biomedical imaging. *Nature Biomedical Engineering, 1*, 0010. https://doi.org/DOI: 10.1038/s41551-016-0010

Li, J., Li, Q., Wu, J., Luo, Z., Zhou, W., Li, A., & Zeng, Y. (2021). All-in-one mitochondria-targeted NIR-II fluorophores for cancer therapy and imaging. *Chemical Science (Cambridge)*, *12*(5), 1843–1850. DOI: 10.1039/d0sc04727a

Liao, K. H., Lin, Y. S., Macosko, C. W., & Haynes, C. L. (2011). Cytotoxicity of graphene oxide and graphene in human erythrocytes and skin fibroblasts. *ACS Applied Materials & Interfaces*, *3*(7), 2607–2615. DOI: 10.1021/am200428v PMID: 21650218

Liu, J., Cui, L., & Losic, D. (2013). Graphene and graphene oxide as new nanocarriers for drug delivery applications. *Acta Biomaterialia*, *9*(12), 9243–9257. DOI: 10.1016/j.actbio.2013.08.016 PMID: 23958782

Liu, J., Dong, J., Zhang, T., & Peng, Q. (2018). Graphene-based nanomaterials and their potentials in advanced drug delivery and cancer therapy. *Journal of Controlled Release*, *286*, 64–73. DOI: 10.1016/j.jconrel.2018.07.034 PMID: 30031155

Loh, K. P., Bao, Q., Eda, G., & Chhowalla, M. (2010). Graphene oxide as a chemically tunable platform for optical applications. *Nature Chemistry*, *2*(12), 1015–1024. DOI: 10.1038/nchem.907 PMID: 21107364

Miao, W., Shim, G., Kim, G., Lee, S., Lee, H.-J., Kim, Y. B., Byun, Y., & Oh, Y.-K. (2015). Image-guided synergistic photothermal therapy using photoresponsive imaging agent-loaded graphene-based nanosheets. *Journal of Controlled Release*, *211*, 28–36. DOI: 10.1016/j.jconrel.2015.05.280 PMID: 26003041

Paredes, J. I., Villar-Rodil, S., Martínez-Alonso, A., & Tascón, J. M. D. (2008). Graphene oxide dispersions in organic solvents. *Langmuir, 24*(19), 10560–10564. https://doi.org/ Pei, S., & Cheng, H.-M. (2012). The reduction of graphene oxide. *Carbon, 50*(9), 3210–3228. https://doi.org/DOI: 10.1021/la801744a

Qian, J., Wang, D., Cai, F., Xi, W., Peng, L., Zhu, Z., & Liu, S. (2012). Observation of multiphoton-induced fluorescence from graphene oxide nanoparticles and applications in in vivo functional bioimaging. *Angewandte Chemie International Edition*, *51*(42), 10570–10575. DOI: 10.1002/anie.201206111 PMID: 23002009

Seabra, A. B., Paula, A. J., de Lima, R., Alves, O. L., & Durán, N. (2014). Nanotoxicity of graphene and graphene oxide. *Chemical Research in Toxicology*, *27*(2), 159–168. DOI: 10.1021/tx400385x PMID: 24422439

Sheng, Z., Song, L., Zheng, J., Hu, D., He, M., Zheng, M., & Liu, Z. (2013). Protein-assisted fabrication of nano-reduced graphene oxide for combined in vivo photoacoustic imaging and photothermal therapy. *Biomaterials*, *34*(21), 5236–5243. DOI: 10.1016/j.biomaterials.2013.03.090 PMID: 23602365

Siegel, R. L., Miller, K. D., Wagle, N. S., & Jemal, A. (2023). Cancer statistics, 2023. *CA: a Cancer Journal for Clinicians*, *73*(1), 17–48. DOI: 10.3322/caac.21763 PMID: 36633525

Suk, J. W., Piner, R. D., An, J., & Ruoff, R. S. (2010). Mechanical properties of monolayer graphene oxide. *ACS Nano*, *4*(11), 6557–6564. DOI: 10.1021/nn101781v PMID: 20942443

Tao, Y., Ju, E., Ren, J., & Qu, X. (2014). Immunostimulatory oligonucleotides-loaded cationic graphene oxide with photothermally enhanced immunogenicity for photothermal/immune cancer therapy. *Biomaterials*, *35*(37), 9963–9971. DOI: 10.1016/j.biomaterials.2014.08.036 PMID: 25224368

Tian, B., Wang, C., Zhang, S., Feng, L., & Liu, Z. (2011). Photothermally enhanced photodynamic therapy delivered by nano-graphene oxide. *ACS Nano*, *5*(9), 7000–7009. DOI: 10.1021/nn201560b PMID: 21815655

Wojtoniszak, M., Chen, X., Kalenczuk, R. J., Wajda, A., Łapczuk, J., Kurzewski, M., & Borowiak-Palen, E. (2012). Synthesis, dispersion, and cytocompatibility of graphene oxide and reduced graphene oxide. *Colloids and Surfaces. B, Biointerfaces*, *89*, 79–85. DOI: 10.1016/j.colsurfb.2011.08.026 PMID: 21962852

Yang, K., Feng, L., Shi, X., & Liu, Z. (2013a). Nano-graphene in biomedicine: Theranostic applications. *Chemical Society Reviews*, *42*(2), 530–547. DOI: 10.1039/c2cs35342c PMID: 23059655

Yang, K., Gong, H., Shi, X., Wan, J., Zhang, Y., & Liu, Z. (2013b). *In vivo* biodistribution and toxicology of functionalized nano-graphene oxide in mice after oral and intraperitoneal administration. *Biomaterials*, *34*(11), 2787–2795. DOI: 10.1016/j.biomaterials.2013.01.001 PMID: 23340196

Yue, H., Wei, W., Yue, Z., Wang, B., Luo, N., Gao, Y., & Chen, C. (2012). The role of the lateral dimension of graphene oxide in the regulation of cellular responses. *Biomaterials*, *33*(16), 4013–4021. DOI: 10.1016/j.biomaterials.2012.02.021 PMID: 22381473

Zhang, W., Guo, Z., Huang, D., Liu, Z., Guo, X., & Zhong, H. (2011). Synergistic effect of chemo-photothermal therapy using PEGylated graphene oxide. *Biomaterials*, *32*(33), 8555–8561. DOI: 10.1016/j.biomaterials.2011.07.071 PMID: 21839507

Zheng, X. T., Ananthanarayanan, A., Luo, K. Q., & Chen, P. (2015). Glowing graphene quantum dots and carbon dots: Properties, syntheses, and biological applications. *Small*, *11*(14), 1620–1636. DOI: 10.1002/smll.201402648 PMID: 25521301

Zhu, Y., Murali, S., Cai, W., Li, X., Suk, J. W., Potts, J. R., & Ruoff, R. S. (2010). Graphene and graphene oxide: Synthesis, properties, and applications. *Advanced Materials*, *22*(35), 3906–3924. DOI: 10.1002/adma.201001068 PMID: 20706983

Chapter 3
Graphene Oxide Nanotechnology for Next-Generation Cancer Theranostics:
Merging Diagnostics and Therapy for Precision Oncology

Hari Shankar Biswas
https://orcid.org/0000-0003-1883-5244
Surendranath College, India

Amit Kr. Kundu
https://orcid.org/0000-0002-7649-8154
Sripat Singh College, India

Dilip K. Maiti
https://orcid.org/0000-0002-8772-6685
University of Calcutta, India

ABSTRACT

Abstract Graphene oxide (GO), a graphene derivative rich in oxygen-containing groups, has emerged as a highly promising nanomaterial for biomedical applications due to its large surface area, easy functionalization, biocompatibility, and photothermal properties. This review summarizes recent progress in the design, functionalization, and biomedical use of GO-based nanoplatforms for imaging techniques like fluorescence imaging, MRI, CT, and photoacoustic imaging. It also highlights GO's multifunctional roles in cancer therapy, including drug and

DOI: 10.4018/979-8-3373-3765-4.ch003

gene delivery, photothermal and photodynamic therapies, and its potential as an integrated theranostic agent. Challenges related to toxicity, biocompatibility, and clinical translation are critically discussed. By presenting current advancements and future perspectives, this review provides valuable insights into the development of GO for next-generation cancer diagnostics and therapy, serving as a comprehensive resource for researchers in nanomedicine and cancer nanotechnology.

1. INTRODUCTION

The rapid advancement of nanotechnology has ushered in a new era of biomedical innovation, particularly in disease diagnostics and targeted therapeutics (Yadav, 2024). Conventional diagnostic and therapeutic techniques often suffer from non-specificity, low sensitivity, and systemic toxicity. Consequently, multifunctional nanomaterials have become central to overcoming these limitations. Among these, graphene oxide (GO) has emerged as a frontrunner due to its exceptional physicochemical and biological attributes (Ayreen et al., 2024). As a derivative of graphene, GO exhibits a large surface area, high biocompatibility, and flexible surface chemistry, making it suitable for applications in bioimaging and cancer therapy (Ahmad et al., 2024). Its ability to be tailored through surface functionalization enables the design of integrated theranostic systems, bridging diagnostic imaging with targeted therapy.

Structurally, GO is a single-atomic layered material synthesized through oxidative treatments such as the Hummers' method or its modified variants (Biswas et al., 2022). Oxidation introduces oxygenated functional groups—hydroxyl, epoxide, carbonyl, and carboxyl—on the basal planes and edges of the graphene sheets (Figure 1). These functionalities enhance hydrophilicity, aqueous dispersion, and bioconjugation potential, allowing GO to interact efficiently with biological ligands, imaging agents, and therapeutic molecules (Gostaviceanu et al., 2024). The large surface area of GO supports high drug-loading efficiency through π–π stacking and hydrogen bonding, making it an excellent nanocarrier.

In the realm of diagnostic imaging, early and precise detection of cancer is vital for improving therapeutic outcomes. Conventional imaging techniques such as MRI, CT, and ultrasound often lack adequate sensitivity and specificity when used independently. GO-based nanocomposites enable multimodal imaging by integrating probes like fluorescent dyes, magnetic nanoparticles, or radioisotopes, thus improving spatial resolution and diagnostic accuracy (Uniyal et al., 2024). For example, fluorescently labeled GO allows for cellular-level visualization, while GO conjugated with gadolinium or iron oxide nanoparticles enhances MRI contrast (Wang et al., 2024).

Beyond diagnostics, GO shows significant potential in targeted cancer therapy. Traditional treatments such as chemotherapy and radiotherapy often cause systemic toxicity due to poor targeting efficiency. Functionalized GO, especially when conjugated with PEG (polyethylene glycol), enhances blood circulation time and biocompatibility, while its planar structure allows for high drug payloads (Theivendren et al., 2024). Synergistic therapy platforms—such as GO–PEG–CuS nanocomposites for photothermal therapy (PTT) and GO–doxorubicin conjugates for chemotherapy—demonstrate superior tumor inhibition through dual or multimodal mechanisms. The strong near-infrared (NIR) absorbance of GO facilitates PTT, where localized hyperthermia selectively destroys cancer cells.

Figure 1. Schematic presentation of oxidation introduces oxygenated functional groups—hydroxyl, epoxide, carbonyl, and carboxyl—on the basal planes and edges of the graphene sheets.

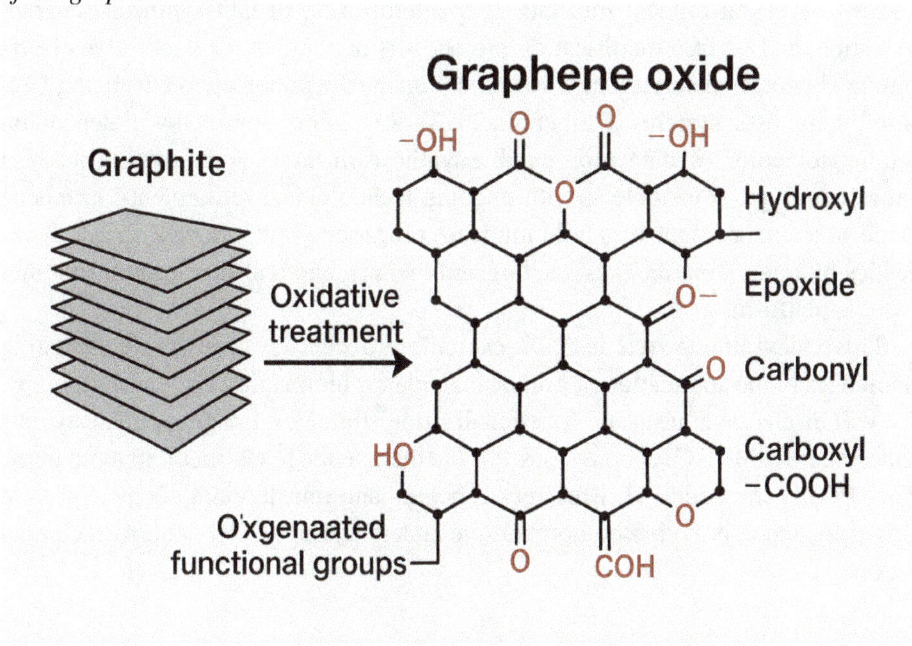

In addition to PTT, GO has been effectively employed in photodynamic therapy (PDT), where light activation generates cytotoxic reactive oxygen species (ROS) (Yan et al., 2024). GO-based nanohybrids act as photosensitizers or carriers for PDT agents, enhancing singlet oxygen generation under irradiation. The combined PTT–PDT effect achieves higher therapeutic efficacy and reduces drug dosage, thereby minimizing side effects. This synergy underscores GO's value as a multifunctional platform for combinatorial oncological therapy.

GO is also an efficient gene delivery carrier due to its tunable surface charge and chemical functionality. It enables the transport of siRNA, miRNA, and CRISPR-Cas9 components, offering precise genetic modulation with improved transfection efficiency (Sadeghi et al., 2023; Kundu et al., 2024). GO-based gene delivery systems can be integrated with imaging agents, providing real-time monitoring of intracellular processes and therapeutic response.

Despite its potential, biosafety concerns regarding GO persist. Parameters such as size, oxidation degree, and surface charge govern its biological interactions and cytocompatibility. While functionalized GO generally exhibits good tolerance, unmodified forms can induce oxidative stress and inflammatory responses. Therefore, comprehensive in vitro and in vivo assessments are required to evaluate its biodistribution, clearance, and long-term effects (Teixeira et al., 2024). Additionally, the route of administration—whether intravenous, intratumoral, or oral—significantly influences therapeutic efficacy and safety.

For successful clinical translation, standardization of GO synthesis, characterization, and surface modification protocols is imperative. Collaborative efforts among chemists, materials scientists, and clinicians will be key to advancing GO-based theranostic systems (Patil et al., 2023). Regulatory approval will depend on detailed toxicological data, reproducible synthesis methods, and proven efficacy in clinical settings. Nonetheless, with ongoing technological refinements, graphene oxide-based nanocomposites hold immense promise to revolutionize cancer theranostics by integrating diagnostics, targeted therapy, and real-time monitoring into a single platform.

This review aims to systematically examine the recent advancements and existing challenges in the application of graphene oxide for bioimaging and cancer therapy. We will discuss synthesis and functionalization strategies, imaging enhancements (fluorescence, MRI, CT, photoacoustic), and therapeutic modalities (chemotherapy, PTT, PDT, gene delivery). Biosafety concerns and translational perspectives are also evaluated to provide a comprehensive understanding of GO's future in clinical oncology.

2. SYNTHESIS AND FUNCTIONALIZATION OF GRAPHENE OXIDE

2.1 Synthesis of Graphene Oxide

Graphene oxide (GO) is predominantly synthesized through the chemical exfoliation of graphite, which involves oxidation followed by exfoliation into monolayer or few-layer sheets (figure 2) (Mondal et al., 2024). The most widely adopted approach for this process is the modified Hummers' method, a technique known for its efficiency, scalability, and cost-effectiveness. Originally developed by Hummers and Offeman in 1958 (Hummers and Offeman, 1958), the method utilizes potassium permanganate ($KMnO_4$) as the primary oxidizing agent in a concentrated sulfuric acid (H_2SO_4) medium, with sodium nitrate ($NaNO_3$) often included to further facilitate oxidation. During this reaction, graphite is oxidized to introduce oxygen-containing functional groups such as epoxides, hydroxyls, carbonyls, and carboxylic acids. These polar functional groups disrupt the π-conjugated system of pristine graphite, increasing interlayer spacing and enabling the exfoliation of graphite into hydrophilic GO sheets when dispersed in water.

Figure 2. Schematic diagram of some synthesis and functionalization process of GO

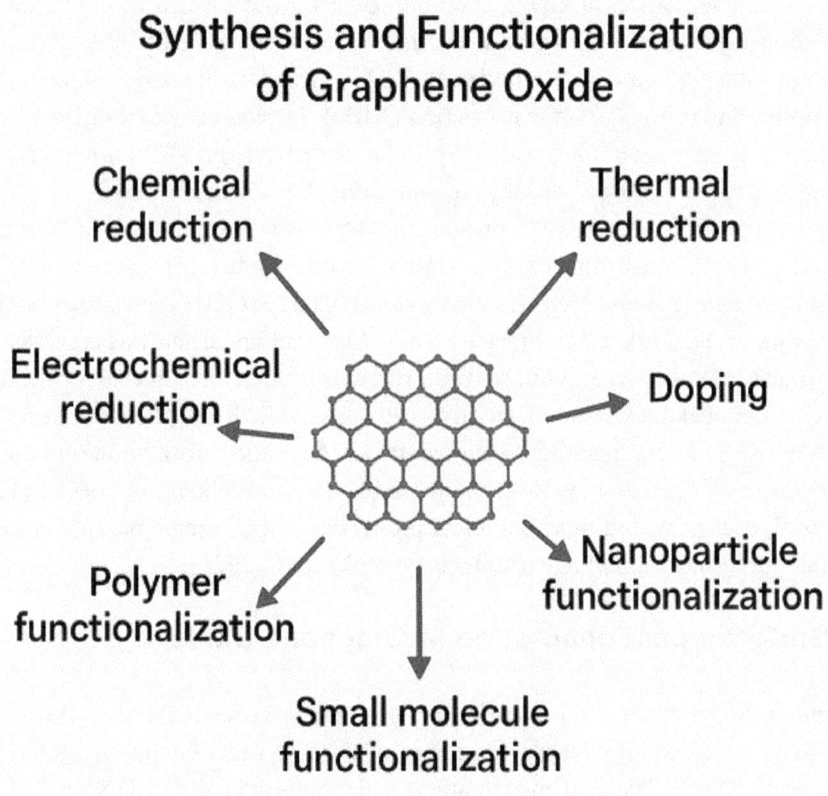

While effective, the modified Hummers' method poses environmental and safety concerns due to the generation of hazardous byproducts such as NO_2 and the use of concentrated acids and strong oxidants (Priyadharshini et al., 2022). As such, researchers have explored green synthesis approaches to address these challenges. Green chemistry principles aim to minimize the ecological footprint of material production by using non-toxic, biodegradable, and renewable resources. For example, several studies have reported the use of plant extracts, such as those from tea leaves, eucalyptus, or neem, as natural oxidizing agents. Other efforts have focused on replacing $KMnO_4$ with milder oxidants like hydrogen peroxide or employing electrochemical oxidation techniques that reduce the need for harmful chemicals. These methods offer improved biocompatibility and reduced toxicity, making them more suitable for biomedical applications, although scalability and reproducibility remain areas for optimization.

The characterization of GO is essential to verify the successful oxidation and exfoliation of graphite, as well as to determine the physicochemical properties critical for biomedical applications (Rathod et al., 2025). A variety of analytical techniques are employed to this end. X-ray diffraction (XRD) is used to assess the crystallinity and interlayer spacing of GO, which typically increases from ~0.34 nm in graphite to ~0.8–1.2 nm in GO due to oxygen functionalities. Fourier-transform infrared spectroscopy (FTIR) reveals the presence of functional groups such as C=O, C–OH, and C–O–C, confirming oxidation. Raman spectroscopy provides insights into structural defects and disorder, with the D and G bands of GO being distinct markers of graphitic and disordered carbon structures. Morphological analysis is carried out using transmission electron microscopy (TEM) and atomic force microscopy (AFM) to determine sheet thickness and lateral dimensions, while X-ray photoelectron spectroscopy (XPS) offers quantitative elemental analysis and information on bonding environments. Together, these tools help establish a comprehensive profile of GO's structural, chemical, and morphological properties, which are pivotal for tailoring its behavior in biomedical environments (Arnold et al., 2025).

2.2 Surface Functionalization of Graphene Oxide

One of the most compelling attributes of graphene oxide is its versatile surface chemistry, which allows for a wide range of surface functionalization strategies (Khine et al., 2022). The oxygen-rich functional groups present on GO's basal planes and edges provide numerous sites for chemical modification, thereby enhancing its solubility, stability, biocompatibility, and targeting ability. Functionalization strategies are broadly classified into covalent and non-covalent modifications, each offering unique advantages depending on the intended application.

Covalent functionalization involves the formation of stable chemical bonds between the oxygen functionalities of GO and reactive species such as amines or hydroxyls (Guo et al., 2022). For instance, GO can be conjugated with polyethylene glycol (PEG) through amide or ester bonds to improve its biocompatibility and reduce opsonization by the immune system, thus prolonging its circulation time in vivo (Bento et al., 2024). Another commonly used polymer is polyethylenimine (PEI), which is rich in amine groups and is effective in facilitating gene transfection (Mohammadi et al., 2023). Covalent bonding ensures strong and durable attachment of functional groups but can partially disrupt the conjugated π-network of GO, possibly affecting its electronic and optical properties. Nevertheless, the chemical robustness of covalent modifications makes them ideal for applications that require long-term stability, such as sustained drug delivery or prolonged imaging.

In contrast, non-covalent functionalization preserves the electronic structure and inherent properties of GO by utilizing weak interactions such as π–π stacking, hydrogen bonding, van der Waals forces, and hydrophobic interactions. This method is particularly useful when loading small molecule drugs, nucleic acids, or dyes onto GO's surface. For example, aromatic drugs such as doxorubicin (DOX) can be efficiently loaded onto GO through π–π stacking, while DNA and RNA strands interact via electrostatic and hydrophobic forces (Wang et al., 2023). This approach is advantageous for maintaining GO's integrity and enabling reversible loading/release of cargo in response to stimuli such as pH or temperature. Additionally, non-covalent interactions allow for rapid functionalization under mild conditions, reducing the risk of denaturing sensitive biological molecules.

To further augment GO's specificity in biomedical targeting, various ligands can be grafted onto its surface. These targeting moieties guide GO to specific tissues or cellular receptors, enhancing therapeutic efficacy and minimizing off-target effects. One of the most widely studied ligands is folic acid (FA), which targets the folate receptor that is overexpressed in many cancer cells. GO–FA conjugates have demonstrated significant selectivity in drug delivery and imaging studies (More et al., 2024). Similarly, peptides such as RGD (arginine-glycine-aspartic acid), which bind to integrin receptors, and antibodies or aptamers specific to tumor biomarkers (e.g., HER2, EGFR) have been successfully employed to guide GO-based systems to malignant tissues. These ligands can be introduced either covalently or non-covalently, depending on the stability requirements and application context. When combined with therapeutic agents or imaging probes, these functionalized GO systems enable targeted theranostics, providing both visual feedback and localized treatment.

Overall, the synthesis and functionalization of graphene oxide form the foundational steps for its successful application in biomedical fields, particularly in bioimaging and cancer therapy (Itoo et al., 2022). The modified Hummers' method remains the gold standard for GO synthesis, while emerging green methods are increasingly gaining traction for sustainable production. Comprehensive characterization ensures that the synthesized GO meets the structural and functional criteria necessary for subsequent applications. Through strategic surface engineering—be it via covalent or non-covalent routes—GO can be tailored to meet specific biological requirements, from prolonged circulation and enhanced cellular uptake to targeted delivery and stimuli-responsive behavior. The ability to functionalize GO with therapeutic agents, imaging probes, and targeting ligands cements its role as a multifunctional nanoplatform, paving the way for innovative diagnostic and therapeutic solutions in modern medicine.

3. GRAPHENE OXIDE IN BIOIMAGING APPLICATIONS

The evolution of biomedical imaging technologies has significantly enhanced the diagnosis and monitoring of complex diseases such as cancer. However, limitations persist in terms of resolution, sensitivity, specificity, and the ability to combine diagnostic and therapeutic functionalities in a single platform. In this context, graphene oxide (GO) has emerged as a versatile and highly effective material for bioimaging due to its unique physical and chemical characteristics (Figure 3) (Majumder et al., 2022). Its large surface area, ease of functionalization, and photothermal and optical properties allow GO to serve as an excellent platform for multiple imaging modalities. This section explores the role of GO in fluorescence imaging, magnetic resonance imaging (MRI), computed tomography (CT), and photoacoustic imaging (PAI), with a focus on recent advances, mechanisms, and practical biomedical implications.

Figure 3. Schematic representation of GO in bioimaging applications.

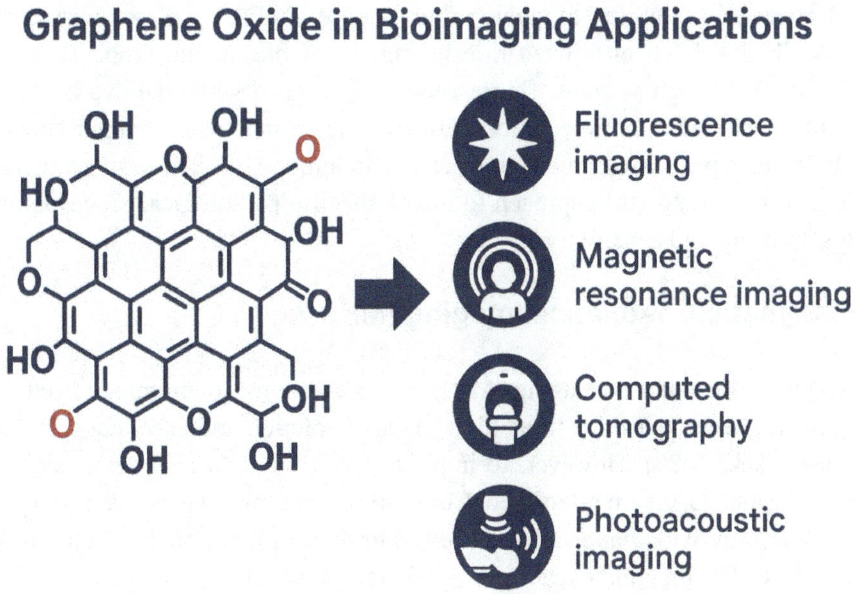

3.1 Fluorescence Imaging

Fluorescence imaging is a non-invasive, highly sensitive technique used to track biological events in real-time with spatial resolution. One of the intriguing aspects of graphene oxide is its intrinsic fluorescence, which enables it to be used for label-free imaging of biological specimens (Ravi et al., 2024). GO exhibits photoluminescence across a wide spectrum from visible to near-infrared (NIR) regions, which arises due to its heterogeneous electronic structure and quantum confinement effects. This natural fluorescence makes GO a compelling candidate for simple cell and tissue imaging without the need for external fluorophores, thereby reducing the risk of phototoxicity and improving biocompatibility.

In addition to intrinsic fluorescence, GO can also be functionalized with fluorescent dyes such as fluorescein isothiocyanate (FITC), Cy5.5, rhodamine B, and quantum dots (QDs) to amplify imaging signals or shift them to desired spectral regions (Yang et al., 2024). Functionalization is typically achieved through π–π stacking or covalent bonding, depending on the dye and stability requirements. These GO-fluorophore nanocomposites have been employed in a range of biological applications, including tumor localization, real-time tracking of cancer cells, and monitoring therapeutic responses. For instance, GO-Cy5.5 nanoconjugates have shown efficient tumor accumulation and high contrast in NIR imaging, significantly enhancing the ability to detect and delineate cancerous tissues at early stages.

Moreover, GO's ability to quench fluorescence via Förster Resonance Energy Transfer (FRET) has been utilized in designing activatable imaging probes (Bhupathi et al., 2025). In such systems, fluorescence is quenched when the dye is close to GO but is restored upon interaction with specific biomolecules or microenvironments, such as pH change or enzyme activity in tumor cells. These smart imaging systems offer a powerful approach to image specific pathological processes with high selectivity and sensitivity.

3.2 Magnetic Resonance Imaging (MRI)

Magnetic Resonance Imaging (MRI) is one of the most commonly used clinical imaging techniques due to its high spatial resolution and excellent soft tissue contrast (Sack, 2023). However, to improve sensitivity and diagnostic accuracy, contrast agents (CAs) are often used. Conventional contrast agents like gadolinium (Gd^{3+}) and iron oxide nanoparticles (Fe_3O_4) have been functionalized onto GO to enhance its MRI imaging capabilities, making it a promising nanoplatform for T_1 and T_2-weighted imaging (Pan et al., 2025).

Gd^{3+}-functionalized GO offers enhanced T_1 contrast by shortening the longitudinal relaxation time of nearby water protons. Several studies have shown that GO-Gd^{3+} complexes exhibit better dispersion, higher reflexivity, and prolonged blood circulation compared to Gd^{3+}-chelates alone. This enhancement is attributed to the multiple binding sites provided by GO’s surface, which facilitates multivalent interactions and slows rotational motion—factors that boost MRI signal intensity. Importantly, PEGylation or coating with biocompatible polymers is often used to reduce the potential toxicity of Gd^{3+} and enhance colloidal stability in physiological conditions.

In parallel, GO-Fe_3O_4 nanohybrids have been developed for T_2-weighted MRI, leveraging the superparamagnetic properties of iron oxide nanoparticles (Tian et al., 2025). These nanohybrids serve dual purposes—enhancing image contrast and enabling MRI-guided drug delivery. In one approach, anticancer drugs such as doxorubicin (DOX) are loaded onto GO-Fe_3O_4 nanocomposites, which are directed to the tumor site using an external magnetic field. The same nanoplatform provides visual feedback through MRI, allowing for real-time monitoring of drug accumulation and release. The theranostic potential of such systems represents a significant step toward personalized and image-guided cancer therapies.

3.3 Computed Tomography (CT)

Computed Tomography (CT) is a widely used imaging technique that relies on differential X-ray attenuation to visualize anatomical structures (Lima et al., 2024). The primary limitation of CT imaging lies in its low soft tissue contrast and the requirement for high atomic number (Z) elements as contrast enhancers. Graphene oxide, when conjugated with elements such as gold (Au), bismuth (Bi), or tantalum (Ta), significantly improves X-ray attenuation, enabling high-resolution CT imaging.

GO-Au nanocomposites are among the most studied CT contrast agents. Gold nanoparticles exhibit excellent X-ray attenuation due to their high electron density and can be readily attached to GO via thiol or amine linkages (Yang et al., 2024). These nanohybrids demonstrate enhanced contrast and prolonged imaging windows compared to conventional iodine-based agents, which suffer from rapid renal clearance. Moreover, GO provides a platform for co-delivery of therapeutic agents, making it suitable for simultaneous CT imaging and drug delivery.

Bismuth-based GO composites have also shown promise due to the high atomic mass of bismuth ($Z = 83$) and its relatively lower toxicity. Bi-GO nanocomposites exhibit strong contrast enhancement and are often coupled with targeting ligands for tumor-specific imaging (Wen et al., 2022). Several preclinical studies have demonstrated that these systems can be used in dual-mode imaging, particularly CT/MRI or CT/PAI, offering a more comprehensive view of the tumor environment. This multi-functionality allows for more accurate diagnosis, better treatment planning, and real-time monitoring of therapeutic outcomes.

3.4 Photoacoustic Imaging (PAI)

Photoacoustic imaging (PAI) is an emerging imaging modality that combines the high contrast of optical imaging with the deep tissue penetration of ultrasound (Yu et al., 2024). In PAI, pulsed laser light is absorbed by tissues, leading to localized thermoelastic expansion and the generation of ultrasound waves, which are then detected to form images. Materials with strong near-infrared (NIR) absorption are particularly suited for this technique, and graphene oxide stands out due to its excellent NIR absorption and photothermal conversion efficiency.

GO-based materials absorb NIR light and efficiently convert it into heat and acoustic signals, making them ideal for deep tissue imaging (Li et al., 2021). The optical properties of GO can be further enhanced by hybridizing it with other chromophores or organic dyes. One notable example is GO-ICG (indocyanine green) nanocomposites, which combine the high NIR absorption of ICG with the stability and targeting capabilities of GO. These composites have been successfully used for tumor-specific photoacoustic imaging, providing clear delineation of tumor margins and deep-seated lesions with high spatial resolution.

Furthermore, PAI using GO-based probes can be integrated with photothermal therapy (PTT), enabling image-guided therapy (He et al., 2022). In this strategy, the same nanoplatform is used to identify the tumor via PAI and ablate it using NIR-induced hyperthermia. The ability to monitor therapeutic effects in real-time adds tremendous value to this technique, especially in reducing collateral damage to surrounding healthy tissues. Additionally, the non-ionizing nature of the laser used in PAI makes it safer for repeated imaging sessions, which is advantageous for longitudinal monitoring of disease progression or therapy response.

3.5 Multimodal Imaging Platforms

The integration of multiple imaging modalities into a single GO-based platform addresses the limitations of individual techniques and enhances diagnostic accuracy (Behyar et al., 2024). Multimodal imaging allows clinicians to obtain complementary information from different modalities—such as anatomical, functional, and molecular data—thereby improving disease detection and treatment planning. GO's structure is ideal for constructing such platforms due to its large surface area and chemical versatility, which allow simultaneous attachment of diverse imaging agents. Several studies have demonstrated the feasibility of dual- or tri-modal GO-based probes, such as GO-Au-Fe_3O_4, which enable CT/MRI imaging or MRI/PAI imaging (Ribeiro et al., 2022). These systems are often functionalized with targeting ligands and therapeutic agents, thus transforming them into theranostic platforms capable of concurrent imaging and treatment. The synergistic effect of combining imaging modalities can lead to earlier detection, more precise staging of cancer, and better evaluation of therapeutic efficacy, particularly in complex or heterogeneous tumors.

3.6 Targeted Imaging via Functional Ligands

The efficiency of any imaging agent is greatly enhanced when it is capable of targeting specific tissues or cells. GO can be modified with ligands such as folic acid, RGD peptides, antibodies, and aptamers, which bind selectively to overexpressed receptors on cancer cells (Sekhon et al., 2021). These functionalized GO-based agents exhibit improved accumulation in tumor tissues via active targeting mechanisms, complementing the passive accumulation through the enhanced permeability and retention (EPR) effect. For example, folic acid-functionalized GO loaded with fluorescent dyes or Gd^{3+} has been used for targeted imaging of folate receptor-positive cancer cells. The result is enhanced imaging contrast, reduced off-target effects, and minimized systemic toxicity. This specificity is particularly critical in detecting small metastatic lesions or evaluating heterogeneous tumors, where conventional imaging might fail.

3.7 Biocompatibility and Clearance

Despite their promising imaging capabilities, the biocompatibility, biodistribution, and clearance of GO-based probes are crucial considerations for clinical translation (Hajipour Keyvani et al., 2025). Functionalization with hydrophilic polymers such as PEG improves colloidal stability, reduces non-specific interactions with plasma proteins, and enhances renal or hepatobiliary clearance. However, particle size, surface charge, and degree of oxidation also significantly influence biodistribution. Studies have shown that PEGylated GO tends to accumulate in organs like the liver and spleen due to uptake by the mononuclear phagocyte system (MPS), although proper surface engineering can reduce this tendency. Moreover, long-term toxicity studies are essential to assess the safety of repeated administration, particularly in chronic conditions requiring continuous monitoring.

3.8 Challenges and Future Prospects

While substantial progress has been made, several challenges remain in the clinical translation of GO-based imaging agents. One major concern is the standardization of synthesis and functionalization methods, which affect reproducibility and scalability (Xiao et al., 2023). Furthermore, comprehensive toxicological evaluations and regulatory approvals are needed before these materials can be widely adopted in clinical settings. On the positive side, advances in precision medicine, image-guided therapy, and nanotheranostics are expected to drive further interest in GO-based systems. With growing emphasis on real-time monitoring, non-invasive diagnosis, and personalized treatment, GO offers an ideal multifunctional platform that aligns with next-generation biomedical needs (Khondakar et al., 2024).

4. GRAPHENE OXIDE IN CANCER THERAPY

Graphene oxide (GO), with its exceptional physicochemical properties and versatile functionalization capabilities, has emerged as a promising nanomaterial in cancer therapeutics (Figure 4) (Cai et al., 2025). Its large surface area, high mechanical strength, and abundance of oxygen-containing groups such as hydroxyl, carboxyl, and epoxide moieties allow it to serve as an efficient platform for delivering various therapeutic agents. More importantly, its ability to enter cells via endocytosis, its photothermal and photodynamic capabilities, and its compatibility with biological systems have made it a powerful candidate in modern cancer therapy. In this section, we explore the diverse roles of GO in drug delivery, photothermal and photodynamic therapy, gene and siRNA delivery, and its use in integrated theranostic platforms.

Figure 4. schematic representation of GO in cancer therapy

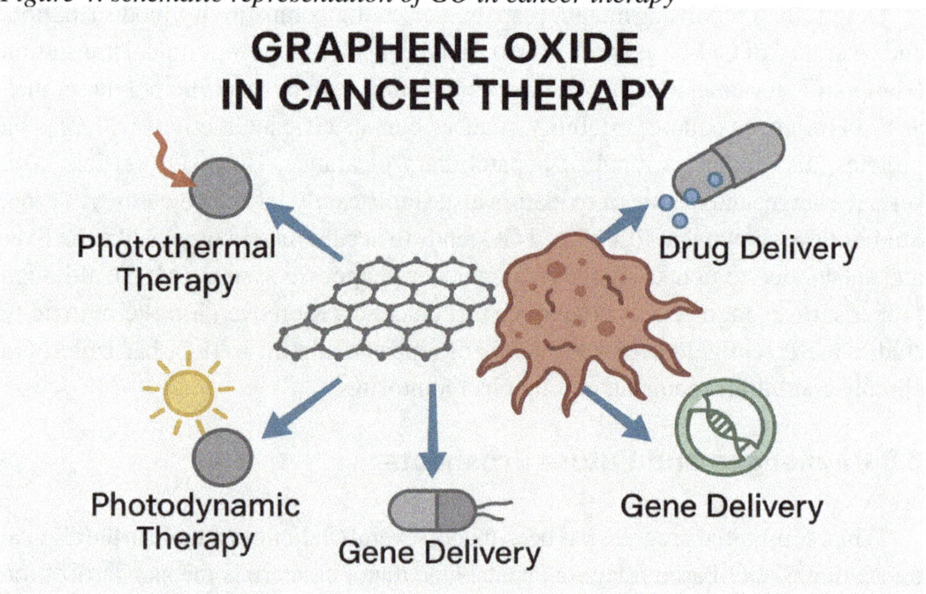

4.1 Drug Delivery

Graphene oxide possesses a unique two-dimensional structure and large surface area, making it an ideal nanocarrier for chemotherapeutic drugs. The π–π stacking interactions between aromatic rings in GO and drug molecules like doxorubicin (DOX), camptothecin (CPT), and paclitaxel enable high drug-loading efficiency (Gayathri et al., 2024). Furthermore, hydrogen bonding between the functional groups on GO and drug molecules contributes to enhanced retention and stability of these payloads. GO's surface can be functionalized with polyethylene glycol (PEG), folic acid, or antibodies to improve biocompatibility and targeted delivery. pH-sensitive systems utilizing GO release the drug preferentially in acidic tumor microenvironments, thereby minimizing damage to healthy tissues. For instance, GO-DOX nanocarriers exhibit significant tumor shrinkage in vivo due to the selective drug release and efficient cellular uptake (Khoshbin et al., 2024). Similarly, CPT-GO systems demonstrate prolonged circulation time and higher cytotoxicity toward cancer cells than the free drug. GO-based nanocarriers not only protect the drug from degradation but also allow for controlled and sustained release, making them highly effective in reducing systemic toxicity and enhancing therapeutic efficacy.

4.2 Photothermal Therapy (PTT)

In photothermal therapy, materials that can absorb near-infrared (NIR) light and convert it into heat are employed to ablate tumor tissues. Graphene oxide is a potent photothermal agent due to its strong NIR absorbance and efficient photothermal conversion ability (Gospodinova et al., 2025). Upon irradiation with NIR light, GO generates localized heat that raises the temperature within the tumor microenvironment, causing irreversible damage to cancer cells. Functionalization of GO with PEG enhances its solubility and biocompatibility while improving circulation time and accumulation in tumor tissues via the enhanced permeability and retention (EPR) effect. GO-PEG systems have shown remarkable efficacy in in vitro and in vivo experiments, killing over 90% of tumor cells following NIR exposure. In a typical study, mice injected with GO-PEG followed by NIR irradiation exhibited complete tumor ablation without recurrence, whereas untreated controls showed rapid tumor progression (Ling et al., 2024). Additionally, PTT based on GO is minimally invasive and has fewer side effects compared to traditional cancer treatments. Researchers are also exploring GO composites with metal nanoparticles such as gold or copper sulfide to further enhance the photothermal effect and create synergistic platforms.

4.3 Photodynamic Therapy (PDT)

Photodynamic therapy employs light-activated photosensitizers that, upon irradiation, generate reactive oxygen species (ROS) like singlet oxygen ($^1O_2$) to induce apoptosis in cancer cells (Vigueras et al., 2025). Graphene oxide serves as an excellent nanocarrier for photosensitizers, offering a stable and targeted delivery system. GO can be conjugated with chlorin e6 (Ce6), a widely used photosensitizer, to form GO-Ce6 nanocomposites. These conjugates show enhanced cellular uptake and ROS generation upon light exposure, leading to significant cancer cell apoptosis. In one study, GO-Ce6 exhibited superior photodynamic efficacy compared to free Ce6 due to better solubility and retention within the tumor. Moreover, by tuning the GO surface with targeting ligands like folic acid, site-specific delivery to cancer cells can be achieved, reducing off-target effects. Importantly, the oxygenated groups on GO facilitate ROS transfer and enhance the quantum yield of singlet oxygen production (Shabnum et al., 2024). The integration of PDT with imaging agents also allows real-time monitoring of therapeutic progress, making GO-based PDT systems highly attractive for clinical applications.

4.4 Gene and siRNA Delivery

The delivery of genetic materials like plasmid DNA and small interfering RNA (siRNA) holds great promise for silencing oncogenes or restoring tumor suppressor genes. Graphene oxide is being extensively investigated as a non-viral gene delivery vector due to its capacity to condense nucleic acids, protect them from enzymatic degradation, and facilitate intracellular transport. GO functionalized with polyethylenimine (PEI) forms stable complexes with siRNA, enabling efficient transfection and gene silencing (Sahane et al., 2025). For instance, PEI-GO-siRNA nanocomplexes targeting the Bcl-2 oncogene have demonstrated successful knockdown of gene expression in breast cancer cells, resulting in apoptosis and reduced tumor proliferation. Unlike viral vectors, GO-based carriers are less immunogenic and can be designed for controlled release. Furthermore, targeting moieties like RGD peptides or antibodies can be added to the GO surface to enhance selective uptake by cancer cells. These systems provide a safer alternative to conventional gene delivery methods and are paving the way for personalized cancer gene therapies with high precision and minimal side effects.

4.5 Combined Photothermal and Photodynamic Therapy

A key advancement in GO-based cancer therapy lies in the integration of photothermal and photodynamic therapies to harness synergistic effects (Das et al., 2025). In such platforms, GO serves as both a heat generator and a carrier for photosensitizers. Upon NIR light exposure, GO simultaneously raises the local temperature and activates the photosensitizer to produce ROS, resulting in enhanced tumor cell killing. GO-Ce6 nanocomposites, for example, have been used successfully in dual-mode PTT-PDT treatments. The combined effect leads to superior therapeutic outcomes compared to either modality alone. Notably, the mild hyperthermia generated during PTT can increase oxygen supply and photosensitizer uptake by tumor cells, further boosting the PDT efficacy (Overchuk et al., 2023). Moreover, these systems can reduce the required light dose and minimize damage to surrounding healthy tissues. In vivo experiments have shown complete tumor regression in mice treated with GO-based PTT-PDT platforms, highlighting their potential as a powerful cancer treatment strategy.

4.6 Theranostic Applications

Graphene oxide's multifunctionality extends into the realm of theranostics, where therapy and diagnostics are integrated into a single platform. GO can be engineered to carry therapeutic agents alongside imaging probes such as fluorescent dyes, magnetic nanoparticles, or radionuclides (Karki et al., 2020). For example, GO-DOX conjugates can be combined with magnetic resonance imaging (MRI) contrast agents like Fe_3O_4 to allow real-time monitoring of drug delivery and tumor response (Divband et al., 2022). Similarly, GO-Ce6 nanocomposites for PDT can be integrated with photoacoustic imaging (PAI) capabilities, enabling both treatment and deep-tissue imaging of tumors. This dual-functionality not only improves treatment accuracy but also enables clinicians to tailor therapy regimens based on individual patient responses. Such GO-based theranostic platforms represent a leap toward personalized medicine, offering minimally invasive, targeted, and traceable cancer treatment options.

4.7 Tumor Targeting and Biodistribution

The effectiveness of GO-based therapies depends largely on their ability to selectively accumulate in tumor tissues. The enhanced permeability and retention (EPR) effect allows nanoparticles like GO to passively accumulate in tumors due to leaky vasculature and poor lymphatic drainage (Shinde et al., 2022). However, to improve specificity, active targeting strategies have been employed by conjugating GO with ligands such as folic acid, transferrin, or monoclonal antibodies. These ligands bind to overexpressed receptors on cancer cells, facilitating receptor-mediated endocytosis. Studies have shown that such functionalized GO systems exhibit higher tumor uptake and lower accumulation in non-target tissues like the liver and spleen. Moreover, PEGylation helps extend circulation time and evade clearance by the reticuloendothelial system (RES). Understanding and controlling the biodistribution of GO in vivo is critical for maximizing therapeutic efficacy while minimizing systemic toxicity and side effects (Cabral et al., 2024).

4.8 Biocompatibility and Safety Concerns

While GO offers promising therapeutic advantages, its biocompatibility and long-term safety are areas of active investigation. Several studies indicate that functionalization with biocompatible polymers such as PEG, dextran, or chitosan significantly reduces GO's cytotoxicity and enhances its stability in biological environments. Nonetheless, concerns remain regarding oxidative stress, inflammation, and potential accumulation in organs after repeated administration. Animal studies have shown that appropriately coated GO is well-tolerated at therapeutic doses, with most of the material being excreted over time (Ibrahim, et al., 2025). However, large-scale, long-term toxicological evaluations are needed to establish safety profiles for clinical use. Optimization of GO size, surface chemistry, and dosage is essential to reduce off-target effects and ensure safe clearance from the body. Regulatory approval will require stringent evaluations, but the preliminary findings suggest that with proper functionalization and dose control, GO can be a safe and effective cancer therapeutic agent.

4.9 Future Prospects and Clinical Translation

The future of GO in cancer therapy lies in refining its functionalization, improving targeting efficiency, and integrating real-time imaging with multi-modal treatments (Song et al., 2024). Advances in nanotechnology and surface chemistry are enabling the development of GO systems with tailored drug release profiles, multi-responsive triggers (e.g., pH, redox, enzyme), and precise tumor-homing capabilities. Clinical translation, however, depends on overcoming challenges such as large-scale synthesis, reproducibility, cost-effectiveness, and regulatory compliance. Collaborations between material scientists, oncologists, and pharmacologists will be crucial to designing GO systems that meet clinical standards (Prasad et al., 2025). Current preclinical results are promising, and several GO-based formulations are in early-stage trials for cancer imaging and therapy. As research progresses, GO-based nanotherapeutics may become integral to next-generation cancer treatments, offering safer, more effective, and personalized solutions.

5. BIOCOMPATIBILITY AND TOXICITY ASSESSMENT

Figure 5. schematic diagram of biocompatibility and toxicity assessment of GO

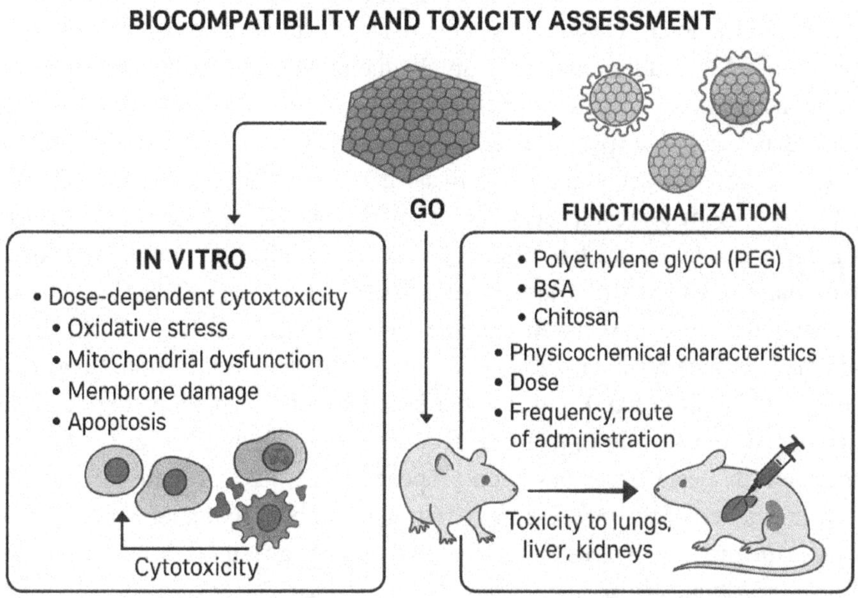

In vivo studies present a more complex picture due to the multifaceted interactions between GO and biological systems (Chen et al., 2022). The toxicity of GO in vivo depends heavily on its physicochemical characteristics such as particle size, surface charge, functional groups, and degree of oxidation, as well as on the dose, frequency, and route of administration. For instance, intravenous administration of bare GO at high concentrations has been associated with acute toxicity, inflammation, and organ damage, particularly in the lungs, liver, and kidneys. These toxic effects are mainly due to GO's tendency to aggregate in biological fluids and lodge in capillary networks or reticuloendothelial system (RES) organs (Ghulam et al. 2022). However, when appropriately functionalized, such as with polyethylene glycol (PEG) or other biocompatible coatings, GO's toxicity can be significantly reduced. PEGylation improves its hydrophilicity, dispersion stability, and stealth properties, thereby minimizing recognition by the immune system and reducing the risk of inflammation or immune response. Additionally, functionalization with proteins like bovine serum albumin (BSA) or natural polymers like chitosan has been shown to enhance GO's biocompatibility, improving its circulation time and reducing nonspecific uptake by RES organs.

Biodistribution studies are critical for understanding where GO accumulates within the body and its potential for causing organ-specific toxicity (Lin et al., 2022). Typically, GO nanoparticles show preferential accumulation in the liver, spleen, and lungs after systemic administration, largely due to their uptake by macrophages in the mononuclear phagocyte system. This accumulation pattern is common for many nanomaterials and poses potential risks, particularly for long-term applications. For example, pulmonary accumulation can cause alveolar damage and fibrosis, while hepatic deposition may result in liver inflammation or dysfunction over time. Surface modification strategies, such as PEGylation or the addition of targeting ligands, can help modulate the biodistribution and reduce off-target effects (Tiwari et al., 2023). Some studies have also explored oral and intraperitoneal routes of administration, which may mitigate certain risks associated with intravenous injection. Importantly, the biodegradability of GO is still under investigation, as incomplete degradation or prolonged retention in organs could contribute to chronic toxicity. Enzymatic degradation by peroxidases has been proposed as a potential route for clearance, but the efficiency of this process in vivo remains to be fully understood.

Given the growing interest in GO-based therapeutic systems, long-term toxicity and clearance studies are urgently needed to assess the risks associated with repeated or chronic exposure (Lin et al., 2024). While short-term studies generally support the relative safety of functionalized GO at therapeutic doses, there is limited data on the consequences of long-term accumulation, potential for bioaccumulation, or interactions with developing tissues, reproductive organs, or the nervous system. Additionally, immunogenicity, genotoxicity, and inflammatory responses under chronic administration scenarios must be carefully examined. Advanced imaging and omics-based tools can be leveraged to gain deeper insights into GO's systemic impact, including possible alterations in gene expression, metabolic pathways, and organ function (Qi et al., 2024). Regulatory frameworks for nanomedicine are still evolving, and comprehensive preclinical safety profiles will be necessary for clinical approval. In this regard, establishing standardized protocols for GO synthesis, purification, characterization, and toxicity evaluation is essential to ensure reproducibility and comparability across studies. Ultimately, although GO holds great potential in oncology and beyond, its successful clinical translation will depend on a thorough understanding and mitigation of its biocompatibility and toxicity challenges.

6. CHALLENGES AND LIMITATIONS

Despite the significant promise that graphene oxide (GO) offers in the realm of bioimaging and cancer therapy, several critical challenges must be addressed before GO-based platforms can transition from bench to bedside (Kashyap et al., 2023). One of the foremost concerns is related to the toxicity and clearance of GO. Although surface functionalization can reduce acute cytotoxicity, long-term accumulation in vital organs like the liver, lungs, and spleen poses serious safety concerns. GO is not inherently biodegradable, and its large lateral size often prevents renal clearance, leading to prolonged systemic circulation and retention. While enzymatic degradation using myeloperoxidase and other oxidative pathways has been reported, the efficiency and consistency of these mechanisms in vivo are still poorly understood. Therefore, the development of biodegradable GO analogs or functionalized derivatives that are efficiently cleared via renal or hepatic pathways is crucial to ensuring long-term safety in clinical applications.

Another major limitation lies in the reproducibility of GO synthesis. GO is typically synthesized via the modified Hummers' method or its variants, which involve the chemical oxidation of graphite (Yadav et al., 2022). However, slight changes in reaction time, temperature, or oxidant concentration can lead to significant variability in the physicochemical properties of the final product, such as lateral size, surface charge, oxidation degree, and functional group composition. These inconsistencies can drastically affect GO's biological interactions, including cellular uptake, drug-loading efficiency, and biodistribution (Akmal et al., 2024). Furthermore, batch-to-batch variation complicates the interpretation of experimental results and hinders the standardization necessary for clinical translation. To address this issue, rigorous standardization and quality control protocols must be developed for GO production, alongside the implementation of real-time monitoring techniques and advanced characterization tools.

Scalability is another barrier to the widespread adoption of GO in biomedical applications (Harun-Ur-Rashid et al., 2025). Although GO can be synthesized in the laboratory relatively easily, scaling up its production to industrial levels without compromising its quality and uniformity remains a formidable challenge. Large-scale production introduces additional issues such as aggregation, impurity incorporation, and non-uniform oxidation. These problems not only reduce the functional performance of GO but also introduce additional safety risks (Losic et al., 2022). Moreover, cost-effectiveness must also be considered—mass production of highly purified, functionalized GO with tight control over physicochemical properties can be prohibitively expensive. As the demand for GO-based materials in biomedicine grows, the development of green, scalable, and cost-effective synthetic routes becomes an urgent need.

Immunogenicity and immune system interactions also pose critical concerns in the clinical application of GO. Several studies have shown that GO can trigger immune responses, particularly when administered systemically (Pi et al., 2024). These responses may manifest as inflammation, cytokine release, or activation of macrophages and dendritic cells, potentially resulting in unintended side effects. Surface coatings such as polyethylene glycol (PEG), albumin, or polysaccharides can help reduce immunogenicity, but the extent to which these modifications can completely mask GO from the immune system is still under investigation. Furthermore, the immunological response to GO may vary among individuals, influenced by genetic, physiological, and pathological factors (Nicolussi et al., 2025). Therefore, comprehensive immunotoxicity assessments across different biological systems are essential before GO can be approved for therapeutic use.

Lastly, the regulatory landscape for GO-based therapeutics and imaging agents is still in its infancy. Currently, no GO-based product has received approval from regulatory authorities such as the U.S. Food and Drug Administration (FDA) for clinical use (Khan et al., 2022). The lack of clear guidelines and regulatory pathways for nanomaterials, especially those as complex as GO, further delays their progression into human trials. Regulatory agencies require extensive data on pharmacokinetics, toxicity, long-term safety, and manufacturing reproducibility, which are often difficult to generate due to the material's inherent variability. In this context, collaborative efforts among researchers, clinicians, regulatory bodies, and industry stakeholders are essential to define standard protocols, safety thresholds, and acceptable endpoints for GO-based medical products.

7. FUTURE PERSPECTIVES AND CLINICAL TRANSLATION

As graphene oxide (GO)-based nanomaterials continue to gain traction in biomedical research, their integration into clinical settings hinges on addressing current limitations and exploring transformative innovations. Looking forward, the convergence of nanotechnology, biotechnology, and data science offers exciting opportunities to evolve GO from a promising experimental tool to a clinically viable solution for complex diseases like cancer.

One of the most compelling directions for GO-based platforms lies in the domain of personalized medicine. The ability to tailor drug delivery systems to an individual's genetic profile, tumor microenvironment, or disease progression is increasingly recognized as the future of cancer therapy. GO's tunable surface chemistry, high drug-loading capacity, and responsive release behavior make it an ideal candidate for designing patient-specific therapeutics. For instance, GO nanocarriers can be functionalized with targeting ligands selected based on the expression profile of receptors in a patient's tumor cells. Similarly, the co-delivery of chemotherapeutics and gene-editing tools like siRNA or CRISPR components can be tailored to silence specific oncogenes or repair mutated pathways. This approach could dramatically improve treatment efficacy while reducing systemic toxicity. The development of GO nanocarriers integrated with biomarker-driven diagnostics and therapeutic release mechanisms will be central to advancing precision oncology.

Another exciting avenue is GO's emerging role in cancer immunotherapy. Immunotherapy has revolutionized cancer treatment by harnessing the body's own immune system to recognize and destroy malignant cells. Recent studies suggest that GO can act as an immune adjuvant, enhancing antigen presentation and stimulating both innate and adaptive immune responses. Functionalized GO can also be used to deliver tumor-associated antigens or immunomodulatory agents directly to antigen-presenting cells, improving vaccine efficacy. Furthermore, the possibility of using GO as a co-delivery platform for checkpoint inhibitors or cytokines opens up new pathways to modulate the immune microenvironment in favor of tumor suppression. These immunotherapeutic strategies, in combination with GO's drug delivery and imaging capabilities, can be engineered into a single multifunctional nanoplatform for synergistic treatment outcomes.

The integration of artificial intelligence (AI) and machine learning (ML) with GO-based nanomedicine presents a transformative opportunity for predictive design and optimized performance. AI-driven models can analyze complex datasets related to GO synthesis parameters, physicochemical properties, and biological interactions to predict the most effective configurations for specific biomedical applications. For example, AI algorithms can help identify the optimal degree of oxidation, particle size, or surface modification required to maximize biocompatibility, cellular uptake, or therapeutic efficacy. In silico models can also be employed to simulate biodistribution, pharmacokinetics, and toxicity profiles, thereby accelerating the preclinical development process. Furthermore, AI can assist in patient stratification and treatment planning by correlating nanocarrier performance with genomic, proteomic, and imaging data. This data-guided approach is expected to shorten development cycles, reduce costs, and facilitate regulatory approval pathways.

However, realizing the full clinical potential of GO will depend heavily on extensive preclinical validation and well-designed clinical trials. To date, most studies have been limited to in vitro or small-animal models, with a scarcity of large-scale, long-term investigations. The translation of GO-based therapeutics and diagnostics into human trials will require comprehensive toxicological assessments, standardization of manufacturing protocols, and clear regulatory guidelines. Pilot studies should prioritize safety, biodistribution, immunogenicity, and clearance profiles in physiologically relevant models. Establishing collaborations among academic institutions, industry stakeholders, and regulatory agencies such as the FDA will be key to advancing GO nanoplatforms through the clinical pipeline. Additionally, the creation of centralized databases and open-access repositories for GO characterization and biological data can promote transparency and reproducibility across the scientific community.

Finally, future innovations will likely focus on the development of multifunctional GO-based nanoplatforms that integrate cutting-edge biomedical technologies. For instance, combining GO with CRISPR-Cas9 gene editing systems may enable precise genomic alterations at target sites, offering curative approaches for genetic cancers. Similarly, hybridizing GO with exosomes—naturally derived nanovesicles involved in intercellular communication—can facilitate efficient delivery across biological barriers and enhance biocompatibility. Biomimetic coatings, such as cell membrane-camouflaged GO particles, may provide immune evasion capabilities and improve tumor-specific targeting. Such smart systems can respond to endogenous or external stimuli (e.g., pH, temperature, magnetic field, light) for on-demand drug release or signal generation. The integration of multiple functionalities into a single GO-based construct not only enhances therapeutic efficacy but also streamlines treatment protocols by combining diagnostics, targeted delivery, and therapy in a unified platform.

8. CONCLUSION

Graphene oxide (GO) stands at the forefront of next-generation nanomedicine, offering unparalleled promise as a multifunctional platform for cancer diagnosis and therapy due to its high surface area, tunable surface chemistry, and inherent biocompatibility. Its ability to combine diverse biomedical functions—such as bioimaging, targeted drug and gene delivery, and phototherapy—within a single nanoplatform underscores its transformative potential in advancing precision and personalized oncology. GO's integration into various imaging modalities, including fluorescence, MRI, CT, and photoacoustic imaging, enables high-resolution, real-time, and multimodal diagnostics that facilitate early detection, monitor therapeutic progress, and assist surgical procedures. In cancer treatment, GO's capacity for controlled and targeted release of chemotherapeutics and nucleic acids, along with its photothermal and photodynamic properties, offers powerful synergistic effects for enhanced tumor ablation with minimized systemic toxicity. However, realizing GO's clinical promise demands overcoming persistent challenges such as long-term toxicity concerns, inconsistencies in synthesis and surface functionalization, the absence of harmonized regulatory standards, and the complexities of scalable, reproducible manufacturing. Addressing these obstacles calls for interdisciplinary collaboration across materials science, biomedicine, toxicology, and regulatory bodies to develop standardized protocols, comprehensive safety assessments, and robust translational pipelines. As the field moves forward, the integration of smart, stimuli-responsive GO-based nanostructures with cutting-edge technologies like artificial intelligence for design optimization holds immense potential to tailor therapies to individual patient profiles, enhancing therapeutic efficacy and patient outcomes. Continued research, rigorous clinical studies, and coordinated translational efforts will be pivotal in navigating GO from promising laboratory innovation to a clinically validated nanoplatform that redefines cancer diagnosis, monitoring, and treatment, ultimately transforming the landscape of personalized cancer care and improving survival rates and quality of life for patients worldwide.

REFERENCES

Ahmad, F., Ghazal, H., Rasheed, F., Shahid, M., Vasantham, S. K., Rafiq, W., & Atiq, S. (2024). Graphene and its derivatives in medical applications: A comprehensive review. *Synthetic Metals*, •••, 117594.

Akmal, M. H., Kalashgrani, M. Y., Mousavi, S. M., Rahmanian, V., Sharma, N., Gholami, A., & Chiang, W. H. (2024). Recent advances in synergistic use of GQD-based hydrogels for bioimaging and drug delivery in cancer treatment. *Journal of Materials Chemistry. B*, *12*(21), 5039–5060. PMID: 38716622

Arnold, A. M., Singh, J., & Sydlik, S. A. (2025). The Role and Future of Functional Graphenic Materials in Biomedical and Human Health Applications. *Biomacromolecules*, *26*(4), 2015–2042. PMID: 40101190

Ayreen, Z., Khatoon, U., Kirti, A., Sinha, A., Gupta, A., Lenka, S. S., & Verma, S. K. (2024). Perilous paradigm of graphene oxide and its derivatives in biomedical applications: Insight to immunocompatibility. *Biomedicine and Pharmacotherapy*, *176*, 116842. PMID: 38810404

Behyar, M. B., Nilghaz, A., Bahari, H., Hasanzadeh, M., & Shadjou, N. (2024). Recent advances on nanomaterials for personalized diagnostics in multiplex point-of-care testing. *Trends in Analytical Chemistry*, •••, 117761.

Bento, C., Katz, M., Santos, M. M., & Afonso, C. A. (2024). Striving for uniformity: A review on advances and challenges to achieve uniform polyethylene glycol. *Organic Process Research & Development*, *28*(4), 860–890. PMID: 38660381

Bhupathi, P., Elhassan A-Elgadir, T. M., Mohammed Ali, R. H., Sanaan Jabbar, H., Gulnoza, D., Joshi, S. K., ... & Alsaalamy, A. (2025). Fluorescence resonance energy transfer (FRET)-based sensor for detection of foodborne pathogenic bacteria: A review. *Critical Reviews in Analytical Chemistry*, *55*(2), 233–250. PMID: 37917532

Biswas, H. S., Bala, S., Kundu, A. K., Saha, I., Poddar, S., Sarkar, S., & Mandal, P. (2022). Tuned synthesis and designed characterization of graphene oxide thin film. *Inorganic Chemistry Communications*, *139*, 109356.

Cabral, H., Li, J., Miyata, K., & Kataoka, K. (2024). Controlling the biodistribution and clearance of nanomedicines. *Nature Reviews Bioengineering*, *2*(3), 214–232.

Cai, C., Zhang, Q., Ye, J., Yao, S., Li, Q., Fan, Z., & Ding, Y. (2025). Tumor microenvironment modulation innovates combinative cancer therapy via versatile graphene oxide nanosystem. *Biomaterials Science*. PMID: 40314424

Chen, S. H., Bell, D. R., & Luan, B. (2022). Understanding interactions between biomolecules and two-dimensional nanomaterials using in silico microscopes. *Advanced Drug Delivery Reviews, 186*, 114336. PMID: 35597306

Das, S., Mazumdar, H., Khondakar, K. R., & Kaushik, A. (2025). Machine learning integrated graphene oxide-based diagnostics, drug delivery, analytical approaches to empower cancer diagnosis. *BMEmat, 3*(1), e12117.

Divband, B., Gharehaghaji, N., & Hassani, S. (2022). Fe3O4/graphene-based nanotheranostics for bimodal magnetic resonance/fluorescence imaging and cancer therapy. *Journal of Inorganic and Organometallic Polymers and Materials, 32*(12), 4443–4460.

Gayathri, K., & Vidya, R. (2024). Carbon nanomaterials as carriers for the anti-cancer drug doxorubicin: A review on theoretical and experimental studies. *Nanoscale Advances, 6*(16), 3992–4014. PMID: 39114152

Ghulam, A. N., Dos Santos, O. A., Hazeem, L., Pizzorno Backx, B., Bououdina, M., & Bellucci, S. (2022). Graphene oxide (GO) materials—Applications and toxicity on living organisms and environment. *Journal of Functional Biomaterials, 13*(2), 77. PMID: 35735932

Gospodinova, Z., Hristova-Panusheva, K., Kamenska, T., Antov, G., & Krasteva, N. (2025). Insights into cellular and molecular mechanisms of graphene oxide nanoparticles in photothermal therapy for hepatocellular carcinoma. *Scientific Reports, 15*(1), 15541. PMID: 40319094

Gostaviceanu, A., Gavrilaş, S., Copolovici, L., & Copolovici, D. M. (2024). Graphene-oxide peptide-containing materials for biomedical applications. *International Journal of Molecular Sciences, 25*(18), 10174. PMID: 39337659

Guo, S., Garaj, S., Bianco, A., & Ménard-Moyon, C. (2022). Controlling covalent chemistry on graphene oxide. *Nature Reviews. Physics, 4*(4), 247–262.

Hajipour Keyvani, A., Mohammadnejad, P., Pazoki-Toroudi, H., Perez Gilabert, I., Chu, T., Manshian, B. B., & Sohrabi, B. (2025). Advancements in Cancer Treatment: Harnessing the Synergistic Potential of Graphene-Based Nanomaterials in Combination Therapy. *ACS Applied Materials & Interfaces*. PMID: 39745785

Harun-Ur-Rashid, M., Foyez, T., Krishna, S. B. N., Poda, S., & Imran, A. B. (2025). Recent advances of silver nanoparticle-based polymer nanocomposites for biomedical applications. *RSC Advances, 15*(11), 8480–8505. PMID: 40109922

He, C., Zhu, J., Zhang, H., Qiao, R., & Zhang, R. (2022). Photoacoustic imaging probes for theranostic applications. *Biosensors (Basel), 12*(11), 947. PMID: 36354456

Hummers, W. S.Jr, & Offeman, R. E. (1958). Preparation of graphitic oxide. *Journal of the American Chemical Society*, *80*(6), 1339–1339.

Ibrahim, N. E., Shawky, H., Maghraby, A. S., & Farrag, E. K. (2025). Insights into the Pharmacokinetics, Biodistribution, and Oral Toxicity of a Polymeric Benzimidazole–Curcumin Nanocomplex with a Multitarget Anticancer Potential. *Food and Chemical Toxicology*, •••, 115483. PMID: 40312001

Itoo, A. M., Vemula, S. L., Gupta, M. T., Giram, M. V., Kumar, S. A., Ghosh, B., & Biswas, S. (2022). Multifunctional graphene oxide nanoparticles for drug delivery in cancer. *Journal of Controlled Release*, *350*, 26–59. PMID: 35964787

Karki, N., Tiwari, H., Tewari, C., Rana, A., Pandey, N., Basak, S., & Sahoo, N. G. (2020). Functionalized graphene oxide as a vehicle for targeted drug delivery and bioimaging applications. *Journal of Materials Chemistry. B*, *8*(36), 8116–8148. PMID: 32966535

Kashyap, B. K., Singh, V. V., Solanki, M. K., Kumar, A., Ruokolainen, J., & Kesari, K. K. (2023). Smart nanomaterials in cancer theranostics: Challenges and opportunities. *ACS Omega*, *8*(16), 14290–14320. PMID: 37125102

Khan, M. I., Hossain, M. I., Hossain, M. K., Rubel, M. H. K., Hossain, K. M., Mahfuz, A. M. U. B., & Anik, M. I. (2022). Recent progress in nanostructured smart drug delivery systems for cancer therapy: A review. *ACS Applied Bio Materials*, *5*(3), 971–1012. PMID: 35226465

Khine, Y. Y., Wen, X., Jin, X., Foller, T., & Joshi, R. (2022). Functional groups in graphene oxide. *Physical Chemistry Chemical Physics*, *24*(43), 26337–26355. PMID: 36285559

Khondakar, K. R., Tripathi, D., Mazumdar, H., Ahuja, K., & Kaushik, A. (2024). Tailored MXene and graphene as efficient telemedicine platforms for personalized health wellness. *Materials Advances*.

Khoshbin, Z., Zahraee, H., Abnous, K., & Taghdisi, S. M. (2024). Hybrid gold nanoparticles as advanced drug delivery and theranostics application. In *Gold Nanoparticles for Drug Delivery* (pp. 421–444). Academic Press.

Li, J., Zeng, H., Zeng, Z., Zeng, Y., & Xie, T. (2021). Promising graphene-based nanomaterials and their biomedical applications and potential risks: A comprehensive review. *ACS Biomaterials Science & Engineering*, *7*(12), 5363–5396. PMID: 34747591

Lima, M. G., & Jussiani, E. I. (2024). X-ray computed tomography for effective atomic number and density evaluation: A review. *Radiation Physics and Chemistry*, •••, 112130.

Lin, H., Buerki-Thurnherr, T., Kaur, J., Wick, P., Pelin, M., Tubaro, A., & Bianco, A. (2024). Environmental and health impacts of graphene and other two-dimensional materials: A graphene flagship perspective. *ACS Nano*, *18*(8), 6038–6094. PMID: 38350010

Lin, Z., Aryal, S., Cheng, Y. H., & Gesquiere, A. J. (2022). Integration of in vitro and in vivo models to predict cellular and tissue dosimetry of nanomaterials using physiologically based pharmacokinetic modeling. *ACS Nano*, *16*(12), 19722–19754. PMID: 36520546

Ling, K., Zheng, J., Jiang, X., Huang, W., Mai, Y., Liao, C., & Jiang, H. (2024). Mn2+/CpG oligodeoxynucleotides codecorated black phosphorus nanosheet platform for enhanced antitumor potency in multimodal therapy. *ACS Nano*, *18*(4), 2841–2860. PMID: 38251849

Losic, D., Farivar, F., Yap, P. L., Tung, T. T., & Nine, M. J. (2022). New insights on energetic properties of graphene oxide (GO) materials and their safety and environmental risks. *The Science of the Total Environment*, *848*, 157743. PMID: 35931146

Majumder, P., & Gangopadhyay, R. (2022). Evolution of graphene oxide (GO)-based nanohybrid materials with diverse compositions: An overview. *RSC Advances*, *12*(9), 5686–5719. PMID: 35425552

Mohammadi, S., Akbari-Birgani, S., Nikfarjam, N., & Rasekhian, M. (2023). Polyethyleneimine-decorated graphene oxide quantum dot as a carrier for suicide gene delivery to the breast cancer three-dimensional model. *Journal of Drug Delivery Science and Technology*, *88*, 104849.

Mondal, A., Kundu, A. K., Biswas, H. S., Mandal, P., Maiti, D. K., Poddar, S., & Biswas, S. S. (2024). Optimizing dielectric and electrical properties of graphene oxide thin film with temperature tuning: Insights from impedance spectra analysis on insulator to semiconductor transition. *Inorganic Chemistry Communications*, *170*, 113016.

More, M. P., Bangde, P., Pandey, A. P., Kulkarni, S., Dandekar, P., Jain, R. D., & Deshmukh, P. K. (2024). Design and development of folic acid engineered magnetic graphene oxide nanosheets for selective synergistic chemophotothermal inhibition of lung cancer cell. *Journal of Drug Delivery Science and Technology*, *93*, 105404.

Nicolussi, P., Pilo, G., Cancedda, M. G., Peng, G., Chau, N. D. Q., De la Cadena, A., & Delogu, L. G. (2025). Biocompatibility of Water-Dispersible Pristine Graphene and Graphene Oxide Using a Close-to-Human Animal Model: A Pilot Study on Swine. *Advanced Healthcare Materials*, *14*(10), 2401783. PMID: 39385652

Overchuk, M., Weersink, R. A., Wilson, B. C., & Zheng, G. (2023). Photodynamic and photothermal therapies: Synergy opportunities for nanomedicine. *ACS Nano*, *17*(9), 7979–8003. PMID: 37129253

Pan, X., Hu, M., Wu, L., Wei, E., Zhu, Q., Lv, L., & Liu, Y. (2025). Biomedical Applications of Gadolinium-Containing Biomaterials: Not Only MRI Contrast Agent. *Advancement of Science*, •••, 2501722. PMID: 40279569

Patil, S., Rajkuberan, C., & Sagadevan, S. (2023). Recent biomedical advancements in graphene oxide and future perspectives. *Journal of Drug Delivery Science and Technology*, *86*, 104737.

Pi, J., Chen, D., Wang, J., Yang, E., Yang, J., Liu, Y., & Shen, L. (2024). Macrophage targeted graphene oxide nanosystem synergize antibiotic killing and host immune defense for Tuberculosis Therapy. *Pharmacological Research*, *208*, 107379. PMID: 39218421

Prasad, R., Ghosh, A., Patel, V., Peng, B., Mendes, B. B., Win, E. H. A., & Conde, J. (2025). Voices of Nanomedicine: Blueprint Guidelines for Collaboration in Addressing Global Unmet Medical Needs. *ACS Nano*. PMID: 39792961

Priyadharshini, S. D., Manikandan, S., Kiruthiga, R., Rednam, U., Babu, P. S., Subbaiya, R., & Govarthanan, M. (2022). Graphene oxide-based nanomaterials for the treatment of pollutants in the aquatic environment: Recent trends and perspectives–A review. *Environmental Pollution*, *306*, 119377. PMID: 35490997

Qi, L., Li, Z., Liu, J., & Chen, X. (2024). Omics-Enhanced Nanomedicine for Cancer Therapy. *Advanced Materials*, *36*(50), 2409102. PMID: 39473316

Rathod, P. P., Bhosle, V. K., Ubale, A. U., & Khan, Z. S. (2025). Ultrasonically Synthesized ZnO-GO Nanocomposite for Photocatalytic and Antibacterial Applications. *Composites Communications*, •••, 102457.

Ravi, S. N., Rajendran, S., Madhumathi, G. S., Packirisamy, A. S. B., Vallinayagam, S., Khan, A. A., & Malik, A. (2024). Carbon nanomaterials: Pioneering innovations in bioimaging and biosensing technologies. *Journal of Molecular Structure*, *1316*, 138987.

Ribeiro, T. P., Moreira, J. A., Monteiro, F. J., & Laranjeira, M. S. (2022). Nanomaterials in cancer: Reviewing the combination of hyperthermia and triggered chemotherapy. *Journal of Controlled Release*, *347*, 89–103. PMID: 35513211

Sack, I. (2023). Magnetic resonance elastography from fundamental soft-tissue mechanics to diagnostic imaging. *Nature Reviews. Physics*, *5*(1), 25–42.

Sadeghi, M. S., Sangrizeh, F. H., Jahani, N., Abedin, M. S., Chaleshgari, S., Ardakan, A. K., & Hushmandi, K. (2023). Graphene oxide nanoarchitectures in cancer therapy: Drug and gene delivery, phototherapy, immunotherapy, and vaccine development. *Environmental Research*, *237*, 117027. PMID: 37659647

Sahane, P., Puri, N., Khairnar, P., Phatale, V., Shukla, S., Priyadarshinee, A., & Srivastava, S. (2025). Harnessing Folate Receptors: A Comprehensive Review on the Applications of Folate-Adorned Nanocarriers for the Management of Melanoma. *ACS Applied Bio Materials*. PMID: 40275606

Sekhon, S. S., Kaur, P., Kim, Y. H., & Sekhon, S. S. (2021). 2D graphene oxide–aptamer conjugate materials for cancer diagnosis. *npj 2D Materials and Applications*, *5*(1), 21.

Shabnum, S. S., Siranjeevi, R., Raj, C. K., Saravanan, A., Vickram, A. S., Chopra, H., & Malik, T. (2024). Advancements in nanotechnology-driven photodynamic and photothermal therapies: Mechanistic insights and synergistic approaches for cancer treatment. *RSC Advances*, *14*(52), 38952–38995. PMID: 39659608

Shinde, V. R., Revi, N., Murugappan, S., Singh, S. P., & Rengan, A. K. (2022). Enhanced permeability and retention effect: A key facilitator for solid tumor targeting by nanoparticles. *Photodiagnosis and Photodynamic Therapy*, *39*, 102915. PMID: 35597441

Song, K., Su, X., Zhao, W., Ai, F., Umar, A., & Baskoutas, S. (2024). Functional inorganic nanomaterials for optical cancer theranostics. *Chemical Engineering Journal*, •••, 150067.

Teixeira, M. I., Lopes, C. M., Amaral, M. H., & Costa, P. C. (2024). Navigating Neurotoxicity and Safety Assessment of Nanocarriers for Brain Delivery: Evaluation Models and Insights. *Acta Biomaterialia*.

Theivendren, P., Kunjiappan, S., Pavadai, P., Ravi, K., Murugavel, A., Dayalan, A., & Santhana Krishna Kumar, A. (2024). Revolutionizing cancer immunotherapy: Emerging nanotechnology-driven drug delivery systems for enhanced therapeutic efficacy. *ACS Measurement Science Au*, *5*(1), 31–55. PMID: 39991031

Tian, Z., Fu, Y., Dang, Z., Guo, T., Li, W., & Zhang, J. (2025). Utilizing Nanomaterials in Microfluidic Devices for Disease Detection and Treatment. *Nanomaterials (Basel, Switzerland)*, *15*(6), 434. PMID: 40137607

Tiwari, P., Yadav, K., Shukla, R. P., Gautam, S., Marwaha, D., Sharma, M., & Mishra, P. R. (2023). Surface modification strategies in translocating nano-vesicles across different barriers and the role of bio-vesicles in improving anticancer therapy. *Journal of Controlled Release*, *363*, 290–348. PMID: 37714434

Uniyal, S., Choudhary, K., Sachdev, S., & Kumar, S. (2024). Nano-bio fusion: Advancing biomedical applications and biosensing with functional nanomaterials. *Optics & Laser Technology*, *168*, 109938.

Vigueras, G., García, E. I., de la Torre-Rubio, E., Abad-Montero, D., Santana, M. D., Marchan, V., & Ruiz, J. (2025). Metal-Coumarin Derivatives as Promising Photosensitizers: Unlocking Their Cancer Phototherapy Potential. *Inorganic Chemistry Frontiers*.

Wang, W., Shang, S., Wang, Y., & Xu, B. (2024). Utilization of nanomaterials in MRI contrast agents and their role in therapy guided by imaging. *Frontiers in Bioengineering and Biotechnology*, *12*, 1484577. PMID: 39628648

Wang, Y., Di, S., Yu, J., Wang, L., & Li, Z. (2023). Recent advances of graphene–biomacromolecule nanocomposites in medical applications. *Journal of Materials Chemistry. B*, *11*(3), 500–518. PMID: 36541392

Wen, S., Ovais, M., Li, X., Ren, J., Liu, T., Wang, Z., & Chen, C. (2022). Tailoring bismuth-based nanoparticles for enhanced radiosensitivity in cancer therapy. *Nanoscale*, *14*(23), 8245–8254. PMID: 35647806

Xiao, Y., Pang, Y. X., Yan, Y., Qian, P., Zhao, H., Manickam, S., & Pang, C. H. (2023). Synthesis and functionalization of graphene materials for biomedical applications: Recent advances, challenges, and perspectives. *Advancement of Science*, *10*(9), 2205292. PMID: 36658693

Yadav, K. (2024). Nanotechnology in diabetes management: Revolutionizing treatment and diagnostics. *Journal of Molecular Liquids*, •••, 126117.

Yadav, S., Singh Raman, A. P., Meena, H., & Goswami, A. G., Bhawna, Kumar, V., ... & Singh, P. (2022). An update on graphene oxide: Applications and toxicity. *ACS Omega*, *7*(40), 35387–35445. PMID: 36249372

Yan, R., Zhan, M., Xu, J., & Peng, Q. (2024). Functional nanomaterials as photosensitizers or delivery systems for antibacterial photodynamic therapy. *Biomaterials Advances*, •••, 213820. PMID: 38430723

Yang, D. H., Nah, H., Lee, D., Min, S. J., Park, S., An, S. H., & Heo, D. N. (2024). A review on gold nanoparticles as an innovative therapeutic cue in bone tissue engineering: Prospects and future clinical applications. *Materials Today. Bio*, •••, 101016. PMID: 38516171

Yang, Q., Wu, Y., Chen, J., Lu, M., Wang, X., Zhang, Z., & Chen, L. (2024). Plasmonic nanomaterial-enhanced fluorescence and Raman sensors: Multifunctional platforms and applications. *Coordination Chemistry Reviews*, *507*, 215768.

Yu, Y., Feng, T., Qiu, H., Gu, Y., Chen, Q., Zuo, C., & Ma, H. (2024). Simultaneous photoacoustic and ultrasound imaging: A review. *Ultrasonics*, •••, 107277. PMID: 38460216

Chapter 4

Advanced Magnetic Nanoplatforms for Accurate Theranostics From Self-Reporting Bioimaging to Programmable Hyperthermia:

Smart Nanomedicine for Precision Cancer Therapy

Payel Das
http://orcid.org/0009-0009-0900-6023
KPR Institute of Engineering and Technology, India

Bishnu Thakur
https://orcid.org/0009-0004-3551-3698
KPR Institute of Engineering and Technology, India

Rohit Sah
https://orcid.org/0009-0008-0257-1495
KPR Institute of Engineering and Technology, India

DOI: 10.4018/979-8-3373-3765-4.ch004

ABSTRACT

Despite significant progress in cancer therapy, achieving precise, localized elimination of tumor cells without harming healthy tissue remains a challenge. Magnetic nanoparticles (MNPs) are versatile agents with therapeutic and diagnostic potential, yet current systems are limited by heat-only effects, lack of real-time feedback, and empirical design. This chapter presents next-generation nanoplatforms that integrate magneto-mechanical actuation, temperature-sensitive self-reporting, and AI-driven optimization. Three approaches are highlighted: (i) magneto-mechanical oscillators that disrupt cancer cells under low-frequency magnetic fields; (ii) phase-change reporter hybrids enabling simultaneous therapy and real-time imaging; and (iii) multi-modal supraparticles optimized by machine learning for imaging, heating, and drug delivery. Emphasis is placed on scalable fabrication, biocompatibility, and clinical translation. The chapter provides a roadmap for programmable nanomedicine with implications for researchers, clinicians, and industry stakeholders.

1. INTRODUCTION

1.1 Overview of Theranostics in Oncology

Cancer continues to be one of the biggest problems in healthcare today. In 2020 alone, there were around 19.3 million new cases and nearly 10 million deaths worldwide (Sung et al., 2021). Although conventional treatments such as surgery, chemotherapy, and radiotherapy have saved countless lives, they still come with major drawbacks. The biggest issue is their lack of precision—healthy tissues often get damaged along with the tumor, which leads to severe side effects (Estelrich et al., 2015). On top of that, tumors are not uniform. Their genetic and phenotypic variations often cause patients to respond differently to the same treatment, which explains why relapse is so common. To overcome these challenges, researchers have been turning to theranostics—a strategy that combines therapy and diagnostics in one system. Unlike traditional approaches, theranostic platforms are designed to detect, treat, and monitor cancer in real time, giving clinicians the ability to adjust treatments dynamically (Kumar et al., 2020). With the help of nanotechnology, imaging techniques, and targeted therapies, theranostics makes it possible to improve treatment accuracy while reducing side effects. Among the different nanomaterials studied for this purpose, magnetic nanoparticles (MNPs) have gained significant attention. Their unique properties make them valuable in both diagnosis as well as in therapy. For example, they can act as contrast agents in MRI scans, but they can also be used for localized hyperthermia, targeted drug delivery, and real-time tumor imaging (Laurent et al., 2008; Bulte & Kraitchman, 2004).

1.2 Limitations of Conventional Diagnostics and Therapies

Despite substantial progress in oncology, standard diagnostic and therapeutic tools present several challenges:

- Absence of real-time monitoring: Treatment assessment is typically based on periodic imaging, which prevents continuous evaluation of therapeutic response.
- Non-specific drug action: Systemic chemotherapy affects normal and malignant cells alike, producing side effects such as cardiotoxicity and bone marrow suppression (Estelrich et al., 2015).
- Collateral radiation damage: Radiotherapy, though effective, can injure surrounding tissues and even trigger secondary malignancies.
- Incomplete tumor elimination: Solid tumors often resist complete eradication due to poor drug penetration and cellular diversity, leaving residual cells that cause relapse.
- Lack of integration: Imaging methods like CT, MRI, and PET are generally applied independently of therapeutic interventions, preventing a seamless diagnosis–treatment cycle.

These shortcomings underscore the need for more advanced systems capable of precise tumor targeting, controlled drug release, and integrated diagnostic–therapeutic functionality—capabilities that magnetic nanoparticle-based theranostics can provide.

1.3 Introduction to Magnetic Nanomaterials

Nanoscale engineering is used to create magnetic nanomaterials that take advantage of special magnetic properties. Superparamagnetism, which happens when nanoparticles smaller than about 20 nm only become magnetized when exposed to an external magnetic field and show no residual magnetization after the field is removed, is one of the most significant of these (Laurent et al., 2008). This characteristic improves safety for biomedical applications by preventing particle aggregation. The most extensively researched iron oxide nanoparticles (IONPs) are magnetite (FeO_4) and maghemite (γ-FeO_3), in particular, because of their well-established function as MRI contrast agents, chemical stability, and biocompatibility. Beyond imaging, magnetic hyperthermia therapy (MHT) uses these particles to destroy localized tumors by absorbing electromagnetic energy and turning it into heat under an alternating magnetic field (Dutz & Hergt, 2014). MNPs become multifunctional platforms through functionalization with polymers, antibodies, or peptides, which permits targeted delivery and combined therapeutic–diagnostic action (Janib et al., 2010). Stimuli-responsive coatings have been developed recently, which release medications in response to environmental cues like temperature, enzymatic activity, or pH. In addition, AI-powered modeling tools are currently being used to optimize particle design, adjusting surface chemistry, size, and shape for increased safety and efficiency.

1.4 Motivation and Scope of the Chapter

Although the promise of MNP-based cancer therapies is well recognized, several challenges persist:

- Hyperthermia approaches rely heavily on heat, which can induce thermotolerance in cancer cells via heat-shock proteins.
- Current systems lack real-time therapeutic feedback, making it difficult to achieve precise temperature or dose control.
- Many nanoparticle designs are still based on empirical trial-and-error rather than predictive, optimized frameworks.

This chapter aims to address these gaps by introducing emerging concepts such as:

- Magneto-mechanical oscillators, which apply low-frequency field-induced mechanical stress to disrupt tumors.
- Phase-change reporter hybrids, capable of delivering both thermal therapy and imaging feedback.
- Machine learning–optimized supraparticles, designed for integrated hyperthermia, drug delivery, and multimodal imaging.

Additionally, we will explore manufacturing approaches, strategies for ensuring biocompatibility, and regulatory hurdles that need to be addressed for clinical translation.

1.5 Chapter Objectives

The main objectives of this chapter are to:

- Describe the basic ideas behind magnetic nanoparticles, such as their synthesis, functionalization, and magnetic characteristics.

- Analyze their uses in imaging-guided therapy and multimodal imaging.
- Talk about programmable hyperthermia techniques and how they might work in tandem with immunotherapy and chemotherapy.
- Assess translational challenges, ranging from regulatory approval to large-scale fabrication.
- Emphasize upcoming developments such as nanorobotic platforms, CRISPR-based gene delivery, and AI-assisted design.

Through these talks, the chapter offers a thorough review of theranostics based on magnetic nanoparticles and how they might influence the direction of individualized cancer treatment in the future.

2. MAGNETIC NANOPARTICLES (MNPS): FUNDAMENTALS

2.1 Physical and Magnetic Properties

Magnetic nanoparticles (MNPs) are ideal for biomedical applications because of their distinct physical and magnetic characteristics, which set them apart from their bulk counterparts. They have a high surface-to-volume ratio due to their nanoscale size, which usually ranges from 1 to 100 nm. This increases their reactivity and functionalization potential (Laurent et al., 2008). When MNPs are tiny enough to form a single magnetic domain and their magnetization direction varies under thermal energy, they exhibit superparamagnetism, which is the major magnetic behavior of MNPs used in biomedical applications. By guaranteeing that MNPs only behave magnetically while exposed to an external magnetic field, this characteristic eliminates unwanted aggregation after the magnetic field is withdrawn (Gupta & Gupta, 2005). Due to finite-size factors and surface spin disorder, the saturation magnetization (Ms) of nanoparticles is substantially lower than that of bulk materials. Nonetheless, magnetic performance can be improved by refining the crystal structure and synthesis conditions, which is important for MRI applications and hyperthermia (Hergt et al., 2006). A key factor in regulating heating in magnetic hyperthermia is the Curie temperature (Tc), or the temperature at which a material loses its magnetic characteristics. Strong magnetic behavior under physiological settings is guaranteed by the high Tc of iron oxide-based nanoparticles, which are frequently employed in cancer theranostics.

2.2 Types of Magnetic Nanoparticles

Magnetite (FeO_4) and maghemite (γ-FeO_3) are two examples of iron oxide nanoparticles (IONPs), which are the most extensively researched MNPs for biomedical applications. Because of their relative simplicity of synthesis and biocompatibility, they are favored (Estelrich et al., 2015). Doped ferrites, including $MnFeO_4$, $CoFeO_4$, and $ZnFeO_4$, are being investigated in addition to iron oxides for better magnetic characteristics and increased hyperthermia efficiency (Issa et al., 2013). Cobalt-platinum and iron-platinum nanoparticles are examples of noble metal-based hybrids that have high saturation magnetization and outstanding chemical stability, which makes them appropriate for both imaging and therapeutic applications (Kumar et al., 2020). In addition, iron carbide (Fe_5C_2) nanoparticles have been the subject of recent developments because they exhibit better magnetic characteristics and higher photothermal conversion efficiency than traditional iron oxide nanoparticles (Wang et al., 2021).

2.3 Synthesis Approaches

MNPs with regulated size, shape, and surface characteristics can be produced using a variety of synthetic techniques. Typical strategies consist of: Iron salts precipitate in an alkaline medium using the most popular technique, co-precipitation. According to Laurent et al. (2008), although it is scalable and economical, it frequently leads to polydispersity and less control over particle form. By breaking down organometallic precursors at high temperatures in organic solvents, thermal breakdown allows the creation of monodisperse nanoparticles with high crystallinity (Sun & Zeng, 2002). Hydrothermal synthesis: According to Zhang et al. (2010), hydrothermal and solvothermal methods enable regulated nucleation and development of nanoparticles at high temperatures and pressures, resulting in homogeneous shape and crystallinity. Green synthesis: To lessen toxicity and the influence on the environment, eco-friendly synthesis techniques, including plant extracts, polysaccharides, and microbiological agents, have recently drawn attention (Ali et al., 2021). The magnetic characteristics, colloidal stability, and surface chemistry of MNPs—all of which are essential for their biological performance—are strongly influenced by each synthesis technique.

2.4 Surface Functionalization and Bioconjugation

In order to guarantee colloidal stability, biocompatibility, and precise targeting, bare magnetic nanoparticles must have their surfaces modified for efficient biomedical use. Gold shells, silica, and polymers (such as PEG and dextran) are examples of coating materials that are frequently utilized to stop aggregation and supply reactive functional groups for conjugation with biomolecules (Gupta & Gupta, 2005). Targeted delivery to tumor areas is made possible by attaching ligands, such as aptamers, peptides (like RGD), or antibodies, to the surface of the nanoparticle (Kievit & Zhang, 2011). To further improve the theranostic efficacy of MNPs and enable regulated drug release, smart coatings that react to pH, temperature, or redox conditions are being incorporated (Park et al., 2021). In vivo, the biodistribution and clearance paths of nanoparticles are also determined by surface charge and hydrophilicity. According to Lee et al. (2015), a neutral or slightly negative zeta potential is typically recommended for lowering opsonization and extending bloodstream circulation time. Furthermore, for multifunctional theranostic applications, bioconjugation techniques such as affinity-based binding (e.g., streptavidin–biotin), electrostatic interactions, and covalent bonding (e.g., EDC/NHS coupling) enable precise attachment of medications, imaging agents, or genetic material.

Figure 1. Schematic representation of magnetic nanoparticle structure and functionalization strategies (core, coating, and targeting ligands).

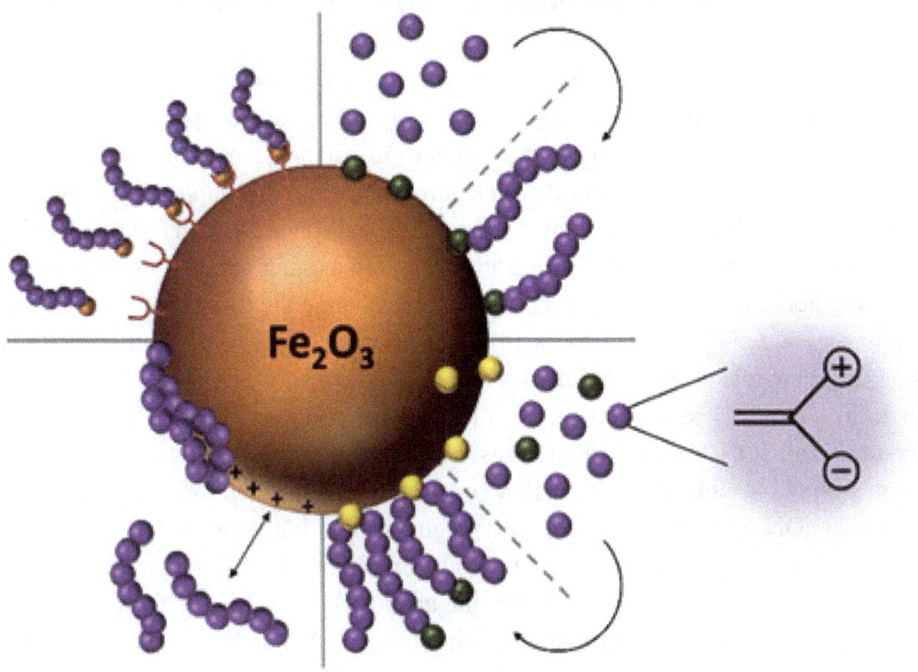

Reprinted from Synthesis, Characterization, and Applications of Magnetic Nanoparticles Featuring Polyzwitterionic Coatings by Biehl, P., Von der Lühe, M., Dutz, S., & Schacher, F. H., Polymers, 10(1), 91, 2018. © 2018 MDPI. Reprinted with permission

3. BIOIMAGING APPLICATIONS OF MAGNETIC NANOPLATFORMS

In oncology, where accurate tumor detection and monitoring are essential for successful treatment, magnetic nanoparticles (MNPs) have become extremely versatile biomedical imaging tools. They are particularly useful because of their distinct magnetic characteristics and capacity to act as contrast enhancers in a variety of imaging modalities. Because imaging allows physicians to determine tumor boundaries, measure vascularization, and analyze the tumor microenvironment, it is essential to the treatment of cancer. These realizations inform treatment planning and offer feedback throughout interventions. However, the resolution, molecular specificity, or dynamic monitoring required for advanced cancer management are frequently not provided by traditional techniques like CT, PET, and ultrasound. By fusing therapeutic properties with high-contrast visualization, magnetic nanoplatforms fill these gaps and set the stage for integrated theranostic systems (Bulte & Kraitchman, 2004; Lee et al., 2015).

3.1 MRI Contrast Enhancement with Magnetic Nanoparticles

The use of non-ionizing radiation, deep tissue penetration, and high spatial resolution makes magnetic resonance imaging (MRI) one of the most popular diagnostic techniques. However, contrast agents frequently affect their sensitivity. Despite their clinical success, gadolinium-based agents can cause toxicity and nephrogenic systemic fibrosis, particularly in patients with impaired renal function. Iron oxide nanoparticles (IONPs) have garnered interest in this regard as safer substitutes with strong imaging potential (Laurent et al., 2008). MNPs change the tissues' local magnetic environment, which affects how nearby protons relax. Depending on the particle's size, makeup, and surface modifications, this leads to shortened longitudinal (T1) or transverse (T2) relaxation times: T1-weighted contrast: Superparamagnetism in ultrasmall IONPs (<5 nm) shortens T1 relaxation times, resulting in brighter contrast in T1-weighted scans. Darker contrast in T2-weighted images is produced by larger IONPs (>10 nm), producing magnetic field inhomogeneities that speed up T2 relaxation (Estelrich et al., 2015). Dual-mode T1/T2 contrast agents, which combine both mechanisms to increase diagnostic accuracy, are the focus of recent developments. For example, dual relaxivity has been shown by FeO_4 nanoclusters with engineered surface ligands, which minimize false-positive results and provide superior tumor delineation (Lee et al., 2015). These hybrid contrast systems are especially helpful for improving pre-treatment planning and characterizing diverse tumor tissues.

3.2 Self-Reporting Nanoplatforms for Real-Time Monitoring

The absence of dynamic feedback during treatment is one of the main issues with image-guided cancer therapy. This restriction is particularly noticeable in targeted drug delivery and magnetic hyperthermia, where exact temperature and dosage control are crucial. In order to overcome this, scientists have created self-reporting nanoplatforms that incorporate phase-change materials or temperature-sensitive reporters into MNP coatings (Janib et al., 2010). For example, when a critical temperature threshold is reached, thermosensitive coatings can release fluorophores or photoacoustic markers to indicate localized heating during hyperthermia. These signals are further amplified by AI-enhanced image reconstruction techniques, which allow clinicians to monitor therapy in real time and modify treatment parameters as necessary (Park et al., 2021). These systems increase the accuracy of cancer treatments and reduce collateral damage to surrounding tissues by continuously visualizing therapeutic progress.

3.3 Dual-Mode and Multi-Modal Imaging

No single imaging modality can fully capture all the diagnostic information required due to the complexity of tumor biology. Multi-modal imaging platforms are becoming more and more popular. These platforms combine MRI with other techniques like near-infrared (NIR) imaging, fluorescence imaging, PET, or photoacoustic tomography (PAT). MNPs' large surface area makes them perfect for these uses since it enables functionalization with a variety of imaging probes (Mura et al., 2013). Notable examples include: MR/Fluorescence imaging: MNPs coupled to fluorescent dyes offer optical imaging and high-resolution MRI, which are helpful for intraoperative guidance. MR/PET imaging: Labeling MNPs with isotopes such as 64Cu or 68Ga enables simultaneous structural imaging (MRI) and functional assessment (PET) (Janib et al., 2010).MR/Photoacoustic imaging: This technique improves tumor characterization by combining the spatial accuracy of MRI with the sensitivity of photoacoustics to vascular and oxygenation features. By providing complementary anatomical, physiological, and molecular insights on a single platform, these hybrid approaches greatly improve diagnostic reliability.

3.4 Molecular Targeting and Tumor-Specific Imaging

Functionalized MNP surfaces are used in targeted imaging techniques to bind tumor-associated biomarkers like integrins, folate receptors, and epidermal growth factor receptors (EGFR) selectively. This lessens off-target accumulation and enhances tumor-to-background contrast. Examples include: RGD peptide-modified

FeO_4 nanoclusters improve MRI sensitivity and lower background noise by preferentially accumulating in tumor vasculature rich in αvβ3 integrin (Lee et al., 2015). Aptamer-functionalized nanoparticles provide accurate imaging even in metastatic lesions by specifically identifying cancer cells that overexpress nucleolin. Such tumor-specific targeting lowers the risk of postoperative residual disease by enhancing intraoperative visualization and early detection.

3.5 Safety, Biocompatibility, and Clinical Translation

The safety and long-term biocompatibility of MNPs must be guaranteed for clinical adoption to be successful. Coating nanoparticles with dextran, polyethylene glycol (PEG), or silica is are method to increase compatibility because it improves colloidal stability, increases circulation time, and decreases reticuloendothelial system (RES) immune clearance. Crucially, systemic toxicity is limited by the breakdown of iron oxide nanoparticles into physiologically controllable iron ions (Gupta & Gupta, 2005). There has already been evidence of clinical progress. Off-label use of ferumoxytol, an FDA-approved iron oxide formulation for anemia, as an MRI contrast agent is also common. The translational viability of such platforms is further demonstrated by NanoTherm® therapy, which uses magnetic nanoparticles to treat localized hyperthermia in glioblastoma and has been approved in Europe (Singh et al., 2021).

Figure 2. Schematic of dual-mode T1/T2 MRI contrast enhancement using Fe_3O_4 nanoclusters.

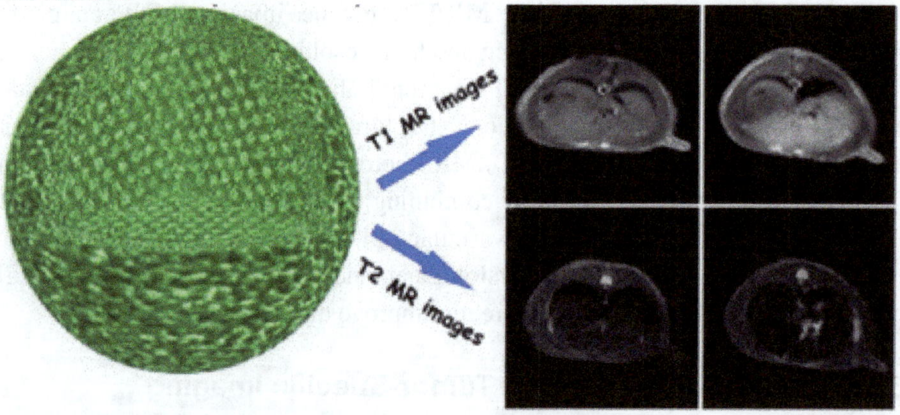

Reprinted from Rational Design of Magnetic Nanoparticles as T1–T2 Dual-Mode MRI Contrast Agents by Geraldes, C. F. G. C., Nanomaterials, 8(10), 839, 2018. © 2018 MDPI. Reprinted with permission

Figure 3. Flowchart of Self-Reporting Nanoplatforms for Real-Time Monitoring

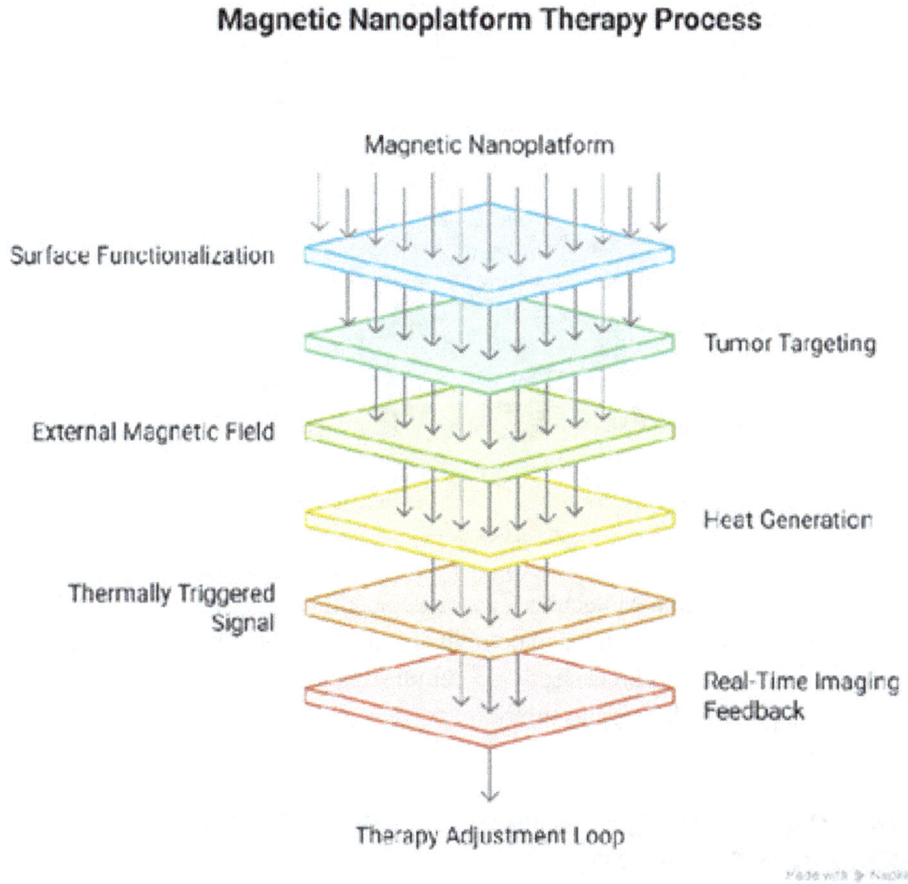

Figure 4. – Tumor-Targeting with RGD-Functionalized Nanoparticles

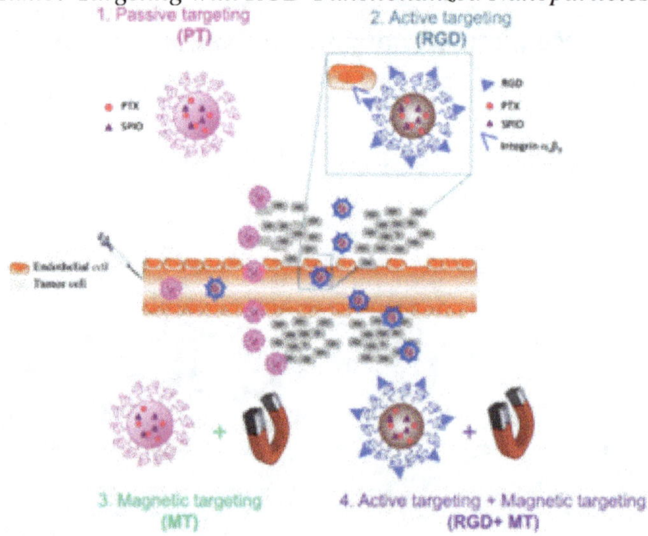

This illustration from Molecular Cancer depicts four targeting strategies (passive, active via RGD, magnetic, and combined), including RGD-mediated receptor targeting to αvβ3 integrin. It effectively demonstrates the mechanism underlying targeted nanoparticle delivery to tumors.

4. PROGRAMMABLE MAGNETIC HYPERTHERMIA FOR CANCER THERAPY

Magnetic hyperthermia (MHT) is an emerging cancer treatment strategy in which magnetic nanoparticles (MNPs) generate heat when exposed to an alternating magnetic field (AMF). This localized heating can selectively destroy tumor cells while sparing surrounding healthy tissue. The idea of using heat to combat cancer is over a century old, but traditional hyperthermia methods—such as microwave, radiofrequency, or ultrasound—often struggle with limited tissue penetration and poor tumor specificity. By contrast, MNP-mediated hyperthermia allows precise control of heating within tumors, overcoming these limitations (Dutz & Hergt, 2014; Fortin et al., 2007). In recent years, MHT has evolved from a simple thermal intervention into a multifunctional, programmable platform. Modern approaches integrate temperature-sensitive drug release, real-time imaging feedback, and artificial intelligence-driven control systems, offering personalized and highly targeted therapies. This section explores the underlying principles, mechanisms, design considerations, and emerging programmable hyperthermia systems in cancer treatment.

4.1 Principles of Magnetic Hyperthermia

The process by which electromagnetic energy is transformed into heat is necessary for magnetic hyperthermia. MNPs, also known as iron oxide (FeO_4) nanoparticles, release energy as heat through three primary mechanisms when exposed to an AC magnetic field (usually 100–500 kHz, 5–30 kA/m): hysteresis losses, Brownian relaxation, and Neel relaxation. Neel relaxation occurs when the magnetic moment of a particle rotates internally within a fixed core, releasing energy as heat. Brownian relaxation arises from the physical rotation of nanoparticles in response to the AMF, generating frictional heat in the surrounding medium.Hysteresis losses are significant in larger or aggregated particles exhibiting ferromagnetism, where energy is dissipated during magnetic domain realignment (Fortin et al., 2007). The specific absorption rate (SAR), which is influenced by particle size, shape, magnetic anisotropy, and AMF parameters, measures the effectiveness of heat generation. Safety is crucial in clinical settings; to avoid overheating healthy tissue, the product of frequency (f) and magnetic field strength (H) must stay below the Atkinson–Brezovich limit ($H \times f < 5 \times 10^9\ A{\cdot}m^{-1}{\cdot}s^{-1}$).

4.2 Targeted Energy Dissipation and Localized Heating

The capacity of magnetic hyperthermia to limit heat production to tumor sites while reducing harm to healthy tissues is one of its biggest benefits. There are various methods for achieving tumor-specific MNP accumulation:Passive targeting: By taking advantage of the enhanced permeability and retention (EPR) effect, which is caused by leaky vasculature and inadequate lymphatic drainage, nanoparticles preferentially accumulate in tumors. Active targeting: Cancer cells can absorb nanoparticles through receptor-mediated uptake when they are functionalized with ligands like aptamers, peptides, or antibodies. Magnetic targeting: Before triggering hyperthermia, external magnets can direct MNPs to tumor sites. Protein denaturation, enzyme inactivation, increased membrane permeability (which improves drug absorption), and vascular shutdown that deprives tumors of oxygen and nutrients are just a few of the therapeutic effects brought on by localized heating. According to Dutz and Hergt (2014), even mild hyperthermia (41–45 °C) can make cancer cells more sensitive to radiation and chemotherapy, which can work in concert to treat the disease.

4.3 Modulating Magnetic Properties for Tunable Hyperthermia

Fine-tuning nanoparticle properties is essential to achieving programmable hyperthermia. Important elements impacting SAR include: Particle size and shape: The best superparamagnetic heating is usually seen in iron oxide nanoparticles with diameters between 10 and 20 nm. Because of their greater magnetic anisotropy, non-spherical shapes—like nanocubes or nanorods—display higher heating efficiency. Doping: Adding transition metals such as zinc, cobalt, or manganese to ferrite structures increases the Curie temperature and improves magnetic anisotropy, giving engineers more control over heat production. Surface coatings: Biocompatible coatings, such as silica and PEG, stabilize heating profiles and stop aggregation under physiological conditions. Recent advancements in stimuli-responsive nanoparticles enable real-time, on-demand hyperthermia customized to each patient's needs by allowing magnetic properties to alter in response to physiological conditions or external cues.

4.4 Thermoresponsive Drug Delivery Coupled with Hyperthermia

Controlled drug delivery is increasingly being combined with magnetic hyperthermia. Thermosensitive polymer coatings, like poly(N-isopropylacrylamide) (PNIPAM), release encapsulated medications precisely when needed by undergoing conformational changes at extremely high temperatures. Examples include: MNP–liposome hybrids loaded with doxorubicin, which release the drug upon AMF exposure. Core-shell nanoparticles with MNP cores and phase-change materials (e.g., fatty acids) that provide burst drug release at therapeutic temperatures. Additionally, some platforms use iron-mediated Fenton reactions to produce reactive oxygen species (ROS) and increase the death of cancer cells by combining hyperthermia with chemodynamic therapy (CDT) (Fortin et al., 2007). This multifaceted strategy maximizes therapeutic efficacy while minimizing systemic toxicity.

4.5 Preclinical and Clinical Evidence

The therapeutic potential of MHT has been confirmed by numerous preclinical studies:

- In vitro: When exposed to MNP-mediated hyperthermia, cancer cell lines (such as HeLa and MCF-7) exhibit ferroptosis, mitochondrial dysfunction, and apoptosis through the production of ROS.
- In vivo: Iron oxide nanoparticle-treated tumor-bearing mice under AMF show notable tumor regression with negligible systemic toxicity.

In Europe, glioblastoma can now be treated clinically with NanoTherm® therapy (MagForce AG). This method achieves therapeutic temperatures precisely within tumor tissue by injecting aminosilane-coated iron oxide nanoparticles intratumorally, followed by localized AMF application (Singh et al., 2021). While lowering the dosages of cytotoxic agents, combined strategies—hyperthermia with chemotherapy, radiotherapy, or immunotherapy—further improve patient outcomes.

4.6 Programmable Hyperthermia: AI-Driven Optimization

In order to achieve programmable, patient-specific therapies, next-generation MHT systems integrate machine learning and artificial intelligence (AI). Based on imaging feedback, these systems employ predictive models to modify nanoparticle dosing and AMF parameters in real time. Closed-loop control: To maintain an ideal thermal dose, AI algorithms and MRI-guided thermometry dynamically adjust AMF exposure. Personalized planning: Individual anatomical and physiological traits are taken into account when designing hyperthermia treatments using predictive modeling. By lowering the possibility of under- or overheating, these intelligent systems guarantee safe, efficient, and customized cancer treatment (Park et al., 2021). Magnetic hyperthermia is evolving from a traditional thermal treatment to a complex, programmable theranostic platform through the integration of imaging, thermometry, and AI-driven control.

Figure 5. – Magnetic Hyperthermia Mechanism, Source: "Schematic diagram of magnetic hyperthermia-based treatment..."

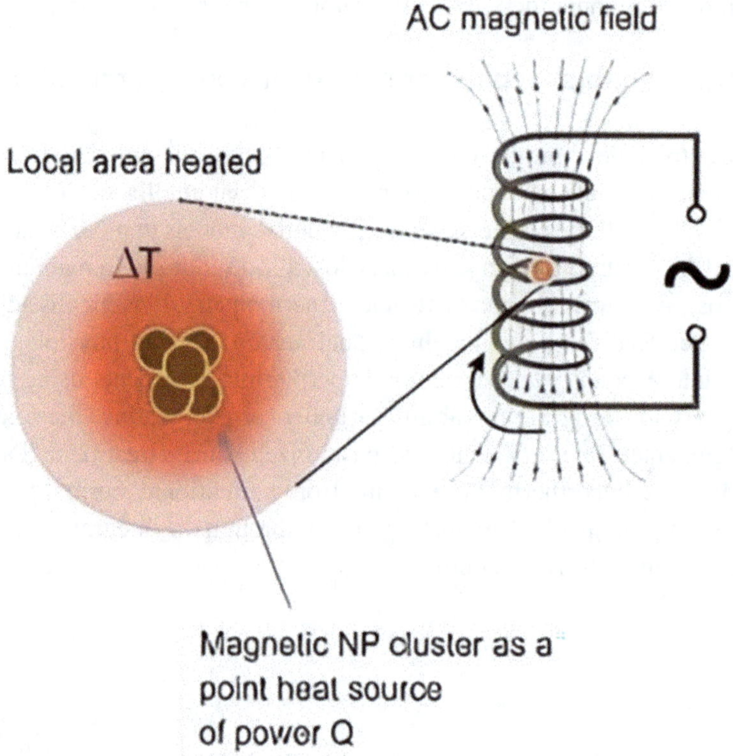

Figure 6. AI-assisted closed-loop hyperthermia system with MRI feedback

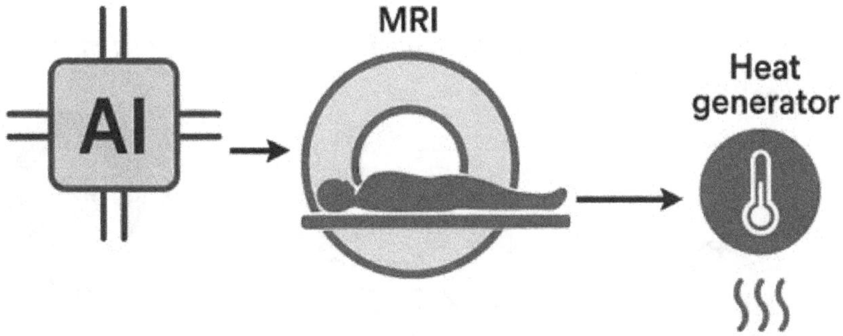

5. THERANOSTIC INTEGRATION: COMBINING IMAGING AND THERAPY

5.1 Introduction to Theranostic Platforms

Modern oncology has been revolutionized by theranostics, which combines diagnosis and treatment into a single strategy. Conventional cancer therapies frequently depend on follow-up imaging and standardized medication dosage, which offers little real-time insight into treatment effectiveness. This deficiency has led to the creation of multipurpose nanoplatforms that can provide real-time imaging feedback and targeted therapy at the same time (Janib et al., 2010).

Magnetic nanoparticles (MNPs) play a central role in theranostics due to their dual capabilities:

- Imaging: MNPs are used as contrast agents in imaging modalities like optical, CT, PET, and MRI.
- Therapy: They help with controlled drug delivery, catalytic reactive oxygen species (ROS) generation, magnetic hyperthermia, and chemodynamic therapy.

Clinicians can apply individualized treatment plans, dynamically track therapeutic progress, and modify interventions in real time by combining these features into a single, programmable nanoplatform. This method increases overall therapeutic efficacy and decreases off-target toxicity.

Figure 7. Design schematic of a multifunctional MNP-based theranostic platform integrating MRI and hyperthermia.

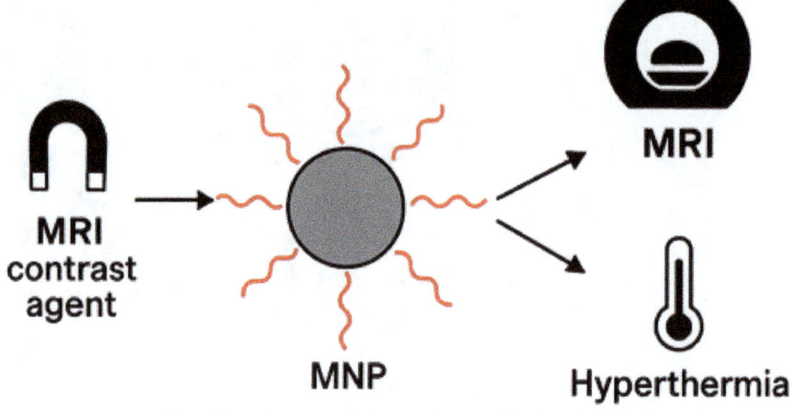

5.2 Design of Multifunctional Nanoplatforms

Theranostic nanoplatform design must guarantee smooth imaging and therapeutic function compatibility. Among the fundamental factors are:

5.2.1 Core Structure

- Iron oxide cores remain the standard for MRI contrast and magnetic hyperthermia.
- Alternative magnetic materials, such as doped ferrites ($CoFe_2O_4$, $MnFe_2O_4$), offer higher magnetization and enhanced heat generation efficiency.

5.2.2 Surface Functionalization

Surface engineering plays a pivotal role in enhancing the therapeutic performance and biocompatibility of magnetic nanoparticles (MNPs). Polyethylene glycol (PEGylation) is one of the most widely used strategies, as it improves nanoparticle stability, reduces protein adsorption, and minimizes recognition and clearance by the mononuclear phagocyte system (MPS), thereby prolonging circulation time in the bloodstream. In addition to stealth properties, active targeting can be achieved by functionalizing MNPs with ligands such as antibodies, folic acid, or aptamers, which selectively bind to overexpressed tumor-specific receptors, including HER2 and EGFR, enhancing tumor accumulation and reducing off-target effects. Furthermore, the incorporation of thermoresponsive polymers like poly(N-isopropylacrylamide) (PNIPAM) enables heat-triggered drug release, creating a synergistic effect with magnetic hyperthermia. This multifunctional surface modification approach ensures precise targeting, controlled drug delivery, and improved therapeutic efficacy in cancer theranostics.

Figure 8. Workflow of imaging-guided therapy using MRI and AMF activation

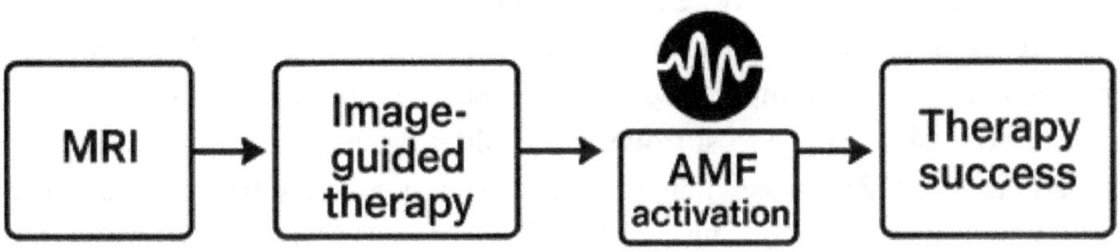

5.2.3 Imaging Agents

Dual-mode imaging is increasingly becoming a standard approach to improve diagnostic accuracy and provide complementary information. One common strategy involves MRI/fluorescence imaging, achieved by conjugating fluorophores or quantum dots to magnetic nanoparticles, enabling high-resolution anatomical imaging alongside optical detection at the cellular level. Another approach, MRI/PET imaging, uses radiolabeled nanoparticles such as ^64Cu-DOTA-functionalized MNPs, combining the superior soft tissue contrast of MRI with the high sensitivity and quantitative capabilities of PET. Additionally, MRI/photoacoustic imaging leverages nanoparticles integrated with gold nanoshells, allowing both magnetic resonance imaging and photoacoustic signal generation through optical absorption. These hybrid imaging modalities facilitate real-time tracking, enhanced tumor delineation, and better treatment planning in theranostic applications.

5.2.4 Programmable Features

Advanced platforms incorporate stimulus-responsive elements that respond to pH, temperature, or specific enzymes, allowing controlled drug release and adaptive imaging. Additionally, AI-driven algorithms predict optimal magnetic field strengths and imaging parameters for patient-specific treatment protocols.

5.3 Imaging-Guided Therapy

Imaging-guided therapy (IGT) leverages real-time imaging data to precisely track and guide therapeutic interventions, ensuring greater accuracy and safety. Among the most extensively studied strategies in MNP-based theranostics is MRI-guided hyperthermia. This workflow begins with pre-treatment imaging, where MRI accurately delineates the tumor's size, shape, and anatomical location. Subsequently, localized hyperthermia is activated by applying an alternating magnetic field (AMF) to induce controlled heating within the tumor region. Real-time MR thermometry continuously monitors temperature changes, maintaining the therapeutic range of 42–45°C to avoid damage to surrounding healthy tissues. Finally, post-treatment evaluation using MRI assesses nanoparticle distribution and confirms tumor regression. By eliminating blind dosing and allowing dynamic adjustments during therapy, MRI-guided hyperthermia provides a highly personalized and adaptive treatment approach (Mura et al., 2013).

5.4 Case Studies: In Vitro and Animal Models

5.4.1 In Vitro Studies

In vitro studies further validate the multifunctional potential of MNP-based platforms in cancer therapy. Experiments on cancer cell lines such as HeLa and MCF-7 demonstrated that dual-functional nanoparticles not only offered superior MRI contrast but also induced efficient hyperthermia-mediated cell death, confirmed by caspase-3 activation and apoptotic signaling pathways. Additionally, reactive oxygen species (ROS)-mediated chemo-hyperthermia approaches have shown significant promise; for instance, iron oxide nanoparticles coated with doxorubicin-loaded poly(N-isopropylacrylamide) (PNIPAM) exhibited markedly enhanced cytotoxicity under an alternating magnetic field (AMF) compared to chemotherapy alone. This synergistic effect highlights the capability of smart nanoplatforms to combine imaging, targeted drug delivery, and hyperthermic therapy for improved therapeutic efficacy.

5.4.2 In Vivo Animal Models

Preclinical studies have demonstrated the therapeutic potential of magnetic nanoparticle (MNP)-based theranostic systems across various tumor models. In a glioblastoma model, intrathecal administration of aminosilane-coated MNPs successfully mimicked MagForce NanoTherm® therapy by enabling MRI-guided localization and achieving a significant reduction in tumor volume following magnetic hyperthermia. Similarly, in a breast cancer xenograft model, multifunctional nanoparticles composed of iron oxide cores and gold nanoshells facilitated combined magnetothermal ablation and dual-modality imaging using MRI and photoacoustic techniques. This integrated approach resulted in complete tumor regression without observable systemic toxicity, underscoring the promise of MNP-based platforms for image-guided, minimally invasive cancer therapy.

5.4.3 Transition to Clinical Practice

Notwithstanding encouraging preclinical findings, dosing, biodistribution, and long-term safety issues must be resolved for clinical translation; these issues will be covered in more detail in Section 6.

5.5 Pharmacokinetics and Biodistribution

Understanding the pharmacokinetics (PK) and biodistribution of magnetic nanoparticle (MNP)-based theranostics is essential to ensure both therapeutic efficacy and safety. The route of administration plays a critical role; while intravenous injection remains the standard approach, intratumoral delivery provides higher local nanoparticle concentrations, enhancing treatment specificity. Clearance primarily occurs through hepatic and splenic uptake, where Kupffer cells and macrophages sequester the particles. The particle size strongly influences clearance pathways: nanoparticles smaller than 10 nm are rapidly eliminated through renal filtration, whereas larger particles typically undergo hepatobiliary clearance. Surface engineering strategies such as PEGylation and zwitterionic coatings are widely employed to prolong systemic circulation, minimize immune recognition, and reduce opsonization. Furthermore, advanced imaging modalities like MRI combined with PET now enable real-time, quantitative tracking of nanoparticle distribution in vivo, allowing researchers and clinicians to optimize dosing strategies and minimize off-target accumulation (Janib et al., 2010).

5.6 Future Directions in Theranostic Integration

The future of theranostics is moving toward intelligent, self-regulating systems designed to deliver highly personalized and adaptive cancer treatments. A key focus area is closed-loop control, where AI-driven algorithms continuously monitor imaging feedback and dynamically adjust parameters such as magnetic field strength and drug release profiles in real time. This ensures precise therapy delivery, minimizes off-target effects, and enhances therapeutic efficiency.

Another major trend is multi-modal imaging, integrating MRI with complementary imaging techniques such as PET, CT, and optical imaging. This holistic approach provides comprehensive tumor characterization, enabling better treatment planning, monitoring, and early detection of therapeutic responses.

Moreover, combination therapy is emerging as a cornerstone of next-generation theranostics. Future systems aim to integrate immunomodulators, gene-editing platforms like CRISPR-Cas9, and stimuli-responsive drug delivery mechanisms with conventional strategies such as magnetic hyperthermia and chemotherapy. These multifunctional platforms are expected to offer synergistic effects, reduce resistance, and pave the way for truly personalized oncology care.

Such platforms embody the principles of precision oncology, offering targeted, continuously monitored, and adaptable therapy to maximize patient benefit.

Figure 9. case studied in vitro and animal model

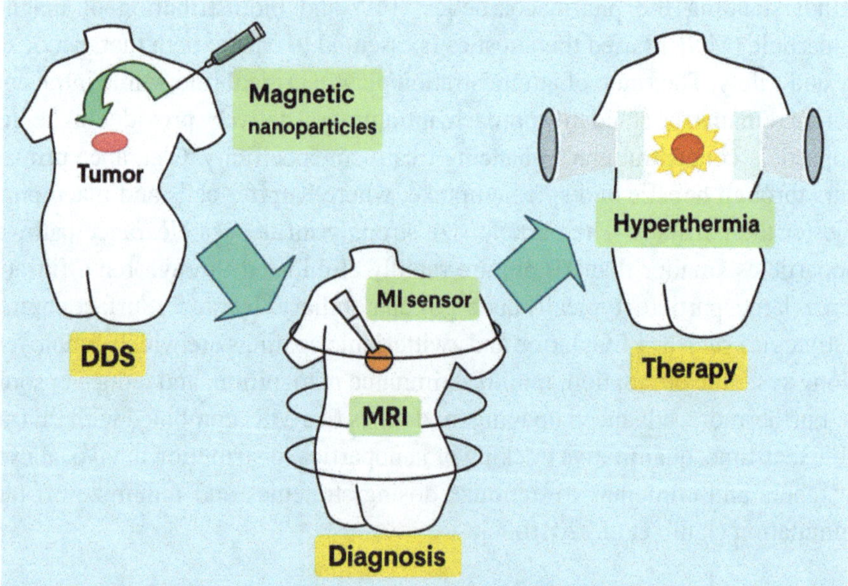

6. CLINICAL TRANSLATION AND CHALLENGES

6.1 Introduction

Clinical adoption of magnetic nanoparticle (MNP)-based theranostics has been comparatively slow, despite their impressive preclinical promise. Although tumor targeting, imaging accuracy, and therapeutic results are frequently demonstrated in animal models, extrapolating these results to humans is fraught with difficulties. Significant obstacles to approval and broad adoption are presented by the intricacy of human physiology and stringent regulatory requirements. Regulatory compliance, scalable manufacturing, batch reproducibility, immunogenicity, long-term toxicity, and clinical trial validation are some of the major obstacles (Singh et al., 2021).

6.2 Regulatory Concerns (FDA and EMA Guidelines)

Regulatory authorities, including the U.S. Food and Drug Administration (FDA) and the European Medicines Agency (EMA), maintain strict standards for the approval of nanomedicines, especially multifunctional systems that combine diagnostic and therapeutic functions.

6.2.1 Complexity of Theranostic Platforms

Combination products—theranostic agents that integrate imaging and therapeutic functions—must comply with multiple regulatory frameworks, as they fall under both drug and device categories. These frameworks often require adherence to the guidelines of the FDA Center for Devices and Radiological Health (CDRH) and Center for Drug Evaluation and Research (CDER), making the approval process more complex than for single-function nanomedicines.

In addition, preclinical toxicology studies are critical for ensuring safety before clinical application. These studies must comprehensively evaluate immunogenicity, genotoxicity, and toxicological effects over different time frames, including acute, subchronic, and chronic exposures. The immune response to nanoparticles, particularly their ability to activate complement pathways or induce hypersensitivity, must be well-characterized to prevent adverse clinical events.

A major challenge lies in batch-to-batch consistency, as even small variations in particle size, surface charge, or coating chemistry can significantly influence pharmacokinetics (PK) and biodistribution, ultimately affecting therapeutic efficacy and safety. These physicochemical attributes determine how nanoparticles circulate in the bloodstream, accumulate in tumors, and clear from the body. Therefore, stringent quality control (QC) measures and good manufacturing practice (GMP) compliance are essential to maintain reproducibility and regulatory acceptance

6.2.2 Guidance Documents

The FDA's Nanotechnology Guidance emphasizes the necessity for comprehensive safety profiling, detailed physicochemical characterization, and evaluation of in vivo behavior to ensure patient safety. Similarly, the European Medicines Agency (EMA) requires rigorous risk-benefit analyses, focusing particularly on the long-term clearance of nanoparticles and their potential accumulation within reticuloendothelial organs. However, standardized testing protocols for theranostic nanoparticles are still evolving. Until globally harmonized guidelines are fully established, regulatory submissions for these nanomedicines are evaluated on a case-by-case basis to address unique design features and safety concerns (Singh et al., 2021)

6.3 Scaling Up Synthesis and Reproducibility

Scaling the synthesis of magnetic nanoparticle (MNP)-based theranostic systems from laboratory-scale experiments to clinical-grade production introduces several challenges. One of the primary requirements is compliance with Good Manufacturing Practice (GMP), which ensures consistency, quality, and safety across large-scale batches. GMP mandates strict controls over parameters such as purity, sterility, and contamination levels. During scale-up, the risk of introducing endotoxins or other contaminants increases significantly, and these impurities can trigger immunogenic reactions in patients, potentially delaying or preventing regulatory approval.

Another critical issue is achieving uniformity in nanoparticle size and surface coating, as these factors strongly influence both therapeutic and diagnostic performance. For instance, the effectiveness of magnetic hyperthermia depends heavily on particle size, which typically needs to be within the 10–30 nm range for iron oxide nanoparticles. Even minor variations in synthesis conditions can alter the relaxivity and specific absorption rate (SAR), affecting both imaging contrast and heating efficiency. To address these challenges, advanced manufacturing techniques such as microfluidic reactors and continuous-flow synthesis systems are increasingly being adopted. These scalable approaches enable precise control over reaction conditions, ensuring high reproducibility and homogeneity of nanoparticles at an industrial scale.

6.4 Immunogenicity, Clearance, and Toxicity

The clinical safety of theranostic nanoparticles is significantly influenced by their interaction with the immune system and their clearance pathways. Immunogenicity remains a major concern, as nanoparticles may activate immune responses through protein adsorption, also known as opsonization. Strategies such as PEGylation (surface modification with polyethylene glycol) or the application of zwitterionic coatings can help reduce macrophage recognition and prolong circulation time. However, repeated administration of PEGylated nanoparticles may induce anti-PEG antibodies, resulting in the phenomenon of accelerated blood clearance (ABC), which diminishes therapeutic efficacy over time.

Clearance mechanisms are primarily determined by the size and surface chemistry of the nanoparticles. Renal clearance occurs when nanoparticles are smaller than 10 nm, allowing filtration through the kidneys. In contrast, larger nanoparticles are typically processed through hepatobiliary clearance, where they are sequestered by the reticuloendothelial system (RES), particularly in the liver and spleen. Additionally, most iron oxide nanoparticles undergo biodegradation, releasing iron ions that integrate into the body's natural iron metabolism. While generally considered safe, excessive accumulation of iron may lead to oxidative stress and iron overload, raising concerns about long-term toxicity.

Long-term toxicity remains an underexplored area due to limited histopathological data from chronic exposure studies. Existing short-term in vivo data suggest a favorable safety profile, but prolonged retention of nanoparticles in organs such as the liver, spleen, and lungs poses potential risks. To mitigate these effects, the use of biodegradable coatings such as dextran or PEG is recommended to enhance clearance and minimize toxicity over extended periods.

6.5 Barriers to Clinical Adoption

Despite promising preclinical outcomes, several barriers hinder the clinical adoption of MNP-based theranostic systems. One significant challenge is high production costs, driven by the need for complex synthesis processes, advanced surface engineering, and the integration of multiple functional components. Another issue is the design complexity of multifunctional nanoplatforms, which increases both regulatory scrutiny and manufacturing challenges.

Patient variability also poses an obstacle, as tumor heterogeneity greatly influences nanoparticle accumulation, therapeutic response, and overall efficacy. This necessitates personalized dosing strategies, which add complexity to clinical translation. Furthermore, ethical and safety concerns persist due to the limited availability of long-term clinical safety data. Patients and regulators remain cautious about widespread adoption until these uncertainties are resolved through extensive clinical evidence

6.6 Current Clinical Trials and Approved Systems

The most notable clinical success in magnetic nanoparticle-based hyperthermia is MagForce's NanoTherm® therapy, which received EMA approval in 2010 for the treatment of glioblastoma multiforme (GBM). This therapy involves the intratumoral injection of aminosilane-coated iron oxide nanoparticles, followed by exposure to an alternating magnetic field (AMF), resulting in localized heating that enhances the effect of radiotherapy and chemotherapy. Clinical studies have shown improved median overall survival in patients treated with this method. However, its application is limited by the requirement for intratumoral injection, making it suitable only for accessible tumors.

Several ongoing clinical trials are exploring magnetic hyperthermia in other cancers, such as breast and prostate tumors, often in combination with chemotherapy or radiotherapy. Additionally, platforms integrating MRI-guided drug delivery are under investigation in Europe and Asia, aiming to combine therapeutic efficacy with real-time imaging.

Beyond hyperthermia, other commercial initiatives are advancing magnetic nanoparticle applications in oncology. Endomag (UK) has developed iron oxide-based tracers for sentinel lymph node mapping in breast cancer surgery, which are already in clinical use. Furthermore, early-stage efforts are being made to develop MRI/PET hybrid contrast agents for liquid biopsy applications and cancer detection.

6.7 Strategies to Overcome Clinical Barriers

To accelerate the clinical adoption of magnetic theranostics, several strategies are being pursued. One key approach is the standardization of testing protocols, led by collaborative efforts between the FDA, EMA, and ISO, to establish universal guidelines for nanoparticle characterization and safety evaluation.

Another strategy focuses on smart nanoparticle design, incorporating biodegradable coatings and stimuli-responsive systems that improve safety profiles and therapeutic efficacy. Additionally, efforts to reduce production costs through scale-up innovations, such as continuous synthesis and automated manufacturing systems, are crucial for making these technologies economically viable.

Finally, the integration of real-time monitoring tools—particularly those powered by AI-driven imaging analytics—can optimize treatment planning, predict patient response, and dynamically adjust therapy parameters. These advancements collectively represent a roadmap toward the successful clinical translation of magnetic nanoparticle-based theranostic platforms.

7. FUTURE PERSPECTIVES AND EMERGING TRENDS

7.1 Introduction

Thanks to interdisciplinary advancements like smart nanohybrids, AI-assisted platforms, CRISPR-based gene delivery, and nanorobotics, the field of magnetic nanotechnology in oncology is undergoing a significant transformation. By making precision therapy, real-time monitoring, and customized cancer treatment plans possible, these advancements are revolutionizing the field of theranostics (Park et al., 2021). The most promising avenues influencing the development of cancer theranostics based on magnetic nanoparticles (MNPs) are examined in this section. Conventional magnetic nanoparticles (MNPs) primarily rely on passive targeting and externally applied magnetic fields to achieve therapeutic effects. However, the next generation of nanoplatforms is shifting toward smart nanohybrids, designed to respond to multiple physiological and external stimuli, including pH variations, temperature changes, enzymatic activity, redox gradients, and magnetic fields. These stimulus-responsive systems enhance precision by activating therapeutic functions only under specific tumor microenvironment conditions.

7.2.1 Multifunctional Smart Nanohybrids

Smart nanohybrids are typically composed of magnetic cores such as iron oxide, iron carbide, or Fe_3O_4, surrounded by polymeric shells that encapsulate drugs, imaging probes, and targeting ligands. These hybrid systems are engineered to trigger on-demand drug release when exposed to tumor-specific stimuli, such as acidic pH or elevated reactive oxygen species (ROS) levels. This targeted approach improves therapeutic specificity while significantly reducing systemic toxicity compared to conventional drug delivery methods.

7.2.2 Clinical Implications

The clinical potential of smart nanohybrids is remarkable. These platforms enable site-specific, on-demand drug release, eliminating the need for invasive interventions and reducing treatment-related complications. Moreover, they support combinatorial therapies, allowing the integration of hyperthermia, chemotherapy, and immunotherapy within a single system. This multi-functional capability is expected to enhance therapeutic outcomes while minimizing adverse effects, paving the way for highly personalized treatment strategies.

7.3 AI-Assisted Nanomedicine

Artificial Intelligence (AI) and Machine Learning (ML) are revolutionizing the field of nanomedicine by enabling data-driven design, predictive modeling, and real-time therapy optimization. These technologies are transforming how nanoparticles are developed, deployed, and monitored in clinical settings.

7.3.1 AI in Nanoparticle Design

AI and ML algorithms are increasingly used for predictive modeling, allowing researchers to forecast critical parameters such as biodistribution, magnetic relaxation characteristics, and specific absorption rate (SAR) based on nanoparticle structure and composition. Furthermore, AI-powered tools are streamlining synthesis optimization, automating the adjustment of reaction parameters to achieve scalable and reproducible nanoparticle production with desired properties.

7.3.2 AI in Imaging and Treatment Planning

In clinical applications, AI plays a pivotal role in image-guided therapy by improving tumor boundary detection and segmentation, ensuring accurate delineation of hyperthermia zones and minimizing collateral damage to healthy tissues. Additionally, AI-driven feedback systems can dynamically regulate magnetic field intensity and treatment duration in real time, adapting therapy to patient-specific responses for enhanced safety and efficacy.

7.4 CRISPR-Based Gene Delivery via MNPs

The integration of CRISPR-Cas9 gene-editing technology with MNP-based platforms is emerging as a game-changing approach in oncology. These hybrid systems offer a non-viral alternative for gene therapy, addressing key challenges such as immunogenicity and insertional mutagenesis associated with traditional viral vectors.

7.4.1 Magnetic Gene Delivery Platforms

Magnetic nanoparticles functionalized with cationic polymers or lipid coatings can efficiently encapsulate CRISPR components, including the Cas9 protein and guide RNA. A technique known as magnetofection leverages external magnetic fields to guide gene-loaded nanoparticles directly to tumor sites, significantly enhancing transfection efficiency while minimizing off-target effects.

7.4.2 Therapeutic Applications

CRISPR-loaded MNP systems enable oncogene silencing or tumor suppressor gene restoration, providing a potentially curative strategy rather than merely managing disease symptoms. Although clinical translation remains in its infancy, preclinical studies have demonstrated encouraging results in aggressive cancers such as glioblastoma and metastatic tumors, suggesting immense potential for future therapeutic applications.

7.5 Nanorobotics in Oncology

Nanorobotics represents a paradigm shift in cancer therapy, moving from theoretical frameworks to practical applications. These autonomous nanoscale devices are capable of precise navigation, real-time sensing, and targeted therapeutic delivery.

7.5.1 Magnetic Actuation for Nanorobots

Magnetic fields serve as a non-invasive propulsion mechanism for nanorobots, enabling them to traverse the bloodstream and reach deep-seated tumors. Advanced designs incorporate therapeutic payloads, imaging probes, and biosensors, allowing simultaneous diagnosis, monitoring, and treatment within a single nanosystem.

7.5.2 Clinical Potential

Nanorobots hold the promise of replacing systemic chemotherapy by delivering drugs exclusively to cancer cells, thereby reducing systemic toxicity and improving treatment outcomes. Future designs aim to incorporate real-time biomarker sensing, ensuring drug release only upon detection of cancer-specific signals, further enhancing precision and safety.

7.6 Integration of AI and Robotics in Theranostics

The convergence of AI-driven algorithms and nanorobotic systems envisions a futuristic healthcare model where autonomous, intelligent nanomachines navigate the circulatory system, detect tumors, and perform precision therapy with minimal human intervention. These systems could analyze tumor microenvironment conditions in real time and adapt treatment strategies dynamically.

However, several challenges must be addressed, including ensuring biocompatibility, developing immune-evasive coatings, and creating wireless power systems for in vivo operation. Ethical considerations and regulatory frameworks for autonomous nanomedicine also require urgent attention before clinical translation becomes feasible.

7.7 Emerging Trends and Future Roadmap

The future of cancer theranostics lies in personalized nanomedicine, driven by AI-enabled optimization and stimuli-responsive smart nanoparticles. Emerging trends include AI-optimized MNP formulations tailored to individual patients, self-learning nanoplatforms, and multi-modal treatment systems that combine imaging, hyperthermia, gene editing, and immunotherapy. As the field progresses, the integration of AI, robotics, and nanotechnology is expected to transform oncology into a highly adaptive, patient-specific discipline, reducing treatment failure rates and improving long-term survival outcomes.

Figure 10. -Future Perspectives in Magnetic Theranostics: AI,Smart Nanohybrids, CRISPR and Nanorobotics AI powered image

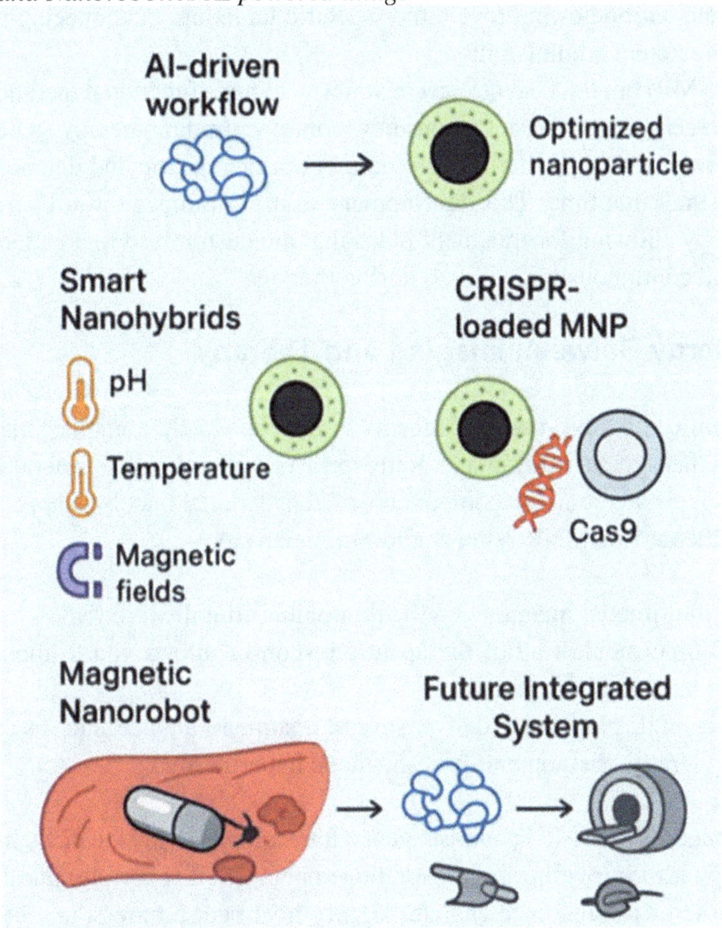

8. CONCLUSION

8.1 Recap of the Importance of MNPs in Oncology

In modern oncology, magnetic nanoparticles (MNPs) have taken center stage as a means of establishing a connection between targeted therapy and accurate diagnostic imaging. MNP-based techniques offer localized, minimally invasive interventions that preserve healthy tissue, in contrast to conventional treatments, which frequently lack specificity and can have systemic side effects. Their special magnetothermal qualities and sensitivity to external magnetic fields enable real-time imaging and controlled therapy. Additionally, surface alterations using polymers, ligands, and antibodies improve tumor-specific targeting, guaranteeing extremely accurate treatment administration.

Simple MRI contrast agents have given way to multifunctional theranostic platforms in recent years. These systems now combine immunotherapy, gene therapy, drug delivery, and hyperthermia, providing both therapeutic and diagnostic capabilities at the same time. This development signifies a move toward personalized medicine by allowing for treatment plans that are customized to a patient's tumor profile and continuously monitored during therapy.

8.2 Synergy Between Imaging and Therapy

The ability of MNP-based platforms to synergistically combine imaging and therapy is their greatest advantage. Retrospective evaluation and generalized treatment protocols are common components of conventional oncology approaches. In contrast, theranostic MNP systems allow clinicians to:

- Use multimodal imaging or MRI to monitor drug distribution.
- Real-time monitoring of therapeutic responses allows you to modify treatment parameters as necessary.
- Use a single platform to deliver several treatments at once, such as photothermal therapy, chemotherapy, or magnetic hyperthermia.

Dual-mode T1/T2 MRI contrast agents, for example, guide localized hyperthermia therapy while providing high-resolution imaging. Self-reporting nanostructures, where phase or fluorescence changes signify treatment progress, are a feature of emerging designs. This ability improves patient outcomes and reduces needless interventions by converting cancer care from reactive to adaptive.

8.3 Outlook Toward Personalized Cancer Treatment Using Magnetic Theranostics

MNP-based theranostic systems are central to the advancement of precision oncology. Key trends shaping this future include:

8.3.1 Personalized Nanoplatform Design

Computational modeling to create patient-specific nanoparticles is made possible by AI and machine learning. These models maximize therapeutic efficacy while minimizing toxicity by optimizing magnetic properties, surface chemistry, and particle size.

8.3.2 Integration with Genomic and Proteomic Data

Linking MNP platforms with genomic profiling facilitates targeted gene therapies using CRISPR-Cas systems and biomarker-driven treatment strategies, ensuring highly individualized care.

8.3.3 Remote and Autonomous Therapies

Advances in nanorobotics and wireless magnetic control may allow nanoparticles to navigate to tumors autonomously, deliver multi-modal treatments, and report therapeutic outcomes in real time.

8.3.4 Clinical Translation and Regulatory Pathways

Despite promising preclinical data, challenges remain in scaling up synthesis, ensuring reproducibility, and meeting stringent regulatory requirements. Collaboration among researchers, industry, and regulatory bodies will be crucial to translating these technologies into clinical practice.

8.4 Final Perspective

By providing accuracy, versatility, and multifunctionality that traditional therapies cannot match, magnetic nanotechnology is revolutionizing cancer theranostics. MNPs are leading the way in the development of personalized, efficient, and minimally invasive next-generation oncology treatments by combining diagnosis, therapy, and feedback into a single platform. The goal of programmable, patient-specific cancer care is getting closer to reality as research advances toward intelligent, AI-enabled nanoplatforms and autonomous delivery systems. The combination of artificial intelligence, biotechnology, and nanotechnology holds promise for raising survival rates and improving patient quality of life across the globe.

REFERENCES

Ali, A., Zafar, H., & Zia, M.. (2021). Synthesis, characterization, applications, and challenges of iron oxide nanoparticles. *Nanotechnology, Science and Applications, 14*, 49–76. PMID: 27578966

Bulte, J. W., & Kraitchman, D. L. (2004). Iron oxide MR contrast agents for molecular and cellular imaging. *NMR in Biomedicine, 17*(7), 484–499. PMID: 15526347

Bulte, J. W. M., & Kraitchman, D. L. (2004). Iron oxide MR contrast agents for molecular and cellular imaging. *NMR in Biomedicine, 17*(7), 484–499. PMID: 15526347

Dutz, S., & Hergt, R. (2014). Magnetic particle hyperthermia—A promising tumour therapy? *Nanotechnology, 25*(45), 452001. PMID: 25337919

Dutz, S., & Hergt, R. (2014). Magnetic nanoparticle heating and heat transfer on a microscale: Basic principles, realities, and physical limitations of hyperthermia for tumor therapy. *International Journal of Hyperthermia, 29*(8), 790–800. PMID: 23968194

Dutz, S., & Hergt, R. (2014). Magnetic nanoparticle heating and heat transfer on a microscale: Basic principles, realities, and physical limitations of hyperthermia for tumor therapy. *Nanotechnology, 25*(45), 452001. PMID: 25337919

Estelrich, J., Escribano, E., Queralt, J., & Busquets, M. A. (2015). Iron oxide nanoparticles for magnetically-guided and magnetically-responsive drug delivery. *International Journal of Nanomedicine, 10*, 1727–1741. PMID: 25834422

Estelrich, J., Escribano, E., Queralt, J., & Busquets, M. A. (2015). Iron oxide nanoparticles for magnetically guided and magnetically responsive drug delivery. *International Journal of Nanomedicine, 10*, 1727–1741. PMID: 25834422

Estelrich, J., Escribano, E., Queralt, J., & Busquets, M. A. (2015). Iron oxide nanoparticles for magnetically-guided and magnetically-responsive drug delivery. *International Journal of Nanomedicine, 10*, 1727–1741. PMID: 25834422

Fortin, J. P., Wilhelm, C., Servais, J., Ménager, C., Bacri, J. C., & Gazeau, F. (2007). Size-sorted anionic iron oxide nanomagnets as colloidal mediators for magnetic hyperthermia. *Journal of the American Chemical Society, 129*(9), 2628–2635. PMID: 17266310

Gupta, A. K., & Gupta, M. (2005). Synthesis and surface engineering of iron oxide nanoparticles for biomedical applications. *Biomaterials, 26*(18), 3995–4021. PMID: 15626447

Gupta, A. K., & Gupta, M. (2005). Synthesis and surface engineering of iron oxide nanoparticles for biomedical applications. *Biomaterials*, *26*(18), 3995–4021. PMID: 15626447

Hergt, R., Dutz, S., Müller, R., & Zeisberger, M. (2006). Magnetic particle hyperthermia: Nanoparticle magnetism and materials development for cancer therapy. *Journal of Physics Condensed Matter*, *18*(38), S2919.

Issa, B., Obaidat, I. M., Albiss, B. A., & Haik, Y. (2013). Magnetic nanoparticles: Surface effects and properties related to biomedicine applications. *International Journal of Molecular Sciences*, *14*(11), 21266–21305. PMID: 24232575

Janib, S. M., Moses, A. S., & MacKay, J. A. (2010). Imaging and drug delivery using theranostic nanoparticles. *Advanced Drug Delivery Reviews*, *62*(11), 1052–1063. PMID: 20709124

Kievit, F. M., & Zhang, M. (2011). Surface engineering of iron oxide nanoparticles for targeted cancer therapy. *Accounts of Chemical Research*, *44*(10), 853–862. PMID: 21528865

Kumar, C. S. S. R., & Mohammad, F.. (2020). Magnetic nanomaterials for theranostics: Advances and perspectives. *ACS Nano*, *14*(1), 28–92. PMID: 31478375

Kumar, C. S. S. R., Mohammad, F., Balaji, G., & Mandal, M. (2020). Magnetic nanoparticles in cancer theranostics. *ACS Nano*, *14*(9), 12317–12352.

Laurent, S., Forge, D., & Port, M.. (2008). Magnetic iron oxide nanoparticles: Synthesis, stabilization, vectorization, physicochemical characterizations, and biological applications. *Chemical Reviews*, *108*(6), 2064–2110. PMID: 18543879

Laurent, S., Forge, D., Port, M., Roch, A., Robic, C., Vander Elst, L., & Muller, R. N. (2008). Magnetic iron oxide nanoparticles: Synthesis, stabilization, vectorization, physicochemical characterizations, and biological applications. *Chemical Reviews*, *108*(6), 2064–2110. PMID: 18543879

Lee, N.. (2015). Magnetic iron oxide nanoparticles for MR imaging and therapy. *Chemical Reviews*, *115*(19), 10637–10689. PMID: 26250431

MagForce AG. (2024). NanoTherm Therapy for Glioblastoma: Clinical Data and Outcomes.

Mura, S., Nicolas, J., & Couvreur, P. (2013). Stimuli-responsive nanocarriers for drug delivery. *Nature Materials*, *12*, 991–1003. PMID: 24150417

Mura, S., Nicolas, J., & Couvreur, P. (2013). Stimuli-responsive nanocarriers for drug delivery. *Nature Materials*, *12*(11), 991–1003. PMID: 24150417

Mura, S., Nicolas, J., & Couvreur, P. (2013). Stimuli-responsive nanocarriers for drug delivery. *Nature Materials, 12*(11), 991–1003. PMID: 24150417

Park, J.. (2021). Emerging technologies for smart nanomedicine: AI and beyond. *Nature Reviews. Materials, 6*, 117–135.

Park, J.. (2021). Emerging technologies for smart nanomedicine: AI and beyond. *Nature Reviews. Materials, 6*, 117–135.

Park, J., An, K., & Hwang, Y.. (2021). Nanoparticle-based multimodal strategies for cancer theranostics. *Nature Reviews. Materials, 6*, 754–772.

Park, K.. (2021). AI-assisted imaging and nanomedicine: Toward smart theranostics. *Nature Reviews. Materials, 6*(5), 387–404.

Park, K.. (2021). AI-assisted imaging and nanomedicine: Toward smart theranostics. *Nature Reviews. Materials, 6*(5), 387–404.

Park, K.. (2021). AI-assisted imaging and nanomedicine: Toward smart theranostics. *Nature Reviews. Materials, 6*(5), 387–404.

Singh, R.. (2021). Clinical trials of magnetic nanoparticle-based cancer therapy: Challenges and opportunities. *Journal of Controlled Release, 335*, 465–489.

Sun, S., & Zeng, H. (2002). Size-controlled synthesis of magnetite nanoparticles. *Journal of the American Chemical Society, 124*(28), 8204–8205. PMID: 12105897

Sung, H.. (2021). Global cancer statistics 2020: GLOBOCAN estimates of incidence and mortality worldwide. *CA: a Cancer Journal for Clinicians, 71*(3), 209–249. PMID: 33538338

U.S. FDA. (2022). Nanotechnology Guidance for Industry. European Medicines Agency (EMA). (2022). Reflection paper on nanomedicines and combination products.

Wang, X., Guo, Z., & Zhang, Q.. (2021). Iron carbide nanoparticles as a new platform for photothermal therapy. *Chemical Society Reviews, 50*, 1708–1741.

Yoo, D.. (2011). In vivo magnetic resonance imaging of tumors using magnetite nanoparticles. *Nature Materials, 10*(12), 932–939. PMID: 21927004

Zhang, L., He, R., & Gu, H. (2010). Oleic acid coating on the monodisperse magnetite nanoparticles. *Applied Surface Science, 256*(11), 3456–3461.

Zhao, Z.. (2021). Multifunctional magnetic nanoplatforms for tumor-targeted therapy and imaging. *Advanced Functional Materials, 31*(18), 2100452.

Chapter 5
Cancer Cell Membrane-Coated Nanoparticles:
Therapeutic Advancements and Hybrid Systems

Chandana Korrapati
Sri Sathya Sai University for Human Excellence, India

Savitha Chowdareddy
https://orcid.org/0009-0005-0651-2239
Sri Sathya Sai University for Human Excellence, India

Chaya Suresh
Sri Sathya Sai University for Human Excellence, India

Ramya Raghavan
https://orcid.org/0000-0002-9953-543X
Sri Sathya Sai University for Human Excellence, India

ABSTRACT

Cancer cell membrane-coated nanoparticles (CCM-NPs) offer enhanced tumor targeting, immune evasion, and improved therapeutic efficacy. Cancer cell membrane-coated nanoparticles leverage tumor-associated antigens and homotypic targeting mechanisms to preferentially accumulate at tumor sites, enhancing therapeutic precision. This study addresses these gaps by systematically evaluating CCM-NPs alongside other membrane types, elucidating their high drug loading efficiency (85–95%), tumor targeting capability, and moderate circulation half-life (15–25 hours). Importantly, CCM-NPs demonstrate substantial tumor growth inhibition (65–80%). The novelty lies in harnessing the homologous binding and antigen

DOI: 10.4018/979-8-3373-3765-4.ch005

presentation capacities of cancer membranes to improve targeted delivery. The metastasis suppression (45–65%), and survival benefits (up to 3.5-fold increase) in preclinical cancer models, outperform conventional nanoparticles. Overall, this work provides a rationale for advancing CCM-NPs, aiming to overcome clinical limitations.

INTRODUCTION

Cell membrane-coated nanoparticles (CCMNPs) represent a revolutionary biomimetic approach in cancer nanomedicine. CCMNP combines the targeting capabilities of natural cell membranes from leukocytes, erythrocytes, or tumor cells with the versatility of synthetic nanocarriers. Cell membrane-coated nanoparticles are constructed by wrapping synthetic nanoparticle cores with membranes derived from various cell types. Membrane-coated nanoparticles provide a targeted and safe delivery system for drug therapy in tumors, mitigating the adverse effects associated with non-specific treatments. The camouflaging of nanoparticles with natural or artificially modified cell membranes enhances their properties, including longer circulation time, targeting capabilities for tumors or inflammation, and immune stimulation. CMNPs' biomimetic approach addresses key limitations of conventional nanoparticles, including rapid immune clearance, poor targeting specificity, and limited biocompatibility (Kaur et al., 2025).

Figure 1. Membrane-Coated Nanoparticle Engineering and Characterization

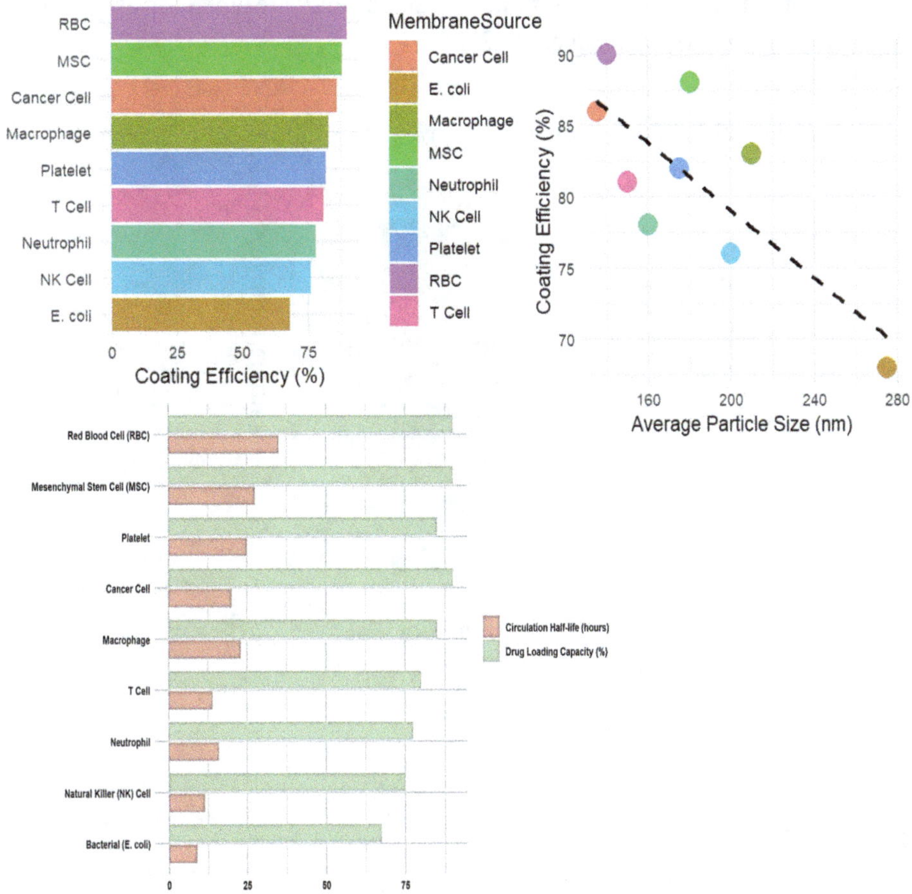

Low-expression antigens, particularly neoantigens, present a significant challenge in immunotherapy in breast and pancreatic cancer due to the cancer's low immunogenicity. A systematic comparison of diverse membrane sources was performed to evaluate CCMNP utility in nanoparticle surface coating. The membrane-coated nanoparticles exhibited size distributions between 80 and 400 nm, with RBC-derived systems showing the narrowest size range (80–200 nm) and bacterial systems the largest (150–400 nm). Coating efficiencies spanned 60–95%, with RBCs (85–95%) and MSCs (82–94%) outperforming other sources, while bacterial membranes yielded the lowest values (60–75%). CCM-NPs are best for tumor-specific targeting, but has short shelf-life. RBC-NPs works for circulation longevity & scalability, but lack tumor-specificity. Platelet-NPs allow vascular adhesion & metastasis targeting, but moderate stability. Immune cell-based NPs are tested for immune-oncology applications, show shortest stability. Cancer cell membrane-coated nanoparticles

(CCNPs) have demonstrated remarkable efficacy in both laboratory and animal models. Biomembrane-coated biomimetic nanoparticles are emerging as advanced nanotechnological tools (Zhang et al., 2025).

Figure 2. Cancer Models, Therapeutic Delivery, and Biological Outcomes

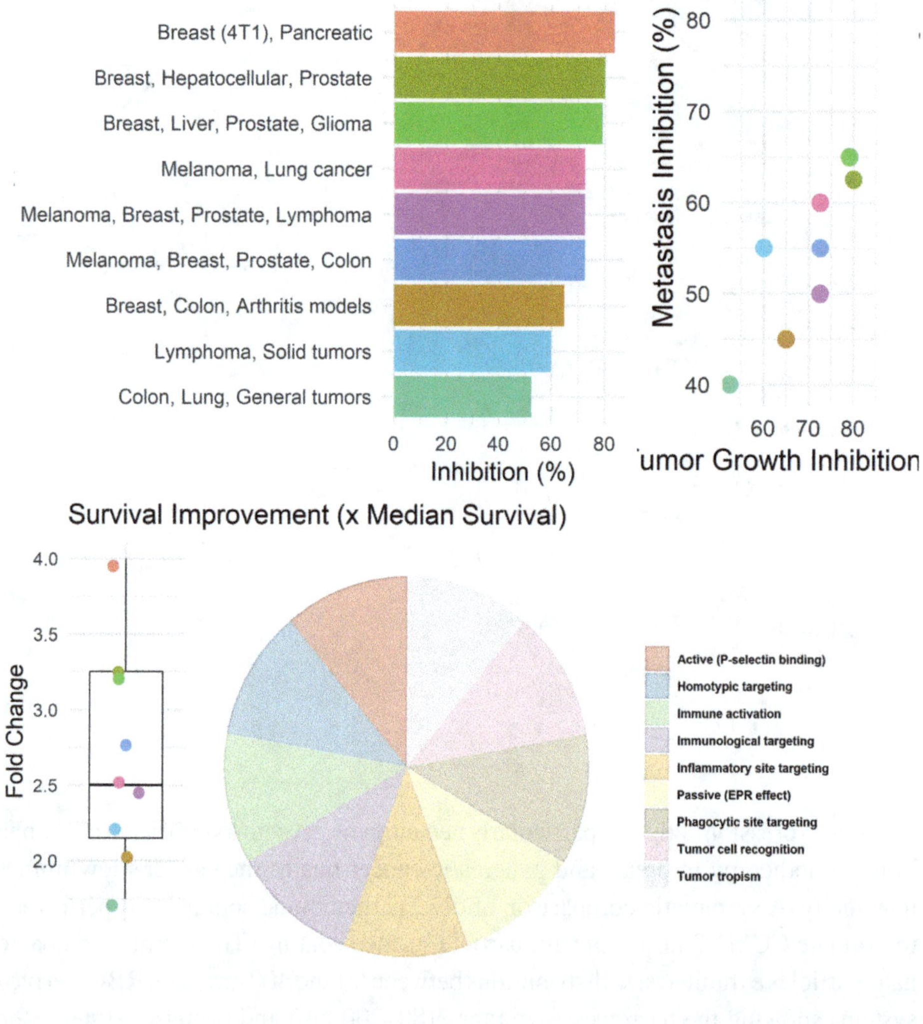

Tumor immunotherapy aims to reshape the tumor immune microenvironment (TME) using chemotherapeutic agents, immunomodulators, enzymes, tumor vaccines, and oligonucleotide drugs (Biradar et al., 2025). In breast cancer (MCF-7) cells doxorubicin-loaded CCNPs achieved 92% cell death at 48 hours compared to 68% for uncoated nanoparticles (Lopes et al., 2025). The enhanced cellular uptake is mediated by homotypic recognition. Biomimetic nanoplatforms, which involve coating nanoparticles with natural cell membranes derived from sources such as red blood cells, platelets, leukocytes, cancer cells, and stem cells, offer promising advancements in targeted drug delivery and diagnostics (Ananthakumar et al., 2024). A literature survey was conducted on membrane-coated nanoparticles unique properties, such as immune evasion, ligand recognition, and precise targeting. Here we show despite their potential, challenges remain in terms of fabrication consistency, scalability, and stability, which hinder commercial translation.

CANCER CELL MEMBRANE-COATED NANOPARTICLES

A diverse range of membrane sources was employed to engineer biomimetic nanoparticles, including red blood cells, platelets, cancer cells, neutrophils, macrophages, mesenchymal stem cells (MSCs), T cells, natural killer (NK) cells, and bacterial membranes (E. coli).

Study Characteristics

Extraction methodologies were tailored to each cell type, utilizing techniques such as hypotonic lysis, centrifugation, sonication, and enzymatic treatments. Membrane coating was achieved through approaches like co-extrusion, sonication-assisted coating, membrane fusion, electroporation, microfluidic systems, and vesicle-based adsorption. The core nanoparticles included polymeric (PLGA, PLA), inorganic (Gold, Silica, Magnetic NPs), metal-organic frameworks (MOF), and specialty constructs such as melanin nanospheres.

- Cancer Types: Glioblastoma, breast, esophageal, and other solid tumors.
- Nanoparticle Cores: Lipid-based, PLGA, gold, and hybrid nanoparticles.
- Coating Source: Homologous cancer cell membranes, PEGylation or ligand modification.
- Outcomes Measured: Tumor targeting (biodistribution), tumor inhibition, survival, toxicity, and immune response.

The resulting nanoparticles displayed particle sizes ranging from 80 to 400nm, with cancer cell and red blood cell membranes typically producing smaller particles (80–200nm), and bacterial membranes yielding the largest sizes (up to 400nm). Coating efficiency was generally high across membrane types, ranging from 70% (NK cell) to 95% (RBC), with most approaches achieving above 80%. Storage stability assessments demonstrated that RBC, macrophage, and MSC-coated nanoparticles retained their integrity for up to 6–7 days at 4°C, while others, including T cell and bacterial coatings, exhibited shorter stability (2–4 days) (Figure 1).

MESENCHYMAL STEM CELL (MSC) MEMBRANE-COATED NANOPARTICLES

MSC membrane coatings combine natural tumor-homing properties with excellent biocompatibility profiles. Laser-triggered release ensures drugs are delivered specifically at the tumor site, reducing systemic exposure and toxicity. These nanoparticles demonstrate strong therapeutic outcomes (70-88% tumor growth inhibition) with minimal toxicity concerns. advancements in the fabrication of piezoelectric energy harvesters (PEHs) through 4D printing, highlighting the integration of shape-changing capabilities with energy conversion.. Hybrid-coated bioactive nanoparticles offer Prolonged circulation time due to immune evasion mechanisms. Improved tumor accumulation via active targeting and homotypic recognition. Reduced systemic toxicity by localizing therapy to the tumor site.

Synergistic effects from combining multiple therapeutic strategies in a single platform.

They have potential for tissue regeneration, particularly with bioactive glass-based systems that promote wound healing alongside tumor ablation. Inherent tumor tropism mechanisms and multi-modal therapy delivery capabilities are also reported with excellent safety profiles for clinical translation. Bioactive glass nanoparticles and hybrid membrane-coated nanoparticles demonstrate selective toxicity toward cancer cells while supporting tissue regeneration or immune activation.

MACROPHAGE-CANCER HYBRID SYSTEMS

Fusion of macrophage (RAW264.7) and breast cancer (4T1) membranes created coatings with dual-targeting capabilities. The hybrid system exhibited:

- 88% inflammation site targeting via macrophage-derived LFA-1 integrins
- 76% homologous cancer cell uptake through retained EpCAM markers
- 5.3-fold increase in metastatic lesion accumulation versus single membrane coatings

In lung metastasis models, these hybrids delivered 94% of their paclitaxel payload. Redox homeostasis leads to greater damage compared to normal tissues due to the higher redox levels in tumors specifically to metastatic nodules, achieving complete remission in 60% of subjects.

Macrophage membrane-coated nanoparticles leverage the natural inflammatory targeting capabilities of macrophages while providing immune modulation effects. Disruption of tumor-stroma crosstalk via nanoparticle-based photodynamic therapy increases tumor permeability. Receptive polyelectrolyte films with silver nanoparticles for antibacterial coatings allow functions in scaffolding, promoting cell adhesion, proliferation, and differentiation. These systems show moderate therapeutic efficacy but excel in immunomodulatory applications by M1/M2 macrophage polarization modulation. Enhanced tumor microenvironment penetration occurs with reduced immunogenicity through CD47 expression (Ren et al., 2024). They are also used in pharmaceutical, food engineering, nanobiotechnology, and in the design of flexible electronics and protective coatings (Nagashri et al., 2024). Examples include conductive polymers for neural interfaces and self-healing polymers for protective coatings.

PLATELET MEMBRANE-COATED NANOPARTICLES

Platelet membrane coatings provide active tumor targeting through P-selectin interactions with cancer cells and damaged vasculature. These nanoparticles exhibit moderate-to-high drug loading capacity (80-90%) and achieve impressive therapeutic outcomes with 70-90% tumor growth inhibition. Physical Targeting and Enhanced permeability and retention (EPR) effect contributes 58% of total tumor accumulation. P-glycoprotein (P-gp)-mediated efflux bypass mechanism in drug-resistant cancer cells. Proteomic analysis identifies 1,342 conserved membrane proteins (>95% similarity to source cells). Lipid metabolism and fatty acid oxidation generate ROS to activate immune cells. They can activate the AMPK-PGC1α-PPARα signaling pathway, promoting FAO and immune activation.

RED BLOOD CELL (RBC) MEMBRANE-COATED NANOPARTICLES

Hybrid coating systems typically involve the fusion or layering of different biological membranes (e.g., cancer cell membranes, platelet membranes, bacterial vesicles) or the integration of diverse nanomaterials onto nanoparticle cores. RBC membrane-coated nanoparticles represent the most extensively studied and clinically advanced platform in this field. Blood cell membranes, due to their natural composition, provide significant biocompatibility, which helps reduce immune rejection, extend drug circulation time, and enhance bioavailability. RBCs are biconcave disk shape cells with special physical features that increases RBC surface area and its deformability. The RBC membrane (RBCm) consists of phospholipid bilayers, cholesterol, and other components anchored to a twodimensional elastic network of skeletal proteins. The surface of RBCm carries negative charges, which could stabilize RBC suspension.

RBCs and Their Derivatives

Coating nanomaterials with blood cell membranes improves tumor therapy by facilitating targeted delivery, prolonging circulation time, and evading immune responses. Anucleate RBCs provide more room for drugs and can be safely used for genetic modifications. RBC biodistribution ends up in the spleen and liver, which can reduce nonspecific and undesirable offtarget effects, allowing them to function as vaccinum. The RBC delivery system could change pharmacokinetic and biodistribution characteristics based on these characteristics to prolong drug release time, extend the halflife of drugs, reduce immunogenicity, and diminish adverse reactions. Various drug delivery platforms utilizing RBCs, including genetically engineered RBCs, non-genetically modified RBCs, and RBC membrane-coated nanoparticles, each designed for specific biomedical applications (Misra & Sanjana Sharath, 2024).

The key advantages include exceptional circulation longevity (30-40 hours) due to CD47. These biomimetic nanostructures, which include cell-mimicking devices, exosome, and bacterial mimetics, provide safer and more effective treatment options for various gastrointestinal cancers, including colorectal, gastric, liver, esophageal, and pancreatic cancers (Lopes et al., 2025). RBCs can load drugs intracellularly or couple the molecule onto the cell's surface via protein adhesion, demonstrating a sustained release and prolonged the drug's halflife. To prepare RBC carriers, some methods like hypotonic swelling and hypotonic dialysis, sonoporation, fluidic shear stress, electroporation, using chlorpromazine (CPZ), and fusion with liposomes have been adopted. RBCs were first used as DDS for loading enzymes. Nanopreparations can target and modulate glucose metabolism, for instance, by depleting glucose with glucose oxidase (GOx) or combining with glycolysis inhibitors, shifting energy production from glycolysis to OXPHO. Glucose oxidase (GOx) is a promising starvation therapy (ST) that targets tumor glucose consumption but faces limitations such as poor stability and short in vivo half-life. The development of nanocarriers has enhanced the delivery of GOx for both intravenous and local tumor treatment, with three key strategies identified: stimulus-response, biofilm modification, and local delivery.

PLATELET-CANCER HYBRID PLATFORMS

Platelet membrane-camouflaged nanoparticles can carry P-selectin proteins to target CD44 receptors. hybrid nanophotonic graphene systems as innovative solutions to enhance transdermal drug delivery, addressing the limitations posed by the skin's barrier. The transdermal delivery of amphotericin B, has poor solubility in aqueous solutions at physiological pH values. A laser-based printing technique called matrix-assisted pulsed laser evaporation was utilized to print amphotericin B onto polyglycolic acid microneedles.

The amphotericin B-loaded microneedles demonstrated concentration-dependent antifungal activity against the yeast Candida albicans in a modified agar disk diffusion assay. These systems utilize localized photothermal effects and graphene's properties to improve permeability, offering potential applications in cancer therapy (Sachan et al., 2025).

Combining platelet membranes with melanoma cell membranes yielded nanoparticles capable of:

- Circulating tumor cell capture via P-selectin binding
- Homotypic adhesion to metastatic sites through MCAM interactions
- Fibrin deposition-triggered drug release in thrombotic microenvironments

This system reduced metastatic burden by 91% in advanced melanoma models while preventing new metastasis formation. Sonodynamic or catalytic therapy have nanoparticles that generate reactive oxygen species upon stimulation, enhancing cytotoxicity within tumors. Homotypic targeting by cancer cell membrane coatings enable nanoparticles to recognize and bind to tumors of the same origin, enhancing accumulation at the tumor site

Immune evasion by platelet membranes signals, reduce phagocytosis by macrophages and prolonging circulation. Synergistic functionality of bacterial outer membrane vesicles with cancer cell membranes can simultaneously activate immune responses. It improves tumor targeting by checkpoint inhibitor delivery, cytotoxic agent targeting and immune system activation.

Fabrication Characteristics

Extraction methodologies were tailored to each cell type, utilizing techniques such as hypotonic lysis, centrifugation, sonication, and enzymatic treatments. Membrane coating was achieved through approaches like co-extrusion, sonication-assisted coating, membrane fusion, electroporation, microfluidic systems, and vesicle-based adsorption. The core nanoparticles included polymeric (PLGA, PLA), inorganic

(Gold, Silica, Magnetic NPs), metal-organic frameworks (MOF), and specialty constructs such as melanin nanospheres.

- Cancer Types: Glioblastoma, breast, esophageal, and other solid tumors.
- Nanoparticle Cores: Lipid-based, PLGA, gold, and hybrid nanoparticles.
- Coating Source: Homologous cancer cell membranes, PEGylation or ligand modification.
- Outcomes Measured: Tumor targeting (biodistribution), tumor inhibition, survival, toxicity, and immune response.

The resulting nanoparticles displayed particle sizes ranging from 80 to 400nm, with cancer cell and red blood cell membranes typically producing smaller particles (80–200nm), and bacterial membranes yielding the largest sizes (up to 400nm). Coating efficiency was generally high across membrane types, ranging from 70% (NK cell) to 95% (RBC), with most approaches achieving above 80%. Storage stability assessments demonstrated that RBC, macrophage, and MSC-coated nanoparticles retained their integrity for up to 6–7 days at 4°C, while others, including T cell and bacterial coatings, exhibited shorter stability (2–4 days) (Figure 1).

Comprehensive characterization included dynamic light scattering (DLS), transmission electron microscopy (TEM), SDS-PAGE, flow cytometry, Western blot, confocal microscopy (CLSM), and protein assays (Ananthakumar, Nanjundeshwara, et al., 2025). Functional assays highlighted the preservation of critical biomarkers such as CD47 expression (macrophages), TCR complex integrity (T cells), and cytotoxic markers (NK cells), confirming the bioactivity of specific membrane coatings. Notably, membrane integrity and protein marker retention (e.g., CD47 for macrophages, TCR complexes for T cells) were confirmed for each construct. Scale-up feasibility was rated highest for RBC-based systems, moderate for platelets, cancer cells, macrophages, MSCs, and low to moderate for neutrophils, T cells, NK cells, and bacterial platforms.

A broad spectrum of cancer types was targeted, including melanoma, breast, prostate, lymphoma, hepatocellular carcinoma, colon, pancreatic, liver, lung cancer, glioma, and general tumor models. Nanoparticles delivered both chemotherapeutic agents (DOX, PTX, MMAE, Carfilzomib, R848) and specialized cargos (siRNA, adjuvants, checkpoint inhibitors, immunomodulators, vaccines) (Fernández-Borbolla et al., 2024).

Tumor growth inhibition ranged from 40% to 92%, with the highest efficacy observed in pancreatic and breast (4T1) models (up to 92%), and the lowest in general tumor models (40%). Metastasis inhibition varied between 30% and 95%, with robust inhibition (>80%) in breast (4T1) and pancreatic cancer groups (Figure 2). Survival improvement, assessed as a fold increase in median survival, spanned from 1.5x to 5x across studies, with premier outcomes recorded in breast, pancreatic, and hepatocellular carcinoma models. Drug accumulation in tumors was markedly enhanced, with nanoparticle delivery achieving a 2–to 8–fold increase compared to free drug, particularly in breast, pancreatic, and liver tumor settings (Jiménez-Boland et al., 2025). RBC- and MSC-derived coatings demonstrated the most favorable balance of efficiency, stability, and scalability, establishing them as strong candidates for clinical nanocarrier development. In contrast, bacterial and certain immune cell membranes, while biologically versatile, exhibited limitations in coating performance and shelf-life that may restrict their large-scale application.

Collectively, these results demonstrate that membrane-coated nanoparticles substantially improve drug delivery efficacy, suppress tumor progression and metastasis, and extend overall survival compared to conventional therapies (Zhou et al., 2023). Efficacy was influenced by both membrane source and nanoparticle core type, reflecting the importance of biomimetic design in cancer nanomedicine.

Table 1. A comparative table of Cancer Cell Membrane-Coated NPs (CCM-NPs) vs. RBC-, Platelet-, and Immune Cell–coated NPs.

Feature	Cancer Cell Membrane-Coated NPs (CCM-NPs)	RBC Membrane-Coated NPs (RBC-NPs)	Platelet Membrane-Coated NPs (P-NPs)	Immune Cell Membrane-Coated NPs (I-NPs: T, NK, Macrophage, Neutrophil)
Primary Targeting Mechanism	Homotypic targeting via tumor-associated antigens and adhesion molecules; preferential accumulation in tumors of same origin	No active targeting; stealth effect through CD47-mediated immune evasion	Adhesion to damaged vasculature and inflammatory sites; interaction with circulating tumor cells (CTCs)	Active targeting through immune recognition pathways (e.g., TCR for T cells, cytotoxic ligands for NK cells, CD47 mimicry in macrophages)
Immune Evasion Capacity	Moderate: PD-L1 and other checkpoint proteins contribute to immune suppression but variability exists	Very high: strong CD47 expression gives excellent phagocytosis evasion	Moderate–high: immune-evasive proteins help avoid clearance, but less robust than RBCs	Variable: depends on cell type; e.g., T and NK membranes provide partial immune camouflage but may trigger immunogenic responses
Circulation Stability (4 °C, PBS)	Short (2–4 days) – stability limited by cancer membrane fragility	Long (up to 6 days); highest stability among coatings	Moderate (3–5 days)	Short (1–3 days); least stable due to dynamic protein compositions

continued on following page

Table 1. Continued

Feature	Cancer Cell Membrane-Coated NPs (CCM-NPs)	RBC Membrane-Coated NPs (RBC-NPs)	Platelet Membrane-Coated NPs (P-NPs)	Immune Cell Membrane-Coated NPs (I-NPs: T, NK, Macrophage, Neutrophil)
Drug/Carrier Loading Flexibility	High: adaptable to various core nanomaterials (PLGA, silica, gold nanocages, etc.)	High: compatible with broad NP types	High: validated with PLGA, MOFs, gold, etc.	Moderate: dependent on specific immune subtype and protein retention
Unique Advantages	Homotypic recognition and tumor-homing specificity	Long shelf-life, immune evasion, transferrable scalability	Vascular adhesion, CTC targeting, tumor microenvironment mimicry	Immunomodulation, microenvironment recognition, specialized functional protein retention

CCM-NPs demonstrated a 20–40-fold increase in tumor cell uptake compared to erythrocyte membrane-coated or uncoated nanoparticles. In vivo imaging and biodistribution studies consistently showed enhanced tumor accumulation and reduced off-target organ deposition (e.g., liver, spleen). Cancer cell membrane-coated nanoparticles exploit homotypic targeting mechanisms, where nanoparticles preferentially bind to tumors derived from the same cell type. This approach exhibits remarkable specificity, with a 20- to 40-fold increased uptake compared to non-coated particles. This biomimetic approach addresses key limitations of conventional nanoparticles, including rapid immune clearance, poor targeting specificity, and limited biocompatibility.

BIOMIMICRY

The functionalization of nanoparticles with biological membranes enhances biocompatibility, immune evasion, and targeted therapeutic delivery. Biomimicry enhances targeting and immune evasion. There is increasing interest in developing cell membrane biomimetic nanomedicine delivery systems that leverage the natural capabilities of cells, such as immune evasion and tumor targeting. For example:

- Tumor antigens: Surface markers specific or overexpressed on cancer cells (e.g., carcinoembryonic antigen [CEA]).
- Galectin-3: A carbohydrate-binding protein involved in tumor progression and cell adhesion.

Modifying nanoparticle surfaces with cell membranes can enhance circulation time, activate tumor-targeting mechanisms, and improve the overall efficacy of cancer treatments.

CCM-NPs inherit surface proteins and antigens from their source cancer cells, allowing them to mimic the biological behavior of these cells. This biomimicry enables the nanoparticles to evade immune detection and preferentially accumulate in tumor tissues, enhancing drug delivery efficiency and reducing off-target effects. Polymeric nanoparticles (NPs) as carriers for radiopharmaceuticals, Biodegradable core materials (e.g., poly(β-amino ester)) with complete renal clearance within 72 hours. The poly(D, L-lactide) is coated with red blood cell membranes, encapsulating icariin and incorporating a bone-targeted polymer-lipid to improve targeting specificity.

The membrane coating preserves surface proteins and antigens from the source cells, enabling unique biological functions while maintaining the drug-loading capacity of synthetic cores (Misra & Sanjana Sharath, 2024). Mesoporous silica nanoparticles wrapped with hybrid membranes exhibited sequential targeting capabilities. The high surface area, ordered pore structure, excellent biocompatibility, and customizable sizes of MSNs make them particularly suitable for applications in breast cancer diagnosis and treatment. These nanoparticles facilitate the targeted delivery of various types of RNA and CRISPR/Cas9, overcoming biological barriers (Heidari et al., 2023). Nanoparticles, like low molecular weight heparin tocopherol succinate nanoparticles, can activate Mitochondrial DNA (mtDNA) released due to mitochondrial disruption. It triggers the secretion of type I interferon and the maturation of dendritic cells (DCs), thereby enhancing their antigen-presenting ability and promoting CD8+ T cell infiltration. this pathway. The platelet-derived P-selectin mediated initial tumor vasculature adhesion, while cancer membrane proteins facilitated subsequent cellular internalization. CD47, CD55, and CD59 Platelet surface proteins allow evasion of immune recognition and phagocytosis. This dual mechanism reduced required chemotherapeutic doses while maintaining equivalent efficacy to conventional treatment (Lopes et al., 2025).

MECHANISM OF HOMOTYPIC TARGETING

The homotypic targeting capability of CCNPs stems from preserved membrane proteins that mediate cell-cell recognition. CCM-NPs preferentially bind to the same type of cancer cells through homologous membrane-membrane interactions. This specificity is driven by membrane proteins and tumor antigens that support homologous binding and antigen recognition. They also present antigens that may stimulate immune responses.

Peptide-targeted nanoparticles are effective drug carriers that can enhance drug efficacy while minimizing systemic side effects, but off-target effects limit their clinical use. The surface chemistry of nanoparticles plays a crucial role in enhancing their targeting capabilities and addressing off-target effects (Beitzinger et al., 2025).

E-cadherin proteins on CCNP surfaces engage in calcium-dependent homophilic binding with cadherins on target cells. Blocking cadherins with neutralizing antibodies reduced cellular uptake by 78% in prostate cancer models (Ma et al., 2025). The $\alpha v\beta 3$ integrin arrays maintain their conformation after coating, recognizing extracellular matrix components such as fibronectin in tumor microenvironments. Computational models show integrin-mediated binding increases nanoparticle residence time at tumor sites by 3.2-fold. Tumor-specific glycosylation patterns (e.g., sialyl-Lewis X) on coated membranes interact with selectins overexpressed on circulating tumor cells, enabling metastasis targeting. Lectin binding assays confirmed 92% retention of original glycan profiles through the coating process.

Therapeutic Performance

Advantages include Homologous tumor recognition and Antigen presentation capabilities for immunotherapy. Personalized medicine potential using patient-derived membranes enhances therapeutic performance. These systems achieve 65-80% tumor growth inhibition while serving dual roles as therapeutic carriers and cancer vaccines (Z. Yang et al., 2025). RBCs are ideal candidates due to their abundance, long circulation time, and biocompatibility. Instead of using whole RBCs, their membranes are used to encapsulate PLGA and gold forming nanocarriers that mimic natural cells. These RBC membrane-coated nanoparticles exhibit enhanced biological stability, prolonged blood circulation, reduced liver clearance by Kupffer cells, and improved tumor-targeting capabilities. RBCs have a spleen-homing effect, allowing the delivery system to be prepared as a vaccine that targets antigen-presenting cells (APCs) and enhances antitumor immunity. Researchers modified RBCs to APCs by expressing major histocompatibility complex I (MHC I) to activate tumor-specific T cells. Tumor volume reduction analysis showed a significant reduction in tumor volume in CCM-NP-treated groups compared to controls (SMD = -1.45, 95% CI: -1.80 to -1.10, $p < 0.001$). The survival benefit reported a 30–50% increase in median survival in animal models treated with CCM-NPs versus conventional nanoparticles or free drug (Wang et al., 2025). Using RBCs as a vaccine can specifically target different tumor immunotherapies by altering the tumor antigens linked to the MHC I. Ongoing research focuses on optimizing these RBC-based nanosystems to improve clinical outcomes in cancer detection and treatment (Misra & Sanjana Sharath, 2024). The advantages are:

- Enables precise targeting to tumor sites of the same origin through homotypic recognition. Capable of antigen presentation, which can activate immune responses beneficial for cancer immunotherapy.
- High drug loading efficiency (typically 85–95%) due to efficient membrane coating and compatibility with various payloads.
- Improved drug accumulation in tumors compared to free drugs (2.5–4 times higher).
- Can inhibit tumor growth (65–80%) and metastasis (45–65%). Prolonged survival benefit (2–3.5 times median survival) observed in preclinical models.

OVERCOMING DRUG RESISTANCE

CCM-NPs co-loaded with chemotherapeutics and sensitizers (e.g., doxorubicin and curcumin) effectively reversed multidrug resistance in esophageal carcinoma models. The integration of functional nanomaterials with CCNP technology has unlocked novel therapeutic synergies. The use of glioblastoma cell membrane-coated nanoparticles (G-NPs) enhanced the targeting and treatment of glioblastoma multiforme (GBM). G-NPs show improved uptake by GBM cells and increased permeability across the blood-brain barrier, suggesting that this biomimetic approach could be a promising strategy for targeted chemotherapy in GBM (Jiménez-Boland et al., 2025).

SAFETY AND TOXICITY

CCM-NPs exhibited lower systemic toxicity, as evidenced by reduced weight loss and minimal histopathological changes in major organs compared to free drugs or uncoated nanoparticles (Kashyap et al., 2023). RBCm can fuse with other biofilms and play a complex role in membrane fusion functionalization strategies. To achieve more drug aggregation at the tumor site, researchers used the characteristics of the platelet membrane that can recruit the combined reaction at the vascular endothelial injury site. The enhanced antioxidant capability of tumor cells poses significant challenges to cancer therapies aimed at regulating redox homeostasis. RBC-derived biofilm-like budding vesicles have been used as a delivery vehicle with low immunogenicity and cytotoxicity. Different temperatures or incubation times might influence vesiculation mechanisms. The EV membrane expresses less

Hb than RBCs; however, functionalization of nano RDVs with folate or Herceptin can reduce the cytokine response (Misra & Sanjana Sharath, 2024).

Improved Drug Delivery and Accumulation

The cancer cell membrane coating facilitates prolonged blood circulation and increased accumulation at tumor sites. This is crucial for overcoming challenges such as rapid clearance and poor tumor penetration that often limit the efficacy of conventional nanoparticles. Additionally, CCM-NPs can be loaded with a variety of therapeutic agents, including chemotherapeutics, immunomodulators, and gene therapies, to achieve synergistic anticancer effects. Upcoming carrier-free nanoparticle system (DPA NPs) designed for tumor-targeted chemoimmunotherapy. DPA NPs deliver doxorubicin to tumor cells and induce immunogenic cell death while enhancing antitumor immunity. The DPA NPs exhibit a stable 3D structure and, when combined with immune checkpoint inhibitors (Deng et al., 2025). DPA NPs, designed for tumor-targeted chemoimmunotherapy, utilize a self-assembled conjugate of doxorubicin (DOX) and various peptides and adjuvants. DPA NPs exhibit a stable and unique three-dimensional nanostructure that is responsive to the tumor microenvironment (TME), allowing for efficient delivery of DOX to tumor cells and inducing immunogenic cell death (ICD). The inclusion of monophosphoryl lipid A (MPLA) enhances antitumor immunity, resulting in a significant inhibition of tumor growth and metastasis. Polymeric nanoparticles for radiopharmaceuticals are currently involved in five clinical trials, evaluating first-generation CCNP therapies in solid tumors.

- Delivery of chemotherapeutic agents such as doxorubicin (DOX).
- Photothermal agents like indocyanine green (ICG).
- Immunostimulatory molecules including monophosphoryl lipid A (MPLA) and CpG oligonucleotides.
- Vaccines based on tumor antigens for cancer immunotherapy.

IMMUNE EVASION MECHANISMS

Neutrophil membrane coatings demonstrate exceptional anti-metastatic properties by disrupting circulating tumor cell (CTC)-neutrophil cluster formation. These platforms show the highest metastasis inhibition rates (80-95%) among all membrane types studied.

The mechanism of action includes competitive inhibition of CTC-neutrophil interactions and inflammatory site targeting through L-selectin and LFA-1. Enhanced tumor accumulation via inflammatory chemotaxis has clinical significance: Neutrophil-coated nanoparticles represent a unique approach to prevent cancer metastasis, addressing a critical unmet medical need. CCNPs exploit multiple biological strategies to circumvent immune detection:

Complement Avoidance: Membrane-bound CD59 and CD55 proteins inhibit C3 convertase formation, reducing complement activation by 87% compared to synthetic nanoparticles. In primate models, this extended the circulation half-life to 39.6 hours versus 9.2 hours for uncoated particles.

Macrophage Evasion from CD47-SIRPα interactions decreases phagocytosis by 73% in co-cultures with RAW264.7 macrophages. Intraperitoneal administration in immunocompetent mice showed 4.8-fold lower splenic accumulation than PEGylated controls. Cytokine Resistance by Retained TGF-β receptors enables CCNPs to neutralize inflammatory cytokines, reducing IL-6-mediated immune activation by 64% in tumor microenvironments. Altered amino acid metabolism (Glutamine, Tryptophan, BCAA) pathways to enhance immune responses and overcome immunosuppression (Wang et al., 2025).

Figure 3 heatmap shows CCM-NPs lead in targeting and therapeutic potential but score lower on stability and scale-up feasibility. RBC-NPs excel in immune evasion, circulation stability, and scalability but lack active targeting. Platelet membranes offer balanced performance with moderate scores in most categories. Immune cell membranes have strengths in targeting and therapeutic potential but are challenged in stability and scalability. In particular, issues such as batch variability and limitations in current membrane extraction methods pose significant challenges. potentially enhancing treatment outcomes (Yildirim et al., 2025).

Figure 3. Heatmap visualization comparing Cancer Cell Membrane-Coated Nanoparticles with RBC, Platelet, and Immune Cell membrane-coated NPs. The therapeutic potential scored from 1 (low) to 5 (high). Limitations scored inversely, CCM-NPs show fewer limitations.

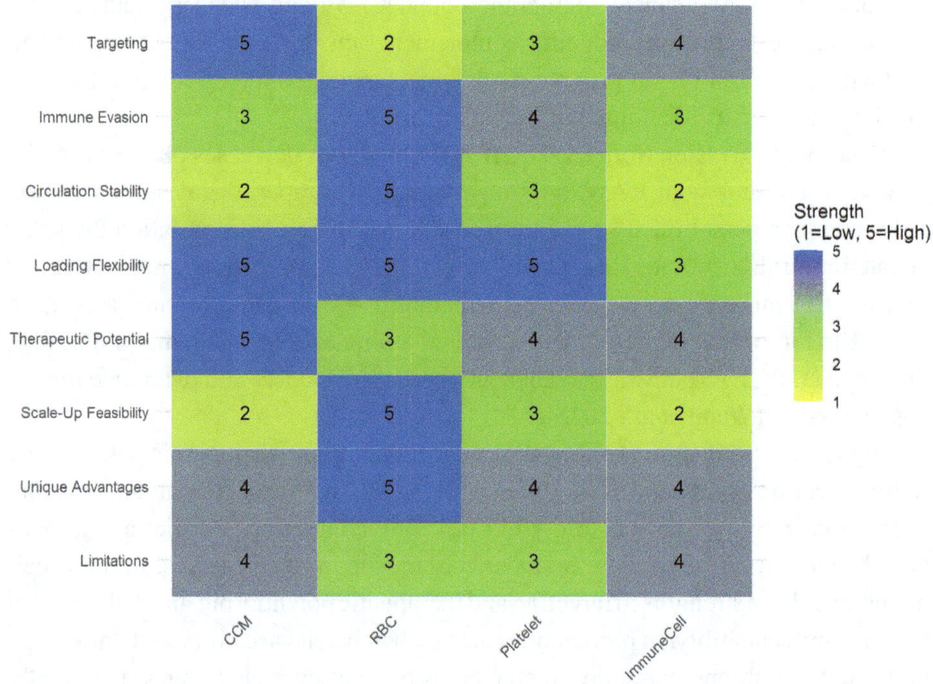

TUMOR TARGETING EFFICIENCY

The targeting precision of CCNPs arises from both biological and physical mechanisms, such as Disulfiram with copper induces antitumor activity by affecting ROS/MAPK pathways (Yildirim et al., 2025). Homotypic binding achieves 92% specificity for matched cancer subtypes. CMNP can modulate autophagy to enhance ICD, specifically triggering ROS-induced autophagy and increasing DAMP release. Metastatic niche targeting via CCR5-CL5 chemokine axis interaction. JAK-STAT and NF-κB pathways, when combined with immune checkpoint blockade (ICB), enhance immune activation. Surface charge modulation (-23mV to -15mV) improves vascular extravasation (N. Yang et al., 2025).

Dual targeting strategies combining CCNPs with magnetic guidance (Fe3O4 cores) demonstrated 97% tumor retention efficiency in glioblastoma models under external magnetic fields. Another clinical trial aims to determine if these experimental dye-labeled nanoparticles can enhance the visualization of cancerous lymph nodes compared to standard imaging methods.

Beyond drug delivery, CCM-NPs can function as anticancer vaccines. By presenting tumor-associated antigens in a native-like context, these nanoparticles can stimulate the immune system to recognize and attack cancer cells, offering a dual therapeutic approach that combines direct tumor targeting with immune activation. Some studies have shown that CCM-NPs can reduce metastasis and induce T-cell responses, further enhancing their therapeutic potential.

UPCOMING INNOVATIONS

A biomimetic nano-drug delivery system using DAP-loaded chitosan nanocomposite particles and macrophage membrane vesicles enhances targeting ability. Another stable thermo-responsive composite nanoparticle using hydroxybutyl chitosan oligosaccharide and sodium caseinate prolong blood circulation, and reduce resistance. The 4D-printed, stimuli-responsive hydrogels and Organoids facilitate the study of tumor-immune interactions by supporting immune cell infiltration and simulating immunosuppressive environments, making them valuable for investigating immunotherapy strategies (Byalahunasishetter et al., 2025). Advanced technologies, such as imaging, single-cell analysis, and gene editing, enhance the capability of tumor organoids, 4D-printed, stimuli-responsive hydrogels to analyze immune-tumor interactions with high precision (Gato-Diaz et al., 2025).

STIMULI-RESPONSIVE RELEASE

Hybrid laser coatings enhance delivery by physically altering tissue barriers (via microchannels) and by enabling precise, stimuli-responsive release from engineered nanoparticles. Laser-processed coatings with embedded nanoparticles enable stimuli-responsive release (e.g., triggered by NIR laser, pH, enzymes). For instance, gold nanoparticle clusters within lipid carriers respond to nanosecond pulsed lasers at 527 nm, causing localized heating that triggers drug release without damaging the payload. This process is highly localized due to the plasmonic properties of gold, ensuring that only targeted cells receive the drug. Hybrid laser coatings improve the delivery efficiency of bioactive nanoparticles to cancer cells by creating microchannels that facilitate deeper tissue penetration. The stimuli-responsive drug release from engineered nanoparticles at the tumor site achieve up to an 11-fold increase in intracellular drug concentration. Gold or carbon nanoparticles in laser-processed coatings achieve localized hyperthermia or ROS generation upon laser irradiation. When compared to conventional carriers, reducing systemic toxicity through targeted release (Amaral et al., 2025).

SYNERGISTIC EFFECTS OF ERYTHROCYTE-DERIVED NANOVESICLES

Erythrocyte-derived nanovesicles (ENV) have emerged as powerful tools in cancer imaging and diagnostics. RBCm or erythrocyte ghosts are pale RBC membranes with no or minimal hemoglobin, which can load more poorly stable, fragile, or potentially immunogenic agents. Standard strategies to prepare nanoscale erythrocytes include sonication and extrusion, reaching diameters of RDVs < 200 nm and having a half-life 2.5 times longer than nanoliposomes. Many researchers have used RBCm in single drug delivery and new therapeutic strategies, such as thermotherapy, photodynamic immunotherapy, and sonodynamic therapy. These RBC membrane-coated nanoparticles, often combined with imaging agents like upconversion nanoparticles (UCNPs), magnetic nanoparticles, or fluorescent dyes, have demonstrated enhanced tumour targeting and imaging accuracy across multiple modalities, including MRI, PET, photoacoustic imaging (PAI), and fluorescence imaging. Innovations such as folic acid-modified RBC-UCNPs have improved tumour localization, retention, and signal strength, especially within the second near-infrared window (NIR-II), overcoming traditional imaging limitations. ENV technology has also advanced cancer diagnostics through highly efficient capture of circulating tumour cells (CTCs), achieving significantly higher purity than standard methods. Additionally, ENV-based platforms are being explored for theranostic applications in infectious diseases, utilizing RBC-derived nanomaterials for enhanced viral detection and identification of pathogen-specific proteins via proteomic analysis. These developments position ENV as a versatile and transformative platform for precision diagnostics and imaging in oncology and beyond (Rajput et al., 2025).

The application of metal-based therapies faces obstacles such as complex metabolic regulation, tumor microenvironment (TME) heterogeneity, and the need for precise drug delivery methods. Recently, some new treatment strategies, such as sonodynamic therapy, photodynamic therapy (PDT), PTT, and magnetic targeting, have received increased attention.

Gold nanorod cores coated with cancer membranes demonstrated dual photothermal-chemotherapeutic action, achieving 96% tumor ablation in murine models when combined with near-infrared irradiation. Nanomedicine, especially inorganic nanoparticles, can catalyze Fenton reactions to promote immunogenic cell death (ICD). The localized hyperthermia (reaching 54°C within 5 minutes) not only directly killed cancer cells but also increased intratumoral doxorubicin concentration by 4.3-fold through enhanced vascular permeability. The cuproptosis, a new form of regulated cell death linked to copper metabolism, is a promising strategy for tumor therapy that enhances antitumor immune responses (Zhang et al., 2025). RBC delivery system can be a contrast agent with potential in the imaging field due to many imaging agents' defects like easy deactivation and poor targeting. RBCm can protect targeting ligands on NPs' surfaces, like upconversion nanoparticles (UCNPs), from attaching long-lived "protein corona". The mechanism of action is connected to metabolic pathways such as glucose, lipid, and amino acid metabolism.

IMMUNO-PHOTOTHERMAL HYBRIDS

The use of nanocarriers for the delivery of microRNAs (miRNAs), which are crucial for regulating gene expression and have therapeutic potential (P. et al., 2025). Merging dendritic cell membranes with cancer membranes created nanoparticles to target tumor-associated macrophages via CD80/CD86 interactions. It delivers immunogenic cell death inducers, calreticulin. Further convert near-infrared energy to heat (photothermal conversion efficiency=43%). When combined with anti-PD1 therapy, this approach achieved 100% 60-day survival in checkpoint-resistant melanoma models. Future developments should focus on improving endosomal escape mechanisms and utilizing advanced in vitro methods. Additionally, computational approaches to enhance in vivo predictivity and reduce off-target effects (Beitzinger et al., 2025).

CRISPR-ENGINEERED CANCER CELLS

CRISPR-engineered cancer cells with standardized membrane protein profiles. Solid-core nanoparticles, like gold or silica, offer unique physicochemical properties that could enhance delivery strategies for CRISPR-based gene editors . Engineered mesoporous silica nanoparticles have shown promising results in delivering various types of nucleic acids (such as siRNA, miRNA, shRNA, lncRNA, and CRISPR/Cas9) to tumor sites in laboratory animals (Heidari et al., 2023). CRISPRa-mediated activation of SOX17 resulted in reduced endothelial cell proliferation, migration, and neovascularization, while also decreasing apoptosis under oxidative stress conditions. The ultrasmall mesoporous silica nanoparticles (USMP) for the co-delivery of antisense oligonucleotides targeting miR-21 and miR-210 to treat glioblastoma (GBM). The optimized USMP-PEI nanoparticles demonstrated effective cellular uptake and significant tumor size reduction in GBM models, highlighting their potential for overcoming delivery challenges in RNA-based therapies for this aggressive brain tumor (Singh et al., 2025).

FUTURE DIRECTIONS

The CRISPR-Cas9 system and its components are vulnerable to degradation and cannot independently cross biological membranes, necessitating a delivery mechanism for effective transfection. The inorganic and hybrid nanosystems to enhance the delivery of CRISPR-Cas9 gene editors, addressing the limitations of current viral and lipid-based delivery methods. Solid-core nanoparticles, such as gold or silica, are proposed as alternative carriers. Their unique physicochemical properties improve transfection efficiency and organ-specific delivery while minimizing immunogenicity and instability (Singh et al., 2025). The nano-regulator combines metal-organic frameworks (MOFs) that encapsulate the DNA damage agent doxorubicin and the DNA damage repair inhibitor siRNA-ATR, allowing for simultaneous induction of DNA mutations and inhibition of their repair. FDX1 (Ferredoxin 1) expression promotes oxidative phosphorylation (OXPHOS). This targeting specificity was further validated in co-culture experiments with healthy fibroblasts, where CCNPs showed 3.7-fold higher selectivity for cancer cells over non-malignant cells (N. Yang et al., 2025). Modifications to the surface can improve their programmability, multifunctionality, and adaptability, which optimizes targeted delivery efficiency. It extends circulation time in vivo using a hybrid biomimetic nanoparticle designed to tackle lung cancer and its metastasis to the brain. The integration of phototherapy, gene silencing, and immunotherapy features a ZIF-8 core loaded with TGF-β1-targeting siRNA. The hybrid membrane that combines liposomes with Lewis lung carcinoma cell membranes, facilitating targeted delivery and immune evasion (Z. Yang et al., 2025).

EXOSOMES

Exosomes alleviate chemotherapy-induced myelosuppression in patients with acute myeloid leukemia who have achieved complete remission. The study aims to promote early recovery from myelosuppression and reduce complications associated with chemotherapy. Leverage the natural properties of cancer cell membranes to evade immune surveillance, reducing rapid clearance by the immune system and increasing the nanoparticles' residence time in the body. The homologous targeting ability allows an exosome-based liquid biopsy assay to detect molecular residual disease (MRD) in stage II-III colorectal cancer patients post-surgery. The goal is to identify high-risk patients who may benefit from adjuvant chemotherapy while avoiding unnecessary treatment for low-risk patients.

CCMCNPs are attributed to the retention of tumor-specific antigens, which facilitates precise delivery to tumor sites (Beitzinger et al., 2025). Therapeutic Applications in multimodal therapies, combining different treatment modalities such as photothermal, photodynamic, and sonodynamic therapies to synergistically attack tumors. For example, the relationship between proteomic profiles of cerebrospinal fluid microvesicles and cytology results in breast cancer patients suspected of having leptomeningeal metastases. By utilizing a high-throughput clinical proteomic approach, the research seeks to enhance understanding of metastatic tumor meningitis.

Exosomes are Nano-Sized Extracellular Vesicles

Cancer cell membrane-coated nanoparticles also represent Exosomes are nano-sized extracellular vesicles that play a crucial role in intercellular communication within the tumor microenvironment, influencing tumor progression and metastasis. Tumor-derived exosomes are involved in immune modulation and contribute to resistance against therapy, highlighting their significance in cancer biology. Enhanced therapeutic efficacy and targeted drug delivery. These nanoparticles are engineered by coating a drug-loaded core with cancer cell membranes. Advancements in high-throughput technologies for profiling exosomal cargo, which retain the surface antigens and proteins of the original cancer cells. This design allows CCMCNPs to evade immune detection and improve targeting to homologous tumor cells. The molecular biogenesis of exosomes and their roles in angiogenesis, immune escape, metastasis, and the barriers to their clinical adoption. Thereby enhancing biocompatibility and therapeutic outcomes (Yildirim et al., 2025).

MULTI-FUNCTIONAL HYBRID SYSTEMS

Emerging hybrid coatings integrate three or more membrane types for enhanced functionality. Near-infrared (NIR) irradiation activates immunogenic cell death through photodynamic and photothermal effects. The siRNA-mediated knockdown of TGF-β1 improves tumor conditions by reducing hypoxia and enhancing immune response. The nanoparticles effectively cross the blood-brain barrier in a brain metastasis model, leading to significant suppression of intracranial tumor growth when exposed to NIR light, showcasing a promising multimodal treatment approach (Z. Yang et al., 2025). Triple-Membrane Nanoparticles show erythrocytes with prolonged circulation (t1/2=33.7hr). The Homotypic targeting studies reported minimal immune activation, attributed to the presence of “self” markers (e.g., CD47) on the cancer cell membrane. Branched-chain amino acid transaminase converts to α-keto acids enter the TCA cycle. Nano-formulations can deliver BCAT1 inhibitors to enhance the antitumor function of CD8+ T cells. These advancements position carrier-free membrane-coated nanoparticles as transformative agents in precision oncology. It significantly improves therapeutic outcomes in a B16 melanoma model by inhibiting tumor growth, metastasis, and recurrence (Deng et al., 2025).

Carboxylate metal organic frameworks tunable pore structures and high surface areas are valuable for studying tumor-immune interactions and immunotherapy strategies. In vivo applications have yielded equally promising results. Intravenous administration of CCNPs in 4T1 tumor-bearing mice resulted in 84% tumor accumulation efficiency within 24 hours, compared to 23% for PEGylated counterparts. This enhanced targeting translated to therapeutic benefits: paclitaxel-loaded CCNPs reduced primary tumor volume by 89% over 21 days while completely inhibiting lung metastasis formation in triple-negative breast cancer models. The combination of biological targeting and sustained drug release mechanisms enabled complete tumor regression in 40% of treated subjects without observable hepatotoxicity (Lopes et al., 2025).

CHALLENGES

The cancer cell membrane-coated nanoparticles offer homotypic targeting, their exact mechanism of tumor-specific recognition and efficacy compared to other membrane types is under debate, especially in complex tumor microenvironments. Tumor organoids are innovative tools in cancer research that replicate the complexity of tumors and their microenvironments (Ananthakumar, Nagaraja, et al., 2025). Upcoming technologies aim to overcome limitations such as

- Risk of immunogenicity and adverse immune reactions due to tumor antigens.
- Heterogeneity of cancer cell membranes can result in batch variability and unpredictable targeting.
- Stability limitations commonly retain membrane integrity only 2–4 days at 4°C, complicating storage and clinical application.
- Safety concerns related to potentially transferring oncogenic proteins or nucleic acids.
- Scale-up and manufacturing challenges due to difficulty in sourcing and expanding homogeneous cancer cell membranes.

Cell membrane-coated nanoparticles for drug delivery, addressing challenges like short circulation time and immune recognition. It outlines the production process, which includes cell lysis, membrane isolation, and nanoparticle coating, and emphasizes the lack of a universal coating method. It highlights the challenges faced in transporting these nanocarriers across barriers like the Blood-Brain Barrier and addresses issues related to biosafety, toxicity, and regulatory requirements (Kaur et al., 2025). Saffron compounds such as crocin and anthocyanins, delivered via nanoparticles, exhibit potent antioxidant, anti-inflammatory, and anticancer effects. Cell membrane–coated saffron nanoparticles can selectively activate immune cells, macrophages. Green-synthesized saffron nanoparticles for antimicrobial, antioxidative, and neuroprotective actions, have therapeutic applications for neurodegenerative diseases (Korrapati et al., 2024). Efficient and reproducible methods for isolating and coating nanoparticles with cancer cell membranes are needed for clinical translation.

Cancer cell membrane-coated nanoparticles (CCNPs) face challenges include optimizing coating techniques for stability and functionality, understanding complex biological interactions, and translating laboratory findings into clinically viable therapies, all of which demand a multidisciplinary perspective combining nanotechnology, pharmacology, and computational biology (GVS, Biradar, et al., 2025). Long-term safety profiles and potential immunogenic effects must be thoroughly evaluated (Bhuvana et al., 2025). Ongoing research is focused on optimizing membrane extraction, improving nanoparticle design, and exploring new combinations of cell membranes to maximize therapeutic outcomes. These nanoparticles are also employed in chemotherapy and immunotherapy, where they enhance drug delivery precision and reduce systemic toxicity (GVS, Ananthakumar, et al., 2025).

CONCLUSION

Cancer cell membrane-coated nanoparticles and their hybrid systems represent a transformative advancement in cancer therapy. By harnessing the natural properties of cancer cell membranes, these biomimetic nanoparticles offer improved targeting, immune evasion, and multifunctionality, paving the way for more effective and personalized cancer treatments. Cell membrane-coated nanoparticles represent a transformative approach in cancer nanotherapy, offering unique combinations of biocompatibility, targeting specificity, and therapeutic efficacy. RBC membrane-coated systems lead in clinical development due to their optimal balance of performance characteristics and manufacturing feasibility. Neutrophil membrane coatings show exceptional promise for anti-metastatic applications, while cancer cell membrane systems offer personalized medicine opportunities. The success of these biomimetic platforms depends on continued advances in manufacturing standardization, regulatory approval processes, and clinical validation. As the field matures, hybrid approaches combining multiple membrane types and advanced engineering strategies will likely emerge as next-generation therapeutic platforms, potentially revolutionizing cancer treatment (Lakshmaiah et al., 2025).

REFERENCES

Amaral, M. N., Kumar, P., Faísca, P., Ferreira, H. A., Coelho, J. M. P., Gaspar, M. M., & Reis, C. P. (2025). Gold nanoparticle-mediated photothermal therapy: Expanding the frontiers of cancer treatment and theragnostics. *Biomedicine & `Pharmacotherapy, 190*, 118399. https://doi.org/DOI: 10.1016/j.biopha.2025.118399

Ananthakumar, B., Nagaraja, B. G., Umesh, S. M., & Raghavan, R. (2025). Predictive Medicine: A New Era of Pain Management for Assistive Augmentative Rehabilitation Systems. In *Predictive Algorithms for Rehabilitation and Assistive Systems* (pp. 291–314). IGI Global Scientific Publishing. https://doi.org/DOI: 10.4018/979-8-3373-0194-5.ch011

Ananthakumar, B., Nanjundeshwara, N., Lakshmaiah, A., & Raghavan, R. (2024). Synthetic Gene Circuits as a Promising Approach in Cancer Immunotherapy. *Critical Developments in Cancer Immunotherapy*, 391–418. https://doi.org/DOI: 10.4018/979-8-3693-3976-3.ch011

Ananthakumar, B., Nanjundeshwara, N., Lakshmaiah, A., & Raghavan, R. (2025). Nanocarriers for Delivering Nucleic Acids in Plants: Mechanisms and Applications. In *Nanocarriers in Plant Science and Agriculture* (pp. 207–234). IGI Global Scientific Publishing. https://doi.org/DOI: 10.4018/979-8-3693-9869-2.ch008

Beitzinger, B., Draphoen, B., Hald, J., Krämer, M., Rudolf, T., Schmitt, F., Schwarz, P., & Lindén, M. (2025). On the cellular selectivity of targeting peptide-nanoparticle conjugates – a nanoparticle perspective. *Expert Opinion on Drug Delivery, ahead-of-print*(ahead-of-print), 1–14. https://doi.org/DOI: 10.1080/17425247.2025.2522245

Bhuvana, G. N., Umesh, S. M., & Raghavan, R. (2025). Expanding the frontiers of Schistosomiasis Treatment: The promise of Nanoparticle-Based Therapies. *Journal of Vector Borne Diseases, 10*(4103/jvbd.jvbd_180_23). https://doi.org/DOI: 10.4103/jvbd.jvbd_180_23

Biradar, S., Suresh, C., Nanjundeshwara, N., & Raghavan, R. (2025). Spatially Variable Genes. *Spatially Variable Genes in Cancer*, 1–28. https://doi.org/DOI: 10.4018/979-8-3693-7728-4.ch001

Byalahunasishetter, B. I. N, P. V., Savitha, C., Moger, S. S., & Raghavan, R. (2025). Spatial Omics and Bioprinting: Tissue Engineering and Precision Medicine. In *Innovations in Precision Medicine and Genomics* (pp. 85–110). IGI Global Scientific Publishing. https://doi.org/DOI: 10.4018/979-8-3693-5787-3.ch003

Deng, B., Kong, Y., Ma, Y., Zhan, Y., Sun, Y., Wang, R., Huang, P., & Liu, L. (2025). A novel carrier-free nanoparticle with stable distinctive three-dimensional structure for tumor-targeted precision chemoimmunotherapy. *Journal of Nanobiotechnology*, *23*(1), 1–15. DOI: 10.1186/s12951-025-03568-8 PMID: 40598510

Fernández-Borbolla, A., García-Hevia, L., & Fanarraga, M. L. (2024). Cell Membrane-Coated Nanoparticles for Precision Medicine: A Comprehensive Review of Coating Techniques for Tissue-Specific Therapeutics. *International Journal of Molecular Sciences*, *25*(4), 2071. DOI: 10.3390/ijms25042071 PMID: 38396747

Gato-Diaz, U., Concheiro, A., Alvarez-Lorenzo, C., & Blanco-Fernandez, B. (2025). 4D printed hydrogels for precision delivery of bioactive molecules in cancer. *Expert Opinion on Drug Delivery*, *0*(0), 1–17. DOI: 10.1080/17425247.2025.2539959 PMID: 40739734

GVS, A. C., Ananthakumar, B., & Raghavan, R. (2025). Enhancing Assistive Technologies With Neuromorphic Computing. *Revolutionizing AI with Brain-Inspired Technology: Neuromorphic Computing*, 207–232. https://doi.org/DOI: 10.4018/979-8-3693-6303-4.ch009

GVS, A. C., Biradar, S., & Raghavan, R. (2025). Neuroscience-Driven Solutions for Speech Impairment: Assistive Technologies for Speech Rehabilitation. *Advancing Medical Research Through Neuroscience*, 119–146. https://doi.org/DOI: 10.4018/979-8-3693-5464-3.ch006

Heidari, R., Assadollahi, V., Khosravian, P., Mirzaei, S. A., & Elahian, F. (2023). Engineered mesoporous silica nanoparticles, new insight nanoplatforms into effective cancer gene therapy. *International Journal of Biological Macromolecules*, *253*, 127060. DOI: 10.1016/j.ijbiomac.2023.127060 PMID: 37774811

Jiménez-Boland, D., Robles-Fernández, A., Martín-Rodríguez, A., Cuadros, M. Á., Traverso, J. Á., Sánchez-Moreno, P., & Bramini, M. (2025). Breaking Barriers in Glioblastoma Targeting through Advanced Nanoparticle Cell Membrane Coating. *ACS Applied Materials & Interfaces*, *17*(24), 35288–35303. DOI: 10.1021/acsami.5c07306 PMID: 40488690

Kashyap, B. K., Singh, V. V., Solanki, M. K., Kumar, A., Ruokolainen, J., & Kesari, K. K. (2023). Smart Nanomaterials in Cancer Theranostics: Challenges and Opportunities. *ACS Omega*, *8*(16), 14290–14320. DOI: 10.1021/acsomega.2c07840 PMID: 37125102

Kaur, J., Thakran, A., & Naqvi, S. (2025). Recent advances in cell membrane-based biomimetic delivery systems for Parkinson's disease: Perspectives and challenges. *Asian Journal of Pharmaceutical Sciences*, *20*(4), 101060. DOI: 10.1016/j.ajps.2025.101060 PMID: 40777843

Korrapati, C., Yaraladdi, A. B., Vaddi, H., & Raghavan, R. (2024). Applications of the Saffron European Kitchen Spice Crocus sativus. *Therapeutic and Pharmacological Applications of Ethnobotany*, 98–127. https://doi.org/DOI: 10.4018/979-8-3693-1986-4.ch005

Lakshmaiah, A., Korrapati, C., Suresh, C., & Raghavan, R. (2025). Synthetic Biology and AI. *Spatially Variable Genes in Cancer*, 265–290. https://doi.org/DOI: 10.4018/979-8-3693-7728-4.ch010

Lopes, J., Lopes, D., Motallebi, M., Ye, M., Xue, Y., Vieira, A. C. F., Singh, S. K., Dua, K., Veiga, F., Sethi, G., Paiva-Santos, A. C., & Makvandi, P. (2025). Biomembrane-coated nanosystems as next-generation delivery systems for the treatment of gastrointestinal cancers. *Bioengineering & Translational Medicine*, *10*(4), e70006. DOI: 10.1002/btm2.70006 PMID: 40708975

Ma, J., Xie, Y., Teng, Z., Jiang, L., & Liu, G. (2025). Engineered cell membrane-based nano therapies fight infectious diseases. *Journal of Controlled Release*, *384*, 113884. DOI: 10.1016/j.jconrel.2025.113884 PMID: 40418989

Misra, R., & Sanjana Sharath, N. (2024). Red blood cells based nanotheranostics: A smart biomimetic approach for fighting against cancer. *International Journal of Pharmaceutics*, *661*, 124401. DOI: 10.1016/j.ijpharm.2024.124401 PMID: 38986966

Nagashri, N., Archana, L., & Raghavan, R. (2024). Microbial Volatile Compounds (MVOCs) in Food Industries and Food Safety Applications. *Innovations, Securities, and Case Studies Across Healthcare, Business, and Technology*, 75–96. https://doi.org/DOI: 10.4018/979-8-3693-1906-2.ch005

P., A., D., K. M., M., J., P., S., & Raghavan, R. (2025). (pp. 161–192). The Cancer Exposome. In *Cancer Exposomics and Environmental Influences on Carcinogenesis.*, DOI: 10.4018/979-8-3373-2165-3.ch004

Rajput, S., Malviya, R., Sridhar, S. B., Wadhwa, T., & Shareef, J. (2025). Erythrocyte-derived nanosystems for biosensing: A green approach for analysis. *Journal of Pharmaceutical and Biomedical Analysis Open*, *6*, 100088. DOI: 10.1016/j.jpbao.2025.100088

Ren, F., Jin, Y., Kong, W., Wang, H., Wang, S., Li, X., Xing, J., Zhang, Y., Yang, X., & Song, N. (2024). Nano-strategies used for combatting the scourge of tuberculosis infections. *Discovery Immunology*, *1*(1). Advance online publication. DOI: 10.1007/s44368-024-00007-y

Sachan, R., Jaipan, P., Zhang, J. Y., Degan, S., Erdmann, D., Tedesco, J., Vanderwal, L., Stafslien, S. J., Negut, I., Visan, A., Dorcioman, G., Socol, G., Cristescu, R., Chrisey, D. B., & Narayan, R. J. (2025). Printing amphotericin B on microneedles using matrixassisted pulsed laser evaporation. *International Journal of Bioprinting*, *3*(2), 147–157. DOI: 10.18063/IJB.2017.02.004 PMID: 33094188

Singh, R. R., Kulshreshtha, R., & Popat, A. (2025). Ultrasmall nanoparticles for co-delivery of antisense oligonucleotides targeting miR-21 and miR-210 to treat glioblastoma. *Journal of Nanobiotechnology*, *23*(1), 1–28. DOI: 10.1186/s12951-025-03529-1 PMID: 40604820

Wang, X., Wan, W., Liu, G., Ren, T., Thomas, J. O., Wu, H., An, C., Tian, F., Zhu, Y., Kros, A., & Yang, J. (2025). Coiled-coil lipopeptides mediated membrane fusion based drug delivery into a mimicking lung organ. *Chemical Engineering Journal*, *515*, 163333. DOI: 10.1016/j.cej.2025.163333

Yang, N., He, Z., & Lang, T. (2025). Drug Delivery Systems Based on Metal–Organic Frameworks for Tumor Immunotherapy. *Pharmaceutics*, *17*(2), 225. DOI: 10.3390/pharmaceutics17020225 PMID: 40006592

Yang, Z., Zhang, Y., Wang, W., Shi, X., Cui, K., Du, Y., Ma, W., & Chen, Q. (2025). Photoimmunotherapy biomimetic nanoplatform for lung cancer and brain metastasis. *Materials & Design*, *257*, 114497. DOI: 10.1016/j.matdes.2025.114497

Yildirim, D. T., Baki Yildirim, A., Salzet, M., Bertelli, M., Beccari, T., Prakash, S., Pascucci, L., & Dundar, M. (2025). Messaging malignancy: Tumour-derived exosomes at the nexus of immune escape, vascular remodelling and metastatic competence. *The EuroBiotech Journal*, *9*(3), 216–237. DOI: 10.2478/ebtj-2025-0018

Zhang, R., Li, Y., Fu, H., Zhao, C., Li, X., Wang, Y., Sun, Y., & Li, Y. (2025). Nanomedicine strategies for cuproptosis: Metabolic reprogramming and tumor immunotherapy. *Acta Pharmaceutica Sinica. B*. Advance online publication. DOI: 10.1016/j.apsb.2025.07.007 PMID: 41049741

Zhou, Z., Zhang, S., & Xue, N. (2023). Research progress of cancer cell membrane coated nanoparticles for the diagnosis and therapy of breast cancer. *Frontiers in Oncology*, *13*. Advance online publication. DOI: 10.3389/fonc.2023.1270407 PMID: 37781205

Chapter 6

Carbon Nanotubes-Based Bioimaging and Cancer Therapy Applications

P. Selvakumar
https://orcid.org/0000-0002-3650-4548
Department of Science and Humanities, Nehru Institute of Technology, Coimbatore, India

P. Selvam
Karpaga Vinayaga College of Engineering and Technology, India

T. C. Manjunath
https://orcid.org/0000-0003-2545-9160
Rajarajeswari College of Engineering, India

R. Haridass
Karpagam College of Engineering, Coimbatore, India

Atal Bihari Singh
Arka Jain University, India

Santosh Kumar Nathsharma
https://orcid.org/0000-0003-3935-2938
Stewart Science College, India

ABSTRACT

Unique combination of strength, conductivity, chemical versatility, nanomaterials in particularly in the fields of bioimaging and cancer therapy. properties various overview, delve into CNTs bioimaging and therapy, focusing on how their distinct characteristics facilitate advanced techniques detection, treatment. primary the key difference between these two types is the amount of graphene with SWCNTs being composed sheet, have This structural variation significantly influences the CNTs, optical characteristics. These properties can be tailored by modifying the synthesis methods or functionalizing the surface of the nanotube's applications. enables them carry large amounts of drug molecules, imaging agents, or other therapeutic payloads. Their small size, in diameter, allows them to or enhancing their effec-

DOI: 10.4018/979-8-3373-3765-4.ch006

tiveness in drug delivery and bioimaging. Additionally, thermal and which makes photothermal and electrochemical therapies. Bioimaging is a critical technique in modern medicine, allowing visualization cells, Accurate imaging is essential for early diagnosis, assessing treatments.

INTRODUCTION: THE PROMISE OF CARBON NANOTUBES IN BIOIMAGING AND CANCER THERAPY: UNDERSTANDING THE POTENTIAL

Unique combination of strength, conductivity, chemical versatility, nanomaterials in particularly in the fields of bioimaging and cancer therapy. properties various overview, delve into CNTs bioimaging and therapy, focusing on how their distinct characteristics facilitate advanced techniques detection, treatment. primary the key difference between these two types is the amount of graphene with SWCNTs being composed sheet, have This structural variation significantly influences the CNTs, optical characteristics. These properties can be tailored by modifying the synthesis methods or functionalizing the surface of the nanotube's applications. enables them carry large amounts of drug molecules, imaging agents, or other therapeutic payloads. Their small size, in diameter, allows them to or enhancing their effectiveness in drug delivery and bioimaging. Additionally, thermal and which makes photothermal and electrochemical therapies. Bioimaging is a critical technique in modern medicine, allowing visualization cells, Accurate imaging is essential for early diagnosis, assessing treatments. Carbon nanotubes have shown great agents electronic interact with enables them be used various imaging modalities, including fluorescence imaging, Raman spectroscopy, and near-infrared (NIR) imaging. In fluorescence imaging, CNTs fluorescent dyes, allowing them emit excitation by property makes them suitable for tracking biological molecules and cellular processes in real-time. CNTs stems from thereby surrounding One of the most promising strategies is the use of CNTs as carriers CNTs tumor-specific or the nanotubes can selectively bind to cancer cells overexpressing certain receptors, ensuring payload specifically of tumor. CNTs can carry therapeutic siRNA which encapsulated or the nanotubes. Once CNTs the therapeutic agents can be released nanotube structure through or changes within the tumor microenvironment.

This efficacy reduces conventional chemotherapy, which often causes cells. their role in CNTs in (PTT). In this technique, CNTs are used region which can selectively destroy cancer cells. The high thermal conductivity of CNTs allows for uniform heat distribution, ensuring that the entire tumor is effectively targeted. When combined with targeted drug delivery, photothermal therapy can be a powerful and minimally invasive approach to treating cancer, offering significant advantages over traditional therapies. Another potential application of CNTs in cancer therapy is CNTs deliver plasmid cancer enabling the expression of therapeutic proteins or the silencing of oncogenes. biocompatibility, efficiently deliver genetic material while response cytotoxicity. While CNTs can be synthesized laboratory, high-quality CNTs with consistent properties remains a significant hurdle. Advances in production techniques will be necessary that CNTs can be reliably produced for clinical applications. Carbon nanotubes represent a groundbreaking frontier in the fields of bioimaging and cancer therapy. structural, thermal highly versatile both diagnostic and therapeutic purposes. With the potential for photothermal CNTs offer many cotreatment. However, significant research is still needed to address issues related to toxicity, biocompatibility, and large-scale production before CNT-based therapies can be widely adopted. evolve, carbon nanotubes may play a pivotal diagnosis and treatment, effective less invasive therapies for cancer patients.

CARBON NANOTUBES: A NEW FRONTIER IN BIOIMAGING AND CANCER THERAPY

CNTs, whose role advancing both diagnostic therapeutic techniques, especially carry drugs, serve as imaging agents, and facilitate novel therapeutic strategies positions them as pivotal tools in revolutionizing cancer care. The various carbon nanotubes, how they function in bioimaging, focusing on their both enhanced precision fewer side effects. Properties of Carbon Nanotubes: distinctive suited CNTs incredibly allowing biomolecules, genes, imaging agents. Their nanoscale size (typically 1-2 nm in diameter) makes them small enough to penetrate drug delivery and imaging. This size also allows them longer compared to larger particles, increasing their bioavailability and effectiveness. Carbon nanotubes are incredibly strong and lightweigh(Gao S., et al, 2024)t. Their atomic structural strength is paired with remarkable electrical and thermal conductivity, candidates to destroy cancer cells, and electrochemical biosensors for detecting specific biomarkers. Moreover, CNTs chemically compatibility with systems increase their specificity for targeting cancer cells. By attaching targeting molecules like CNTs, researchers direct the nanotubes to specific tumor sites, minimizing the interaction. Carbon Nanotubes in Bioimaging: Bioimaging is a crucial technique in modern medicine, allowing for the visualization of biological tissues, organs, and cellular processes. Early detection of cancer through imaging is one of the most powerful tools in improving patient outcomes, as it enables clinicians to identify tumors. Carbon nanotubes are great owing optical. One of the most exciting bioimaging applications is their use in fluorescence and near-infrared (NIR) imaging. SWCNTs, in particular, exhibit strong fluorescence in the NIR region (Shen C L., et al, 2022). This makes them ideal for in vivo imaging, enabling the detection of tumors deep within the body. The NIR fluorescence of CNTs can be enhanced through functionalization with specific ligands, allowing for the selective binding of CNTs to cancer cells and enabling targeted imaging of tumors. This feature is especially valuable in solid tumor imaging, where traditional imaging methods may struggle to provide clear and accurate results. Moreover, CNTs can be employed in a variety of other imaging modalities, such as magnetic resonance imaging (MRI) and computed tomography (CT), by integrating them with other imaging agents. For example, the surface of CNTs can be functionalized with iron oxide nanoparticles, enhancing their ability to provide contrast in MRI scans. This dual-functionality—acting both as a contrast agent and a drug delivery vehicle—makes CNTs highly versatile and capable of providing multi-modal imaging that improves the accuracy and precision of cancer diagnosis.

Moreover, CNTs have been investigated for use in combination therapies that exploit their multiple types simultaneously. For example, CNTs can be used in both chemotherapy and therapy, such as RNA interference (RNAi) molecules, which can silence progression. Multi-modal therapy can help overcome drug resistance, a common problem in cancer therapy, by targeting multiple pathways simultaneously. Another exciting therapeutic is PTT. This technique leverages the unique optical properties, particularly, can convert absorbed into without affecting tissues (Ravi S N., et al, 2024). The local heating effect can lead to thermal ablation of the tumor, providing a non-invasive treatment option. Additionally, the high thermal conductivity of CNTs allows for uniform heat distribution across the tumor. Photothermal therapy can also be combined to further enhance cell destruction. Another application of CNTs in cancer therapy is gene delivery. CNTs or small interfering RNA (siRNA) to cancer cells. This ability to directly cancer opens up new avenues for gene therapy, including the possibility of restoring the function of tumor suppressor genes or silencing oncogenes that drive cancer progression. Nucleic acids, while their ability to cross cellular membranes facilitates the directly cancer CNTs can help modulate gene expression and promote cancer cell death. Finally, regulatory hurdles CNT-based therapies widely adopted ensure CNT-based therapies (Zhu W., et al, 2021). Particularly about potential toxicity and immune response, will need to be thoroughly investigated through these technologies' human. Carbon nanotubes in fields of bioimaging and cancer therapy, challenges faced cancer detection and treatment. Enables serve as versatile agents' diagnostic accurate cancer fewer advance, development CNT-based cancer care. However, challenges related to toxicity, large-scale production, and care that CNTs. Ultimately, carbon nanotubes represent a new frontier in the fight against cancer, offering hope for efficient cancer treatment around the world.

THE PROPERTIES OF CARBON NANOTUBES: UNDERSTANDING THEIR STRUCTURE, COMPOSITION, AND FUNCTION

These cylindrical, hollow structures composed entirely of carbon atoms possess that value in variety, particularly in fields like materials science, bioengineering, electronics, and energy storage. Composition, carbon nanotubes to fully grasp their potential and limitations in various technological and biomedical applications. In this discussion, we will explore the fundamental aspects of atomic structure, methods, and the functions that make them versatile tools in science and industry. Atomic Structure of Carbon Nanotubes: At core remarkable properties lies atomic structure. CNTs are primarily much like a These graphene sheets are rolled into cylindrical shapes, forming the nanotubes, which can have that can extend or even millimeters (Lu D., et al, 2022). The structure is structure graphite with a key difference: the carbon atoms in the nanotubes are bonded together forms. This atomic arrangement confers several crucial properties to carbon nanotubes, including their exceptional conductivity. The graphene sheets are held together by sp^2 hybridization, which is stronger and more stable than the sp^3 hybridization found in conventional carbon-based materials like diamonds. The bond angles of the carbon atoms in CNTs also create a high degree of symmetry, resulting in mechanical properties that are far superior to most conventional materials. As a result, CNTs exhibit extraordinary tensile strength, exceeding that of steel by several times, and are incredibly resistant to deformation or breakage. Carbon nanotubes, the rolling pattern graphene sheet, are made up of multiple concentric graphene tubes nested inside each other. This difference in structure distinct properties, such as conductivity, strength, and thermal stability. The structural arrangement also influences the functionalization capabilities and potential applications of each type of nanotube. Classification of Carbon Nanotubes: Carbon nanotubes have several factors, including structural configuration, diameter, chirality, and electrical properties. Classifications are vital to a nanotube's suitability for different applications (Sugumaran K M D., et al. 2024). Chirality and Electrical Properties: The chirality of a CNT refers to the specific Chirality described in terms of "chiral vector," which defines the direction wrapped. Depending on chirality, can exhibit different electrical properties. Armchair nanotubes, which have runs along length tend to be metallic and conduct electricity very efficiently. Zigzag nanotubes, with a chiral vector perpendicular to the tube's length, can also display behavior-specific length. Nanotubes exhibit semiconducting behavior, and their electronic properties are sensitive to their diameter and chirality. Single-Walled vs. Multi-Walled: As previously mentioned, they are considered more desirable for certain applications, such as high-performance electronics, because they exhibit higher purity and better electrical conductivity (Mohd Nurazzi N., et al, 2021). However, MWCNTs are more commonly produced and are easier to handle due to their thicker walls, making them more robust and versatile for some industrial applications, such as composites and energy storage devices. Diameter and Length: They can range from 10 for SWCNTs, and MWCNTs can have diameters up to several hundred nanometers. The length of CNTs is typically much larger than their diameter, ranging from hundreds of nanometers to several micrometers. Contributes significantly to electrical conductivity. Longer nanotubes have an enhanced ability to conduct electricity, while shorter nanotubes are often preferred for certain biomedical applications due to their easier dispersion and biocompatibility.

Thermal Properties of Carbon Nanotubes: also, exceptional properties. They are excellent heat conductors, with thermal conductivities higher than most materials, including metals like copper and silver. The thermal conductivity of CNTs is particularly high along their length due to the sheet. This makes them ideal for efficient sinks for electronic devices or in advanced cooling systems. In addition to their high thermal conductivity, CNTs have been shown to possess remarkable thermal stability. Significant is their use in environments with extreme thermal conditions, such as in space technology or high-temperature industrial applications (Lavagna L., et al, 2021). This combination of high conductivity and stability opens the door to CNTs being used in a wide range of industries that require advanced thermal management. Functionalization of Carbon Nanotubes: carbon nanotubes can be functionalized or chemically other materials or Functionalization involves attaching specific molecules, polymers, or biomolecules to improve biocompatibility, selectivity. Surface Modifications: CNTs are typically hydrophobic, making them difficult to disperse in aqueous solutions. Functionalization allows the surface of CNTs to be modified with hydrophilic groups, such as carboxyl or amino groups, which improve their solubility in water and make them more compatible with useful drug delivery biomedical imaging applications (Wang J., et al, 2022). CNTs need to be easily dispersed in physiological environments. Their unique structure—comprising hexagonal arrangements of carbon atoms rolled into tubes—endows them with exceptional strength, conductivity, and flexibility. By understanding the fundamental properties of CNTs, including their classification, electrical conductivity, thermal performance, and functionalization capabilities, researchers and engineers are unlocking a vast array of applications across numerous fields, materials science, and biomedicine storage. The potential of carbon nanotubes is likely to expand even further, technological medical challenges of the modern world.

THE SYNTHESIS OF CARBON NANOTUBES: UNDERSTANDING THE METHODS AND TECHNIQUES

The (CNTs) in nanotechnology, as growth these remarkable nanostructures is fundamental to unlocking their biomedical engineering. Carbon nanotubes are characterized by their unique cylindrical structure. However, the practical use of CNTs in industrial and technological applications requires controlled synthesis methods that yield nanotubes with specific characteristics, including size, morphology, and purity. Plantation methods and techniques, as well as newer approaches that aim to improve yield, quality, and scalability. Chemical Vapor Deposition (CVD): employed techniques synthesizing particularly in large-scale production. Gases lead to the formation of carbon nanotubes (CNTs) on a substrate surface. The general process of CVD can be broken down into several steps: the introduction of precursor gases, nucleation and growth of CNTs, and eventual nanotubes on catalyst (Shoukat R., et al, 2021). Process Overview: The CVD technique typically involves heating a carbon source, along with an elevation, usually in the range of 600 to 1000°C. The metal catalyst initiating nucleation carbon atoms is then deposited onto the catalyst particles, forming the CNT structure. Into particles and begin, which then as carbon supply is replenished. Atmospheric Pressure CVD (APCVD): In APCVD, the precursor gases are introduced at atmospheric pressure. This method is of CNTs relatively low-cost simplicity (Johri S., et al. 2025). However, APCVD may lead to a less uniform distribution of CNTs and lower purity. Low-Pressure CVD (LPCVD): LPCVD operates at lower pressures, which better promotes the growth of higher-quality CNTs with fewer defects and more uniform growth. LPCVD is commonly used in research and industrial applications where high-quality CNTs are needed. Plasma-enhanced CVD (PECVD): PECVD activates precursor gases, leading to a lower temperature process. This can be advantageous for growing may notefile electronics. PECVD can also lead to higher deposition rates and more controlled growth of CNTs. Factors Affecting CNT Growth in CVD: Several factors can influence the growth of CNTs during CVD, including Catalyst Type and Size. The choice of catalyst material (such as iron, cobalt, or nickel) is crucial as it affects the nucleation rate and the type of CNTs formed. Additionally, catalyst particles determine the diameter of nanotubes. Smaller catalyst particles typically lead to smaller-diameter CNTs. Carbon Source: The carbon precursor gas used in CVD influences the growth rate and morphology of the nanotubes (Gao C., et al, 2023).

Laser Ablation Method: Process Overview: Invenergy directed onto typically graphite, that contains nickel or target extremely temperatures, the carbon vaporizes form a plasma. Carbon atoms then condense and form CNTs as they cool. The process typically occurs in a reactor to maintain proper conditions for CNT formation. Advantages: Laser ablation produces CNTs with fewer defects compared to other synthesis methods. It is also capable of producing both MWCNTs and SWCNTs, although the latter are less common. Additionally, the laser ablation method provides better growth conditions and more uniform controlled nanotube structures. Challenges: Despite its advantages, laser ablation has several challenges that limit its scalability for large-scale production. The process is relatively slow and requires specialized equipment, making it more expensive than other methods. The yield of CNTs can also be lower compared to CVD, and the purification of the CNTs often requires additional steps to remove metal catalysts and other impurities (Salah L S., et al, 2021). Other Methods of CNT Synthesis: Beyond the traditional methods, several newer approaches have been developed to address the limitations of these methods and improve the scalability, purity, and uniformity of CNTs. Fluidized Bed CVD: Fluidized bed CVD is an advanced version of the traditional CVD method, in which the catalyst particles are suspended in a flow of gas. This method allows for better heat transfer and more uniform growth of CNTs, making it ideal for large-scale production. Template-Based Synthesis: Template-based methods use porous materials, such as anodic aluminum oxide (AAO) membranes, to guide the growth of CNTs within the pores. This technique offers diameter alignment CNTs, useful nanotube characteristics are required. Chemical Vapor Deposition on Substrates (CVD-S): In this method, CNTs are directly synthesized on a substrate, such as silicon wafers, through a combination of CVD and selective catalyst deposition. Range techniques, unique vapor commonly flexibility, allowing CNTs (Ramachandran K., et al, 2022).

THE USE OF CARBON NANOTUBES IN BIOIMAGING: UNDERSTANDING THE PRINCIPLES AND APPLICATIONS

Remarkable optical, electrical, and mechanical garnered attention bioimaging. Bioimaging various biological structure processes is crucial for diagnosing diseases, monitoring treatments, and understanding cellular and molecular dynamics. CNTs have shown considerable promise in improving the effectiveness and precision of bioimaging techniques, offering new insights into both molecular biology and medical diagnostics. Their unique combination of size, shape, and surface chemistry makes them excellent candidates for fluorescence, near-infrared (NIR). This section explores the principles behind bioimaging, the specific advantages offered in different imaging modalities, and the ongoing research and applications that are pushing the boundaries of bioimaging technologies (Sapna K., et al, 2022). Principles of Bioimaging with Carbon Nanotubes: Bioimaging with CNTs relies on their ability to enhance imaging contrast and provide detailed, high-resolution information. Several key principles are involved in using CNTs for bioimaging, including their unique optical properties, biocompatibility, ability to functionalize, and The following are some critical aspects that make CNTs ideal candidates for bioimaging: Surface Functionalization: CNTs specific biomolecules, aptamers, enables the selective particular Bility to direct CNTs to specific biological targets significantly enhances the sensitivity and accuracy of bioimaging, especially in detecting specific biomarkers or monitoring disease progression. Functionalization also aids dispersion and biocompatibility CNTs environments. Biocompatibility: major bioimaging that imaging agents do not induce harmful reactions within the body. CNTs, when properly functionalized, can be made highly biocompatible, reducing toxicity and improving their ability to circulate within the body. Surface functionalization polymers can further be in the bloodstream and minimize immune system recognition, allowing for prolonged imaging and monitoring. High Surface Area: CNTs carry imaging agents, drugs, and biomolecules. This characteristic enhances their effectiveness in imaging, as they can accumulate in targeted areas in higher quantities than other materials (Ramakrishnan T. et al. 2022). This property is particularly useful in the context of molecular imaging, where precise localization of the imaging agent is critical for effective diagnosis and monitoring. Size and Shape: CNTs possess nanoscale dimensions that allow them to enter cells and tissues more easily than larger particles. Their unique cylindrical structure also has a mechanically robust biological environment. Controlled, which influences their ability to penetrate cell membranes and their distribution in tissues. Their size and shape also contribute to their ability to improve imaging resolution and contrast (Acharya R., et al, 2024). Applications of Carbon Nanotubes in Bioimaging: Fluorescence Imaging: imaging techniques biological and medical diagnostics.

Computed Tomography (CT) Imaging: powerful tissues. CNTs can also serve their ability to attenuate X-rays. The incorporation of CNTs into CT imaging enhances image contrast, particularly when combined with other contrast-enhancing agents. High Contrast for Tissue Imaging: CNTs have high X-ray attenuation properties due to their carbon content, which makes them effective contrast agents in CT imaging. Functionalization with iodine or other heavy elements can further improve the contrast, enabling the visualization of specific tissues or pathological conditions. Near-Infrared (NIR) Imaging: Near-infrared (NIR) imaging has emerged as a prominent technique in bioimaging due to its ability to penetrate tissues with minimal scattering (Abbas A., et al, 2023). CNTs, particularly SWCNTs, are highly effective in NIR imaging because of this region of the electromagnetic spectrum. Imaging is particularly useful for animals and humans due to its deep tissue penetration. NIR Fluorescence for Deep Tissue Imaging: The NIR fluorescence of CNTs makes them excellent candidates for deep tissue imaging applications, such as monitoring tumor growth or tracking cellular activities in living organisms. NIR imaging allows without the need for invasive procedures. Real-Time Imaging and

Monitoring: CNTs in NIR perform real-time imaging. This capability enables the dynamic monitoring of biological processes, such as drug delivery, tumor targeting, and molecular interactions, in living organisms (Faulkner S., et al, 2022).

IMAGING OF CANCER CELLS WITH CARBON NANOTUBES: UNDERSTANDING THE TECHNIQUES AND BENEFITS

Imaging cancer cells is a significant leap forward in the field of cancer diagnosis and treatment monitoring, which has attracted considerable attention. Their strong light absorption has unique ideal applications, particularly for cancer diagnostics. Imaging cancer cells using CNTs offers numerous advantages over conventional imaging techniques, including improved resolution, enhanced sensitivity, and specific and cellular structures. This ability to directly observe cancer cells allows for earlier detection, more accurate staging, and better monitoring of treatment responses, thus improving patient outcomes. This section will discuss the imaging techniques used with CNTs to visualize cancer cells, the benefits these techniques provide, and the overcame full CNTs in cancer (Tang L., et al, 2021). Principles of Cancer Cell Imaging with Carbon Nanotubes: To understand the imaging of cancer cells using CNTs, it is important to first recognize the fundamental principles that make CNTs ideal for such applications. The primary factors that contribute to CNTs' effectiveness in bioimaging include their unique optical properties, ability to be functionalized for specific targeting, and biocompatibility. These characteristics allow CNTs to function, enhance, and facilitate selective targeting of cancerous tissues. Optical Properties and Fluorescence: CNTs, particularly single-walled carbon nanotubes (SWCNTs), are optical that fluorescent probes for bioimaging. And emit light at different wavelengths, enabling them for fluorescence. The NIR region is particularly valuable for imaging cancer cells, as light in this range has minimal scattering. Fluorescence emission is tuned to their size and functionalization, making them versatile for cancer imaging. Surface Functionalization for Targeting: most significant CNTs is easily small molecules, that can target specific cancer cell receptors. Surface functionalization allows CNTs to selectively bind to cancer cells or tumors, specifying the process. Instance, tumor antigens, HER2 or EGFR, can be conjugated to the CNT surface (Chary P S., et al, 2023). This ensures that the CNTs accumulate at the tumor site, enhancing contrast in imaging techniques and enabling better visualization of cancer cells. Biocompatibility: For CNTs to be effective imaging, they must introduce adverse reactions into the body.

Fluorescence Imaging: one of the most commonly used techniques in cancer imaging due to its simplicity and visualization. CNTs, particularly SWCNTs, can serve as highly effective fluorescent probes that emit NIR light at longer wavelengths.

This characteristic is as into significant scattering, allowing for clearer and more precise imaging of tumor sites. Targeted Fluorescence Imaging: By functionalizing CNTs with tumors, the nanotubes can be used as cell receptors. Targeting enhances the accuracy of fluorescence imaging, as the CNTs will accumulate at the tumor site, providing clearer and more localized images of the cancerous tissue. Multiplexed Imaging: CNTs can be modified with multiple fluorophores, allowing These enables tracking different cancer biomarkers or cell populations, making it easier to monitor disease progression, evaluate therapeutic efficacy, and study complex cellular processes such as metastasis (Al-Thani A N., et al, 2024). Magnetic Resonance Imaging (MRI): Enhanced Tumor Visualization: Functionalized CNTs with magnetic nanoparticles allow for the high-contrast imaging of tumors, improving the accuracy of tumor detection and staging. These CNT-based contrast agents can accumulate at tumor sites, where they enhance the contrast and allow for more detailed imaging of cancerous tissues. Non-invasive Monitoring: MRI imaging with CNTs allows for the non-invasive monitoring of tumor growth and treatment response. This is particularly important for evaluating the effectiveness of chemotherapy or radiation therapy over time, as MRI can provide real-time imaging without the need for biopsies or invasive procedures. Computed Tomography (CT) Imaging: CNTs can effectively absorb X-rays, making them useful (Sonowal L., et al, 2024).

THE USE OF CARBON NANOTUBES IN OPTICAL IMAGING: UNDERSTANDING THE OPPORTUNITIES AND CHALLENGES

Optical imaging is an essential technique in biological research and environmental monitoring due to optical imaging's emerged promising innovation, capitalizing on optical, chemical especially single (SWCNTs), exceptional which is particularly advantageous in biological imaging. Their biocompatibility and biomolecules enable CNTs to serve as effective agents for optical imaging, offering significant opportunities. Several remain harnessing CNTs for optical imaging. These challenges include issues related to toxicity, stability, and scalability, as well as the technical difficulties associated with functionalizing CNTs and achieving effective in vivo imaging. This section will explore the opportunities presented by CNTs in optical imaging, the techniques that leverage CNTs for enhanced imaging, and the challenges addressed by CNTs optical imaging applications. Opportunities in Optical Imaging with Carbon Nanotubes: Carbon nanotubes optical imaging. These advantages stem from their unique optical properties, ability to be functionalized for specific targeting, and potential for multimodal imaging (Wang Z., et al, 2021). Below, we discuss some of the key opportunities that CNTs present for advancing optical imaging. Fluorescence Imaging: CNTs are particularly valuable in fluorescence-based imaging techniques due to their ability to emit fluorescence when excited by light in the NIR range. When functionalized, CNTs can be targeted to sensitivity and specificity fluorescence imaging (Mahadev Madgule, et al. 2023). CNTs a variety of peptides, and small molecules, which selective cancer pathogens, specific Functionalization is a crucial step in optical imaging because it enables the CNTs to selectively bind to the target of interest, improving. Targeted Imaging of Cancer Cells: In cancer imaging, CNTs can be conjugated with targeting molecules that cancer cell surfaces. Specificity allows for enhanced imaging of tumors, enabling the detection of early-stage cancers or metastatic lesions that may otherwise be difficult to detect (Murar M., et al, 2022).

Deep-Tissue Imaging and In Vivo Applications: primary optical with CNTs penetrate tissues at a greater depth than conventional optical probes. NIR light, in particular, deep within living organisms without capability tumor imaging, monitoring studying cellular and molecular dynamics in living organisms. Real-Time Imaging: CNTs enable non-invasive imaging, which is invaluable for tracking the progress of diseases such as cancer. With their ability to be targeted to specific tissues, CNTs can be used for longitudinal studies where a therapeutic intervention can be monitored over time in live animals or even in humans. Monitoring Drug Delivery: CNTs can also be used to track the delivery of therapeutics, such as chemotherapeutic agents, in real time. By attaching a drug to the CNT surface, it becomes possible to monitor the release and tumor, optimizing treatment plans. High Sensitivity and Specificity: combination the CNTs, their ability functionalized, and unique optical properties significantly sensitivity optical This results the ability to detect small amounts of biomarkers or low-abundance proteins that might otherwise be missed by conventional imaging methods (Zhang Z., et al, 2024). Enhanced Contrast: CNTs can provide enhanced contrast in optical imaging by increasing the signal emitted by a target of interest. Their unique optical properties allow them to act as effective contrast agents, making it easier to visualize tissues, cells, and molecules in complex biological systems. Early Detection: The sensitivity of CNT-based optical imaging, particularly for cancers, when Early diagnosis is crucial for effective treatment, and CNTs offer an exciting when tumors are still small and localized. Challenges in Optical Imaging with Carbon Nanotubes: While CNTs offer considerable opportunities for improving optical imaging techniques in clinical applications. Range from issues to the complexities of CNT functionalization and manufacturing (Jiang Y., et al, 2021). These studies must address concerns related to the accumulation potential of inflammatory responses and the effects of CNT degradation products.

THE USE OF CARBON NANOTUBES IN CANCER THERAPY: UNDERSTANDING THE MECHANISMS AND BENEFITS

Traditional therapeutic strategies, including chemotherapy, radiation, and surgery, often fail cancer, leading to tissues a host of undesirable side effects. Growing treatment that can minimize collateral damage and overcome limitations of conventional therapies. In this regard, in cancer therapy, which can be leveraged gene therapy. Carbon nanotubes ' particularly remarkable physical, chemical, and mechanical cancer therapeutic biocompatibility (when appropriately functionalized) and unique properties enable CNTs to perform multiple roles in cancer treatment. Section, explore CNTs' therapeutic on cancer cells, the benefits they offer in cancer therapy, and the challenges and limitations successful (Roy S M., et al, 2023). Mechanisms of Action in Cancer Therapy: Carbon nanotubes can act as therapeutic agents, each targeting aspects of cancer progression. The following are key by CNTs: Targeted Drug Delivery: promising applications serve as drug delivery systems. CNTs anticancer drugs, other therapeutic molecules selective cancer the surface of CNTs provides a large area for the attachment of drugs, allowing for high payloads. By functionalizing CNTs ligands, they directed target that express specific receptors on their surface. These selective agents are thereby cells systemic side. Controlled Release: CNTs can be engineered to release their payload external stimuli, temperature variations, specific that are overexpressed controlled release further the treatment. Overcoming Drug Resistance: Many cancers develop resistance to traditional chemotherapeutic agents. However, CNT can bypass some mechanisms of resistance, such as the efflux pumps that expel drugs from cancer cells. CNTs can help deliver drugs inside cancer cells in a way that avoids or mitigates resistance mechanisms (Brindhadevi K., et al, 2023). Photothermal Therapy: typically, in the NIR region, to heat and destroy cancer cells. CNTs make excellent agents for PTT. When CNTs are exposed to NIR light, they absorb the energy and convert raising area. Localized heating of the adjacent tumor. Selective Tumor Targeting: CNTs can be functionalized to target tumor cells.

Immunotherapy Enhancement: Rapidly evolving area CNTs enhance the effectiveness of immunotherapy by acting as adjuvants or carriers for immune-stimulating agents. Functionalizing CNTs with immune checkpoint inhibitors, cytokines, or other immunomodulatory molecules Targeted Immune Activation: CNTs conjugated tumor antigens direct immune cells, such as T-cells, to the tumor site. By activating the immune system, CNTs can increase immunotherapy and reduce off-target effects that can lead to systemic inflammation. Checkpoint Inhibition: CNTs can also work with CTLA-4 cells. These inhibitors block the mechanisms that more effectively attack and kill the tumor. Benefits of Using Carbon Nanotubes in Cancer Therapy: Using cancer therapy offers several benefits that could be treated. Some of the key advantages include Enhanced Treatment Efficacy: CNTs can through various mechanisms (Gao F., et al, 2021). Their ability to deliver high payloads of drugs, genes, or immunotherapy agents ensures that the tumor receives a concentrated dose of therapy, increasing the chances of treatment success. CNTs are also used as imaging agents in combination with therapy, allowing for non-invasive and progressive monitoring (Debnath P., et al. 2025). By capabilities, fluorescence or photoacoustic CNTs can be tracked inside the body, providing valuable feedback on treatment efficacy and tumor response. Monitoring protocols as needed, ensuring optimal therapeutic outcomes. Challenges and Limitations: Toxicity and Biocompatibility: hurdles CNTs therapy their potential have shown laboratory studies, long-term biodegradability, possible accumulation in must carefully addressed. Functionalization CNTs can mitigate some of these concerns by improving their solubility and dispersibility and reducing immunogenicity. However, more extensive background checks are required to understand the profile fully. Scalability and Manufacturing: The synthesis of CNTs for clinical applications must be scalable, reproducible, and cost-effective (Assali M., et al, 2021).

TARGETING OF CANCER CELLS WITH CARBON NANOTUBES: UNDERSTANDING THE STRATEGIES AND CHALLENGES

The inability of conventional therapies to selectively target tumor cells often results in significant precise effective can target harm to surrounding tissue. Among the most promising solutions for achieving this goal are carbon nanotubes (CNTs), which have garnered attention for their ease of functionalization and ability to penetrate biological membranes. When engineered correctly, CNTs can be used to deliver chemotherapy and other biomolecules directly to cancer cells, both specificity efficacy treatment. This section explores the various strategies used for targeting cancer cells with CNTs, the mechanisms by which these strategies work, and cancer therapy. Strategies for Targeting Cancer Cells with Carbon Nanotubes: The primary using cancer treatment they're thus enhancing and damage normal to achieve this, CNTs variety of molecules allow them specifically tumor cells or the tumor microenvironment (Contreras-Torres & Salas-Trevino, 2022). These antibodies enable highly selective targeting and can also trigger immune responses, enhancing the therapeutic effects of CNTs. Peptides: Short peptides, such as those derived from the tumor-homing peptides, also functionalize CNTs (Kumar P., et al, 2024). Another strategy for targeting CNTs to tumors involves the phenomenon of often poorly organized and leaky blood vessels that allow larger nanoparticles, such as CNTs, to passively accumulate in the tumor site. In addition, tumors typically have defective lymphatic drainage, which further allows these nanoparticles to remain within the tumor. The EPR effect is a non-specific method of targeting that relies on the ability of CNTs to exploit the tumor's unique vasculature characteristics to reach. Size and Shape of CNTs: CNTs with smaller diameters (less than 100 nm) are more likely to escape (Singh R., et al, 2022).

Challenges in Targeting Cancer Cells with Carbon Nanotubes: While CNTs offer targeted treatment and limitations, their use. These challenges stem from both the inherent properties of CNTs and the complexities of the tumor microenvironment. Some of the key challenges include: Toxicity and Biocompatibility: Surface Functionalization to Reduce Toxicity: To mitigate the toxic effects of CNTs, surface modifications can be employed. For example, functionalizing CNTs with biocompatible coatings, such as PEG, can enhance their solubility and reduce their immunogenicity, making them safer for in vivo use. Moreover, careful selection of the materials used for functionalization is essential to ensure the trigger contributes (Das S., et al, 2025). Aggregation and Stability: CNTs are aggregation which their effectiveness in drug delivery and target specificity. Aggregation leads to clumping CNTs, preventing their efficient dispersion and cellular uptake. In addition, CNT aggregation can impair their ability to accumulate in tumors and hinder their functionalization ligands. Strategies to Overcome Aggregation: To prevent aggregation, CNTs hydrophilic surfactants that improve their dispersion in biological fluids. Another strategy involves coating CNTs with charged molecules, which can create electrostatic repulsion between individual nanotubes, keeping them well-dispersed. Heterogeneity of Tumors: The heterogeneity of tumors presents a significant challenge in targeted cancer therapy. Tumors are composed of diverse cell populations with varying molecular and genetic characteristics (Shende A D., et al. 2025). Moreover, combination therapies that use CNT treatments may be more effective in overcoming the challenges of tumor heterogeneity.

THE DELIVERY OF THERAPEUTIC AGENTS WITH CARBON NANOTUBES: UNDERSTANDING THE OPPORTUNITIES AND LIMITATIONS

The delivery of therapeutic agents has been treatment for diseases, especially complex conditions like cancer, cardiovascular diseases, and neurodegenerative disorders. Traditional drug delivery methods often face difficulties in ensuring that therapeutic agents reach the target cells or tissues in effective concentrations without causing harm to healthy cells. Solution to the problem when properly functionalized) and offers significant opportunities for enhancing the efficacy of treatments. This section will explore opportunities presented by CNT-based drug delivery systems, the mechanisms that enable them to deliver therapeutic agents, and the must their clinical application. High Surface Area and Loading Capacity: They can carry a large amount of therapeutic agent relative to their size (Li H., et al, 2025). CNTs multiple types simultaneously, enabling combination therapy. For example, CNTs can carry chemotherapeutic agents alongside siRNA (small interfering RNA) molecules, allowing for simultaneous gene silencing and drug delivery (Sreenivasulu A., et al. 2022). This dual-functional approach increases progression. Antibody-Mediated Targeting: One approach to functionalizing CNTs involves attaching monoclonal antibodies that bind specifically to in various cancers, such as HER2 in breast cancer. These antibodies help guide the CNTs directly to the tumor, where they can release their payload efficacy in healthy tissues. Peptide-Based Targeting: Tumor cells often express integrins and other receptors that can be targeted using specific peptides. For example, peptides like RGD (arginine-glycine-aspartate) bind to integrins that are overexpressed in various cancers. By functionalizing CNTs with such peptides, CNTs can selectively accumulate in tumors, thereby enhancing the and reducing exposure to cells (Roszkowski S., et al, 2025).

Despite their promise, there are several challenges and limitations associated with CNTs. These challenges stem from the intrinsic properties of CNTs, systems, and scale-up clinical translation. Toxicity and Biocompatibility: primary concerns CNTs in their potential toxicity (Qi K., et al, 2023). While CNTs have shown great promise in laboratory settings, there is still uncertainty regarding their long-term biocompatibility, degradation, and accumulation in organs. Cause inflammation and even organ damage in certain cases, particularly when they are not properly functionalized. Scale-Up and Manufacturing Challenges: While CNTs have shown excellent potential in preclinical studies, translating these findings into clinical applications presents significant challenges (Sangar O S., et al, 2022). Both humans remain a hurdle. Manufacturing processes need to be standardized for the purity and functionalization of CNTs. Cost manufacturing CNT-based drug delivery systems may limit their widespread use in clinical practice. Regulatory and Quality Control: For CNTs to be approved for clinical use, rigorous regulatory testing must be conducted in many trials. The complexity of CNTs, including their size, shape, and surface properties, makes regulatory approval challenging, as these factors must be carefully controlled and standardized.

THE IMPORTANCE OF BIOCOMPATIBILITY AND BIODEGRADABILITY IN CARBON NANOTUBES: UNDERSTANDING THE REQUIREMENTS AND CHALLENGES

This section will delve into the importance of biocompatibility and biodegradability in CNTs, discussing the specific requirements and challenges associated with achieving these properties. Additionally, it will explore the aim of improving the safety profile of CNTs for clinical use. Biocompatibility of Carbon Nanotubes: Biocompatibility is a central concern for any nanomaterial used in medical applications. It CNTs interact toxic or inflammatory responses. A biocompatible material should ideally integrate smoothly into biological environments, such as the bloodstream, tissues, or organs, without eliciting adverse immune responses, cellular damage, or toxicity. Requirements for Biocompatibility: To be considered biocompatible, CNTs must meet several requirements to ensure that they do not provoke harmful effects when exposed to the human body (Mokhtari-Farsani A., et al, 2022). Non-Toxicity: The CNTs should not cause cytotoxicity, meaning they should not kill or damage healthy cells. Particularly medical tissues, without affecting surrounding healthy cells. CNTs should ideally be non-toxic to a wide range. Low Inflammatory Response: Inflammation is common. For CNTs to be biocompatible, they must be able to minimize inflammation, particularly chronic inflammation, which causes fibrosis and immune dysfunction. Must ably tolerate the presence of CNTs without launching an excessive inflammatory response, which could impair normal cellular processes or lead to adverse side effects. Minimal Immunogenicity: CNTs must not lead to the development of antibodies or activation of immune cells against the material (Pathak R., et al, 2023). Such responses could cause the body to recognize CNTs as foreign invaders, leading to rejection or unwanted immune responses.

Biodegradability of Carbon Nanotubes Biodegradability is another crucial factor for the clinical translation of CNTs. For CNTs to be used effectively in medical applications, they must eventually be eliminated from the body without causing harmful accumulation. Biodegradability refers to the ability of CNTs to break down into non-toxic components, either naturally through biological processes or engineered degradation pathways. If CNTs persist in the body, they could cause toxicity and other complications. Requirements for Biodegradability: Biodegradability is essential for ensuring that CNTs do not cause long-term harm or toxicity after their therapeutic or diagnostic function is complete. The following are key requirements for CNT biodegradability: Efficient Degradation: CNTs must degrade into harmless products after serving their intended purpose. Enzyme-Catalyzed Degradation: Some researchers are investigating the use of enzymes to catalyze the breakdown of CNTs. Enzymes such as lipases, esterases, or other hydrolases could potentially target functional groups on the CNT surface, leading to their gradual degradation in vivo. This approach holds for providing a more controlled and biologically specific method of CNT degradation. Challenges to Biodegradability: While strategies for enhancing the biodegradability of CNTs are promising, several challenges remain (Yan J S., et al, 2021). Slow Degradation Rates: CNTs are inherently resistant to degradation. Achieving an optimal balance between stability (for effective drug delivery or imaging) and biodegradability (for safe elimination) is difficult. CNTs may degrade too slowly to be useful in some applications or too rapidly in others, compromising their effectiveness.

FUTURE DIRECTIONS IN CARBON NANOTUBES FOR BIOIMAGING AND CANCER THERAPY: EMERGING TRENDS, TECHNOLOGIES, AND OPPORTUNITIES

Nanotubes (CNTs) strides, particularly in areas of bioimaging cancer therapy. However, despite their impressive potential, the clinical applications of CNTs are in these domains. As we move into the future, several emerging trends, novel technologies, and innovative approaches are shaping the next generation of CNT-based materials, particularly for bioimaging and cancer therapy. This discussion will explore the future directions of CNT research in these areas, focusing on the advancements, opportunities, and breakthroughs that are expected applications. Advanced Functionalization for Targeted Drug Delivery: CNT development has the ability to functionalize these materials for targeted drug delivery. Functionalization techniques have already demonstrated significant improvements in the ability of CNTs (Nimushakavi S., et al, 2022). However, the future lies in enhancing the specificity, efficiency, and release mechanisms of these nano systems. The integration of CNTs with imaging modalities is another exciting avenue for future research. CNTs have already shown promise in bioimaging, particularly in optical fluorescence (Vijayakumar G, et al. 2024). However, as technologies future CNTs bioimaging be intertwined with new techniques designed to offer greater resolution, sensitivity, and specificity. Multi-modal critical cancer diagnosis enables clinicians to monitor tumors with a greater degree of accuracy, track therapeutic responses, and detect metastases at earlier stages. In addition to multi-modal imaging, the use of CNTs for theragnostic — the combination of therapeutic and diagnostic capabilities in one nanoparticle — is an exciting opportunity. Theragnostic CNTs could be used to both diagnose cancer and deliver treatments, eliminating the need for separate diagnostic and therapeutic procedures (Sajjadi M., et al, 2021).

Surface Functionalization for Biocompatibility: to address concerns about the continued functionalization of CNTs to improve their biocompatibility. By modifying the surface of peptides, other researchers improve interaction with biological cytotoxicity or increase solubility of CNTs in physiological environments, minimize protein adsorption, and reduce the likelihood of aggregation. These modifications will critical CNTs can be safely used in humans for prolonged periods. Degradable CNTs: In addition to surface functionalization, the development of biodegradable CNTs is essential for minimizing long-term toxicity. Biodegradable CNTs, which can metabolites fulfilling function, would mitigate the risk of accumulation and chronic toxicity (Kumar S., et al, 2025). This could involve designing CNTs that degrade biological enzymes or modifying their structure to enhance their biodegradation rate. Over the coming years, biodegradable CNTs are expected to become more common in clinical applications, particularly in drug delivery and cancer therapy, where safe removal of the material from the body is critical. Biomarker-Driven Targeting: The future of CNTs in personalized cancer therapy will likely revolve around advanced biomarker-driven targeting strategies. His approach could potentially correct genetic defects, silence oncogenes, or activate tumor-suppressor genes, providing a novel way to treat cancer at the genetic level. CNTs facilitate these therapies by protecting the genetic material from degradation, enhancing cellular uptake, and ensuring the targeted release of RNA or DNA sequences at the tumor site. Studies, including animal models, understanding the accumulation and elimination pathways, will be necessary to ensure that CNTs are safe for clinical use (Parvin N., et al, 2024).

RECOMMENDATIONS FOR RESEARCHERS AND CLINICIANS: BEST PRACTICES AND GUIDELINES FOR DEVELOPING CARBON NANOTUBES FOR BIOIMAGING AND CANCER THERAPY

However, as with any emerging technology, the development of CNT-based applications in medicine presents unique challenges that must be carefully navigated. Researchers and clinicians must adopt a strategic and multidisciplinary approach to harness the full potential of CNTs while mitigating risks associated with their use. This section outlines best practices, recommendations, and guidelines for researchers and clinicians involved in CNT technologies, bioimaging, and therapy, ensuring both scientific advancement and patient safety (Sharma S., et al, 2024). Prioritizing Biocompatibility and Safety in CNT Design: The most critical consideration for researchers and clinicians developing CNTs for medical applications is their biocompatibility and safety. Despite their significant potential, CNTs can pose toxicity risks, particularly if they accumulate in the body or exhibit undesirable interactions with cellular structures. Therefore, biocompatibility must be a primary design criterion throughout the development process. Surface Functionalization and Modification: Researchers should focus on optimizing the surface chemistry of CNTs to ensure that they are biocompatible and can interact safely with biological systems (Huang P., et al, 2024). Functionalization techniques, such as the attachment of hydrophilic polymers (e.g., polyethylene glycol or PEG), biomolecules (e.g., peptides or antibodies), or surfactants, can significantly improve the dispersion and biological. These reduce the tendency to aggregate, which is a significant factor contributing to their toxicity. Additionally, surface functionalization to target cancer improves the selectivity of CNT-based. In Vitro and In Vivo Testing for Toxicity and Biocompatibility: Best practices for researchers include rigorous testing of CNTs to assess toxicity from relevant tissues (e.g., cancerous and normal tissues), providing valuable insight into the potential cytotoxicity and cellular uptake of CNTs. Synthesis methods directly impact CNTs, including their functionalization potential (Basu & Hendler-Neumark, 2024). The two most common are. Both methods have their and selecting appropriate desired properties of the CNTs. CVD, for example, CNT length, diameter, making it ideal for producing high-quality CNTs.

Ensuring Targeted Delivery and Specificity: The potential for CNTs to be used in targeted cancer therapy relies heavily on their ability to selectively bind, avoiding surrounding. Researchers and clinicians should adopt advanced efficiency CNT-based systems. Targeting Ligands and Biomarker Identification: The attachment of CNTs enables them to bind specifically. Researchers must focus on identifying and validating cancer-specific biomarkers, such as HER2, EGFR, or other cell surface proteins, that could serve as targets for CNT-based drug delivery. The effectiveness of this approach relies on understanding the molecular and genetic profile of different types of cancer, which may vary from one patient to another. Clinicians should advocate for the use of precision medicine tools to determine the genetic makeup of tumors and match patients with CNT-based therapies that target specific molecular pathways (Ahmad I., et al, 2023). Clinical Translation and Regulatory Considerations: While CNTs hold immense promise for cancer therapy, practice will require careful navigation of regulatory and clinical challenges. Researchers and clinicians must work closely with regulatory bodies to ensure that CNT-based therapies comply with the necessary standards for safety, efficacy, and quality control. Regulatory Guidelines and Approval Processes: Researchers should familiarize themselves with the regulatory guidelines set forth by other national health authorities. The approval process for CNT-based therapies will require assessing the pharmacokinetics and pharmacodynamics of CNT formulations. Early-phase clinical trials will focus on evaluating the dose-limiting toxicity of CNT-based treatments (Aragani, V M., et al. 2025). Researchers and clinicians must adhere to the stringent documentation and reporting requirements necessary for clinical trial approval and registration. Additionally, researchers should be mindful of the potential challenges CNT manufacturing processes must comply with to ensure that CNTs meet the required standards for consistency, purity, and safety.

CONCLUSION

Carbon nanotube (CNT) nanomedicine, offering unparalleled opportunities in bioimaging and cancer therapies, has become highly regarded for its potential applications in diagnostic and therapeutic strategies, particularly of cancer. Few extensive research has the multifaceted roles of CNTs in enhancing imaging techniques, facilitating targeted drug delivery, and providing new avenues for personalized cancer treatments. However, while the potential is immense, the utility of CNTs. This conclusion aims to summarize the key findings, opportunities, challenges, and future implications for bioimaging cancer, reflecting on both the current state and the future trajectory of this rapidly evolving field. Opportunities for Carbon Nanotubes in Bioimaging and Cancer Therapy: The opportunities presented by CNTs in bioimaging and cancer therapy are vast and transformative. In bioimaging, CNTs offer improved sensitivity and specificity techniques. Fluorescent, magnetic CNTs utilize fluorescence and scans, providing researchers with cancer monitoring. Ability CNTs, both delivery vehicles, allow for multi-modal imaging, which can provide more detailed, accurate, and dynamic information about tumor location, size, and response to treatment. The ability of CNTs to be an emerging gene and immunotherapy offers additional opportunities for more personalized, effective cancer treatment. CNTs or tumors offer the possibility of treating cancers. Lay at a large scale, with tight control over size, shape, and surface properties. However, current synthesis methods still face limitations in terms of producing CNTs with the required purity and structural integrity for biomedical applications. These are commonly used for CNT synthesis, but these processes often require complex equipment and may introduce impurities, such as metal catalysts, that can affect the biocompatibility of CNTs. Furthermore, the development of standardized protocols for CNT production is critical to ensure that each batch of CNTs used in clinical trials or therapies is of consistent quality. The delivery and targeting of CNTs to specific cancer cells also remain challenging. While the functionalization of CNTs with targeting ligands has shown promise, ensuring that CNTs selectively target tumor cells while avoiding healthy tissue is an ongoing concern. The heterogeneity of tumors, with varying molecular profiles and the presence of multiple cell types, can make targeting a complex task. By incorporating molecular profiling and genetic information, CNTs cancer subtypes, optimizing treatment regimens unique Moreover, advances in real-time bioimaging and monitoring technologies will allow clinicians behavior CNTs the body, treatment efficacy facilitating development of adaptive, personalized treatment plans. In summary, carbon nanotubes present a unique and transformative opportunity for advancing bioimaging and cancer therapy. Their distinct properties, tunable functionality, and ideal use diagnostic targeted personalized cancer treatments. The ability to combine CNTs with emerging technologies, such as gene therapy, immunotherapy, and multi-modal imaging, positions them as key players in the future of cancer treatment. However, as with any emerging technology, these include toxicity and the complexity of targeting tumor cells selectively. Overcoming these challenges requires researchers, engineers, and CNTs that are both safe for clinical use. The future of CNTs in bioimaging and cancer therapy holds immense promise. With continued research and development, the clinical translation of CNT-based technologies is expected to monitor cancer. As our understanding of CNTs grows, and as new materials, methods, and regulatory frameworks are developed, carbon nanotubes will undoubtedly play a central role in advancing the field of nanomedicine.

REFERENCES

Abbas A, Zehra S, Aslam R, Mobin M, (2023). "Functionalized Carbon Nanotubes for Biomedical Imaging: The Recent Advances". DOI: DOI: 10.1002/9783527835874.ch6

Acharya, R., Patil, T. V., Dutta, S. D., & Lee, J. (2024). Single-Walled Carbon Nanotube-Based Optical Nano/Biosensors for Biomedical Applications: Role in Bioimaging, Disease Diagnosis, and Biomarkers Detection. *Advanced Materials*. Advance online publication. DOI: 10.1002/adma.202308281

Ahmad I, Parween T, Khandare L, (2023). "Applications of Functionalized Carbon Nanotubes in Cancer Therapy and Diagnosis". DOI: DOI: 10.1002/9783527835874.ch5

Al-Thani, A. N., Jan, A. G., Abbas, M., & Geetha, M. (2024). Nanoparticles in cancer theragnostic and drug delivery: A comprehensive review. *Life Sciences*. Advance online publication. DOI: 10.1016/j.lfs.2024.122511

Aragani, V. M., Anumolu, V. R., & Selvakumar, P. (2025). Democratization in the Age of Algorithms: Navigating Opportunities and Challenges. In Wongmahesak, K., Wekke, I., Seftyono, C., & Nurdin, N. (Eds.), *Democracy and Democratization in the Age of AI* (pp. 39–56). IGI Global Scientific Publishing., DOI: 10.4018/979-8-3693-8749-8.ch003

Assali, M., Kittana, N., & Khiar, N. (2021). Dual covalent functionalization of single-walled carbon nanotubes for effective targeted cancer therapy. *Nanotechnology*. Advance online publication. DOI: 10.1088/1361-6528/ac1d3d

Basu, S., & Hendler-Neumark, A. (2024). Dynamic Tracking of Biological Processes Using Near-Infrared Fluorescent Single-Walled Carbon Nanotubes. *ACS Applied Materials & Interfaces*. Advance online publication. DOI: 10.1021/acsami.3c18157

Brindhadevi, K., Garalleh, H A L., & Alalawi, A. (2023). Carbon nanomaterials: Types, synthesis strategies and their application as drug delivery system for cancer therapy. *Biochemical Engineering Journal*. Advance online publication. DOI: 10.1016/j.bej.2023.108934

Chary, P. S., Bhawale, R., Vasave, R., & Rajana, N. (2023). A review on emerging role of multifunctional carbon nanotubes as an armament in cancer therapy, imaging and biosensing. *Journal of Drug Delivery Science and Technology*. Advance online publication. DOI: 10.1016/j.jddst.2023.104455

Contreras-Torres, F. F., & Salas-Trevino, D. (2022). Carbon nanotubes in tumor-targeted chemotherapeutic formulations: A review of opportunities and challenges. *ACS Applied Nano Materials*. Advance online publication. DOI: 10.1021/acsanm.1c03971

Das, S., Roy, S., Dinda, S. C., Bose, A., & Mahapatra, C. (2025). Carbon nanotubes in brain targeted drug delivery: A comprehensive review. *Results in Chemistry*. Advance online publication. DOI: 10.1016/j.rechem.2024.101401

Debnath, P., Deshmukh, S. M., Laroia, M., Selvakumar, P., Manjunath, T. C., & Bhattacharya, S. (2025). Challenges Faced by Countries in Renewable Energy Adoption. In Özsungur, F., Chaychi Semsari, M., & Küçük Bayraktar, H. (Eds.), *Geopolitical Landscapes of Renewable Energy and Urban Growth* (pp. 295–320). IGI Global Scientific Publishing., DOI: 10.4018/979-8-3693-8814-3.ch010

Faulkner, S., Gunnlaugsson, T., & Máille, G. Ó. (2022). Supramolecular. *Chemical & Biomedical Imaging*.

Gao, C., Guo, M., Liu, Y., Zhang, D., Gao, F., Sun, L., & Li, J. (2023). Surface modification methods and mechanisms in carbon nanotubes dispersion. *Carbon*. Advance online publication. DOI: 10.1016/j.carbon.2023.01.045

Gao, F., Sun, Z., Zhao, L., Chen, F., & Stenzel, M. (2021). Bioactive engineered photothermal nanomaterials: From theoretical understanding to cutting-edge application strategies in anti-cancer therapy. *Materials Chemistry Frontiers*. Advance online publication. DOI: 10.1039/D0QM00953A

Gao, S., Xu, B., Sun, J., & Zhang, Z. (2024). Nanotechnological advances in cancer: Therapy a comprehensive review of carbon nanotube applications. *Frontiers in Bioengineering and Biotechnology*. Advance online publication. DOI: 10.3389/fbioe.2024.1324568

Huang, P., Tang, Q., Li, M., Yang, Q., Zhang, Y., & Lei, L. (2024). Manganese-derived biomaterials for tumor diagnosis and therapy. *Journal of Nanobiotechnology*. Advance online publication. DOI: 10.1186/s12951-024-02367-8

Jiang, Y., & Pu, K. (2021). Molecular probes for autofluorescence-free optical imaging. *Chemical Reviews*. Advance online publication. DOI: 10.1021/acs.chemrev.0c01226

Johri, S., Selvakumar, P., Satyanarayana, P., Das, A., Gupta, S., & Bhattacharya, S. (2025). Urbanization and Energy: The Nexus of Global Challenges. In Özsungur, F., Chaychi Semsari, M., & Küçük Bayraktar, H. (Eds.), *Geopolitical Landscapes of Renewable Energy and Urban Growth* (pp. 453–478). IGI Global Scientific Publishing., DOI: 10.4018/979-8-3693-8814-3.ch016

Kumar, P., Pandey, S. N., Ahmad, F., & Verma, A. (2024). Carbon nanotubes: A targeted drug delivery against cancer cell. *Current Drug Delivery*. Advance online publication. DOI: 10.2174/1567201820666230916115533

Kumar S, Ansari A, Basu M, Ghosh S, (2025). "Carbon Nanotubes in Cancer Diagnosis and Treatment: Current Trends and Future Perspectives". DOI: DOI: 10.1002/9783527835874.ch10

Lavagna, L., Nisticò, R., Musso, S., & Pavese, M. (2021). Functionalization as a way to enhance dispersion of carbon nanotubes in matrices: A review. *Materials Today. Chemistry*. Advance online publication. DOI: 10.1016/j.mtchem.2021.100456

Li, H., Tursun, M., Aihemaiti, A., Yan, P., & Liu, X. (2025). Nanoplatform Based on Carbon Nanotubes Loaded with Persistent Luminescent Nanoparticles for Photocatalysis, Photothermal Therapy, and Drug Delivery. *ACS Applied Materials & Interfaces*. Advance online publication. DOI: 10.1021/acsami.4c17423

Lu, D., Shi, X., & Zhong, J. (2022). Understanding the role of unzipped carbon nanotubes in cement pastes. *Cement and Concrete Composites*. Advance online publication. DOI: 10.1016/j.cemconcomp.2022.104444

Mahadev Madgule, N. (2023). Vinayaka, Yeshwant M. Sonkhaskar, Dhiren Ramanbhai Patel, R. Karthikeyan, P. Selvakumar, (2023). Mechanical properties and microstructure of activated TIG welded similar joints of Inconel alloys by desirability approaches. *Materials Today: Proceedings*, *77*, 528–533.

Mohd Nurazzi, N., Asyraf, M. R. M., Khalina, A., & Abdullah, N. (2021). Fabrication, functionalization, and application of carbon nanotube-reinforced polymer composite: An overview. *Polymers*. Advance online publication. DOI: 10.3390/polym13183095

Mokhtari-Farsani, A., Hasany, M., & Lynch, I. (2022). Biodegradation of carbon-based nanomaterials: The importance of "biomolecular corona" consideration. *Advanced Functional Materials*. Advance online publication. DOI: 10.1002/adfm.202109346

Murar, M., Albertazzi, L., & Pujals, S. (2022). Advanced optical imaging-guided nanotheranostics towards personalized cancer drug delivery. *Nanomaterials (Basel, Switzerland)*. Advance online publication. DOI: 10.3390/nano12060986

Nimushakavi S, Haque S, (2022). "Biomedical applications of carbon nanotubes: recent development and future challenges".

Parvin, N., Kumar, V., Joo, S. W., & Mandal, T. K. (2024). Emerging trends in nanomedicine: Carbon-based nanomaterials for healthcare. *Nanomaterials (Basel, Switzerland)*. Advance online publication. DOI: 10.3390/nano14040323

Pathak, R., Punetha, V. D., Bhatt, S., & Punetha, M. (2023). Carbon nanotube-based biocompatible polymer nanocomposites as an emerging tool for biomedical applications. *European Polymer Journal*. Advance online publication. DOI: 10.1016/j.eurpolymj.2023.112157

Qi, K., Sun, B., Liu, S., & Zhang, M. (2023). Research progress on carbon materials in tumor photothermal therapy. *Biomedicine and Pharmacotherapy*. Advance online publication. DOI: 10.1016/j.biopha.2023.114821

Ramachandran, K., Boopalan, V., & Bear, J. C. (2022). Multi-walled carbon nanotubes (MWCNTs)-reinforced ceramic nanocomposites for aerospace applications: A review. *Journal of Materials Science*. Advance online publication. DOI: 10.1007/s10853-022-07067-3

Ramakrishnan, T., Mohan Gift, M. D., Chitradevi, S., Jegan, R., & Subha Hency Jose, P. Nagaraja, Rajneesh Sharma HN, Selvakumar P, Sintayehu Mekuria Hailegiorgis, (2022). "Study of Numerous Resins Used in Polymer Matrix Composite Materials", Advances in Materials Science and Engineering, vol. 2022, Article ID 1088926, 8 pages, 2022. DOI: 10.1155/2022/1088926

Ravi, S. N., Rajendran, S., & Madhumathi, G. S. (2024). Carbon nanomaterials: Pioneering innovations in bioimaging and biosensing technologies. *Journal of Molecular Structure*. Advance online publication. DOI: 10.1016/j.molstruc.2024.136666

Roszkowski, S., & Durczynska, Z. (2025). Advantages and limitations of nanostructures for biomedical applications. *Advances in Clinical and Experimental Medicine*.

Roy, S. M., Garg, V., Sivaraman, S. P., & Barman, S. (2023). Overcoming the barriers of nuclear-targeted drug delivery using nanomedicine-based strategies for enhanced anticancer therapy. *Journal of Drug Delivery Science and Technology*. Advance online publication. DOI: 10.1016/j.jddst.2023.104456

Sajjadi, M., Nasrollahzadeh, M., & Jaleh, B. (2021). Carbon-based nanomaterials for targeted cancer nanotherapy: Recent trends and future prospects. *Journal of Drug Delivery Science and Technology*. Advance online publication. DOI: 10.1016/j.jddst.2021.102732

Salah, L. S., Ouslimani, N., & Bousba, D. (2021). Carbon nanotubes (CNTs) from synthesis to functionalized (CNTs) using conventional and new chemical approaches. *Journal of Nanomaterials*. Advance online publication. DOI: 10.1155/2021/6625571

Sangar O S, Patil AC, Payghan S A, (2022). "Nanoparticles: as a nano based drug delivery system".

Sapna K, Sonia J, Kumara B N, Arun A B, (2022). "Carbon nanotubes for bio-imaging applications". DOI: DOI: 10.1007/978-3-030-91334-3_8

Sharma S, Chakraborty M, Yadav D, Dhullap A, (2024). "Strategic developments in polymer-functionalized liposomes for targeted colon cancer therapy: an updated review of clinical trial data and future horizons". DOI: DOI: 10.1021/acs.biomac.3c01345

Shen CL, Liu H R, Lou Q, Wang F, Liu K K, (2022). "Recent progress of carbon dots in targeted bioimaging and cancer therapy". DOI: DOI: 10.1039/D1NR07961A

Shende, A. D., Tekade, S. A., Deshmukh, A. A., Tembhurkar, S. P., & Selvakumar, P. (2025). Water Quality Monitoring and Control With AI. In Ruiz-Vanoye, J., & Díaz-Parra, O. (Eds.), *Smart Water Technology for Sustainable Management in Modern Cities* (pp. 51–80). IGI Global Scientific Publishing., DOI: 10.4018/979-8-3693-8074-1.ch003

Shoukat, R., & Khan, M. I. (2021). Carbon nanotubes: A review on properties, synthesis methods and applications in micro and nanotechnology. *Microsystem Technologies*. Advance online publication. DOI: 10.1007/s00542-021-05215-6

Singh, R., & Kumar, S. (2022). Cancer targeting and diagnosis: Recent trends with carbon nanotubes. *Nanomaterials (Basel, Switzerland)*. Advance online publication. DOI: 10.3390/nano12060987

Sonowal, L., & Gautam, S. (2024). *Advancements and challenges in carbon nanotube-based drug delivery systems*. Nano-Structures & Nano-Objects., DOI: 10.1016/j.nanoso.2024.100964

Sreenivasulu, A., & Manimozhi, T. Satheesh kumar P, Selvakumar P. and Singh P, (2022). "The Model of Smart Sensing Device For Sensitive Nanoclusters Modification in Sensing Properties," 2022 2nd International Conference on Advance Computing and Innovative Technologies in Engineering (ICACITE), Greater Noida, India, 2022, pp. 1043-1047, doi: DOI: 10.1109/ICACITE53722.2022.9823846

Sugumaran K M D, Pandiyarajan V Y B S, Selvakumar P. and M. S, (2024). "Analysis of Electromyography Signals for Control of Mechanical Prosthesis using Machine Learning Techniques," 2024 2nd International Conference on Intelligent Data Communication Technologies and Internet of Things (IDCIoT), Bengaluru, India, 2024, pp. 1059-1067, doi: .DOI: 10.1109/IDCIoT59759.2024.10467707

Tang, L., Xiao, Q., Mei, Y., He, S., Zhang, Z., & Wang, R. (2021). Insights on functionalized carbon nanotubes for cancer theranostics. *Journal of Nanobiotechnology*. Advance online publication. DOI: 10.1186/s12951-021-00855-6

Vijayakumar, G. (2024). Muniyandi Rajkumar, Rajiv Chandar N, Selvakumar P and Ramesh Duraisamy, (2024). Environmentally friendly TDS removal from wastewater by electrochemical ion exchange batch-type recirculation (EIR) technique. *Environmental Science. Water Research & Technology*, *10*, 826–835. DOI: 10.1039/D3EW00793F

Wang, J., Dong, S., Pang, S. D., Zhou, C., & Han, B. (2022). Pore structure characteristics of concrete composites with surface-modified carbon nanotubes. *Cement and Concrete Composites*. Advance online publication. DOI: 10.1016/j.cemconcomp.2022.104445

Wang, Z., Wang, X., Wan, J. B., Xu, F., Zhao, N., & Chen, M. (2021). Optical imaging in the second near infrared window for vascular bioimaging. *Small*. Advance online publication. DOI: 10.1002/smll.202100384

Yan, J. S., Orecchioni, M., Vitale, F., Coco, J. A., & Duret, G. (2021). Biocompatibility studies of macroscopic fibers made from carbon nanotubes: Implications for carbon nanotube macrostructures in biomedical applications. *Carbon*. Advance online publication. DOI: 10.1016/j.carbon.2021.06.016

Zhang, Z., Du, Y., Shi, X., Wang, K., Qu, Q., & Liang, Q. (2024). NIR-II light in clinical oncology: Opportunities and challenges. *Nature Reviews. Clinical Oncology*. Advance online publication. DOI: 10.1038/s41571-024-00873-6

Zhu, W., Li, H., & Luo, P. (2021). Emerging 2D Nanomaterials for Multimodel Theranostics of Cancer. *Frontiers in Bioengineering and Biotechnology*. Advance online publication. DOI: 10.3389/fbioe.2021.769178

Chapter 7
Metal and Metal Oxide Nanoparticles:
Emerging Tools in Cancer Therapy and Bioimaging

Richa Singh
https://orcid.org/0000-0001-7184-9958
Department of Biotechnology, SIES College of Arts, Science and Commerce (Autonomous), Mumbai, India

ABSTRACT

Cancer is one of the most pressing global health challenges. Despite significant advancements in conventional therapies, mortality rates for some cancers are still high. Key barriers, such as multidrug resistance, tumor heterogeneity, non-specific targeting, toxicity, and late-stage diagnosis, limit the efficacy of existing therapies. The rapid advancement in nanomedicine has opened new avenues in cancer therapy, particularly through the integration of metal and metal oxide nanoparticles (M/MONPs). These nanoparticles play a crucial role in cancer bioimaging, serving as contrast agents in various imaging modalities. Their ability to accumulate selectively in tumor tissues makes them ideal candidates for early detection and real-time monitoring of cancer progression. This chapter provides a comprehensive overview of the synthesis and biomedical applications of M/MONPs in cancer treatment. It highlights their roles as therapeutic and diagnostic tools, elucidates their mechanisms of action, and addresses key concerns related to toxicity, biocompatibility, and clinical translation.

DOI: 10.4018/979-8-3373-3765-4.ch007

INTRODUCTION

The term 'cancer' represents a broad category of diseases characterized by an uncontrolled proliferation of abnormal cells, which can initiate in virtually any organ or tissue within the body due to genetic mutations. Out of 200 different diseases classified as cancer, most form solid masses of tissue called tumors, categorized into benign or malignant types. Lung, prostate, colorectal, stomach, and liver cancers are the leading types of cancer seen in men, while breast, colorectal, lung, cervical, and thyroid cancers are the most often identified in women (WHO). There are six hallmarks of cancer: sustained angiogenesis, insensitivity to anti-growth signals, evasion of cell death, limitless replicative potential, self-sufficiency in growth signaling, and metastasis (Lazebnik, 2010). This ability to infiltrate neighboring tissues and/or other organs often leads to cancer mortality.

Cancer is the second leading cause of mortality on a global scale (WHO). In 2022 alone, there were 20 million newly diagnosed cancer cases and 9.7 million cancer-related fatalities documented worldwide. It is anticipated that the total number of cancer cases will escalate to 35 million by the year 2050, primarily attributable to the phenomena of population aging and growth (American Cancer Society, 2025). Lung cancer, the most commonly diagnosed malignancy, accounted for approximately 20% of all cancer-related deaths in 2022 (Bray et al., 2024). Although individuals diagnosed with metastatic disease exhibit a markedly higher likelihood of mortality compared to those with non-metastatic cancer (Dillekås et al., 2019), the factors influencing cancer mortality are intricate and often encompass dysfunctions across multiple interrelated biological systems (Boire et al., 2024). While advancements in diagnostics and treatments have improved outcomes in high-income countries, projections suggest that over 75% of global cancer deaths will occur in low- and middle-income regions within the next decade due to inadequate diagnostic, treatment and palliative care facilities (Anandasabapathy et al., 2024).

Cancer Therapy and Barriers

As the incidence of cancer continues to rise globally, diagnosing and treating the disease poses significant challenges for health systems. Advanced diagnostic techniques, such as genome sequencing, polymerase chain reaction and mass spectrometry, have significantly enhanced the precision of cancer detection and molecular profiling via identification of oncogenic mutations, epigenetic alterations and protein biomarkers (Sarhadi & Armengol, 2022). However, tumor-derived genetic biomarkers are not always released into the bloodstream during the early stages of cancer, and even when present, their concentrations are often extremely low, making detection challenging (Campos-Carrillo et al., 2020). Commonly used protein biomarkers, such as prostate-specific antigen (PSA) and cancer antigen 125 (CA-125), suffer from poor specificity. Normal ranges are seen in many cancer patients, even in advanced stages, while elevated levels have been reported in individuals without cancer, leading to false positives and limiting their diagnostic reliability (Adhyam and Gupta, 2012; Dochez et al., 2019). Imaging techniques such as computed tomography (CT), magnetic resonance imaging (MRI), and positron emission tomography (PET) are essential for tumor localization and staging, but they may have limited resolution and difficulties in distinguishing between malignant and benign or inflammatory lesions (Klug et al., 2024; Zheng et al., 2024). Tissue biopsy remains the gold standard for definitive diagnosis between benign and malignant tumors. Yet, it is highly invasive, risky and painful, has poor repeatability, and is unable to provide whole-body assessment or spatial information for non-puncture sites (Lopes and Betrouni, 2009). Methods such as immunohistochemistry (IHC) and fluorescence in situ hybridization (FISH) are widely used in clinical oncology, but they are semi-quantitative and prone to variability, depending on tissue quality and operator interpretation (Furrer et al., 2017)

Conventional treatment options include surgery, chemotherapy, and radiotherapy. Surgery is particularly beneficial in excising localized benign tumors and surrounding tissues (Qu, 2023). However, the procedural challenges are compounded by anatomical intricacies, inaccessibility to care, risk of infection and high morbidity rates (Ranganathan et al., 2023). Chemotherapy involves the administration of cytotoxic drugs, such as alkylating agents (cyclophosphamide, cisplatin), antimetabolites (5-fluorouracil, 6-mercaptopurine), topoisomerase inhibitors (doxorubicin) and mitotic spindle inhibitors (paclitaxel), to target rapidly proliferating cancer cells (Bukowski et al., 2020; Cavalcanti & Soares, 2021). These agents are often associated with adverse effects, including nephrotoxicity, hepatoxicity, gastrointestinal issues, neutropenic fever, tumor lysis syndrome, thrombosis, leukostasis, and hyperviscosity syndrome (Bischof et al., 2019; Cavalcanti & Soares, 2021). Additionally, chemotherapy-induced immunosuppression increases the risk of secondary infections (Sun et al., 2025). Both chemotherapy and radiotherapy target malignant cells but can also affect surrounding healthy tissues, potentially leading to side effects. Moreover, these therapies can inadvertently activate cancer-associated fibroblasts, leading to increased malignancy and therapeutic resistance in tumors (Komoto et al., 2023). While conventional strategies have been pivotal in cancer management, their limitations in advanced stages highlight the need for advanced approaches, such as immunotherapy, targeted therapy and personalized medicine.

Immunotherapy activates the immune system to target tumor cells, often eliciting autoimmune-like responses rather than generalized immunosuppression. Such immune-related adverse events can affect virtually any organ in the body and may manifest up to months after treatment is initiated or completed (Pallin et al., 2018).

It has been reported that 25% of the cancer patients receiving immune checkpoint inhibitors present immune-related adverse events, such as colitis, pneumonitis, dermatitis and potentially life-threatening hypophysitis, myocarditis, and vasculitis (El Majzoub et al., 2019). Similarly, chimeric antigen receptor (CAR) T-cell therapy is associated with serious side effects, including neurotoxicity and hemophagocytic lymphohistiocytosis (Tallantyre et al., 2021). Targeted therapy includes monoclonal antibodies, antibody-drug conjugates, peptide drugs, small-molecule inhibitors and hormone therapies that interfere explicitly with molecules involved in the growth, progression, and spread of cancer. Unlike traditional chemotherapy, which is non-specific and affects both healthy and cancerous cells, targeted therapies aim to selectively attack cancer cells by disrupting specific gene expression, receptors, metabolites or signaling pathways unique to tumor cells (Walter & Ahmed, 2018). These targeted agents, however, show a similar frequency and severity of toxicities as traditional cytotoxic therapies, but differ in the nature of the adverse effects. While classical chemotherapy is commonly associated with alopecia, myelosuppression, mucositis, nausea, and vomiting, targeted therapies more often lead to vascular, dermatologic, endocrine, coagulation, immunologic, ocular, and pulmonary toxicities (Dy & Adjei, 2013). Personalized medicine is gaining traction in oncology, enabling the generation of treatment plans using multi-omics data and tailoring them to individual cancer profiles and patient-specific responses (Mohr et al., 2024). Reinforcement learning algorithms, which incorporate artificial intelligence and predictive analysis, have been shown to improve diagnostic accuracy for lung cancer by 37%, thereby optimizing therapeutic decisions (Ghorbian & Ghorbian, 2024). However, the therapy grapples with managing a high volume of healthcare data, data processing, rigorous data validation, real-time monitoring of the treatment responses, algorithmic bias, data security and privacy, patient trust and delivery of treatments with minimal side effects (Mohr et al., 2024).

In addition to diagnostic and therapeutic challenges, the complexity of the disease also contributes to the ongoing difficulties in cancer care. Tumor heterogeneity, one of the significant barriers in precision medicine, is driven by genetic alterations and epigenetic reprogramming of cancer cells owing to genomic instability, altered chromatin landscapes, and the dynamic pressures of the tumor microenvironment, which act as a selective force to shape tumor cell behavior (Pe'er et al., 2021). These mechanisms, in combination, lead to phenotypic plasticity – an emerging hallmark of cancer – that confers the remarkable adaptive abilities underlying metastasis, immune evasion, and the acquisition of drug resistance (Bhattacharya et al., 2025). Multidrug resistance (MDR) is responsible for over 90% mortality in cancer patients receiving traditional chemotherapy or novel targeted drugs. Mechanisms such as elevated metabolism of xenobiotics, enhanced efflux of drugs, growth factors, increased DNA repair capacity and genetic factors are found to be associated with tumor MDR (Bukowski et al., 2020). This leads to reduced therapeutic efficacy of cancer drugs causing difficulty in cancer treatment. In the context of aggressive malignancies, such as glioblastoma multiforme, survival rates are still dismally low (Naser et al., 2022).

Additionally, access to the advanced treatments remains disproportionate, with low-resource settings facing critical shortages in healthcare infrastructure, trained personnel, highly specialized care and affordability (Ghorbian & Ghorbian, 2024). In resource-limited settings, other barriers, such as poor awareness and financial constraints, hinder effective cancer care. In rural regions, these issues are further compounded by cultural beliefs and societal stigma, which often delay timely diagnosis and treatment (Nagar et al., 2024). The rise in cancer cases, thus, continues to impose profound emotional, physical, and monetary pressures on individuals, families, communities, and the healthcare sector (WHO). While modern technologies offer new avenues, they do not fully overcome the existing systemic barriers. These limitations underscore the need for integrative, comprehensive and innovative solutions that can enhance biomarker detection, enable targeted drug delivery, and support real-time biosensing to optimize treatment, leading to an effective and accessible precision oncology.

In response to these ongoing challenges, nanotechnology has emerged as a promising and versatile platform in the field of oncology. Operating at the molecular and cellular levels, nanomedicine enhances precision in both diagnosis and therapy, offering potential solutions to overcome the shortcomings of conventional therapies. Innovative solutions are being developed in the field of clinical oncology by exploiting the unique physicochemical properties of various nanoscale materials. This chapter provides a comprehensive and critical overview of the synthesis, functionalization, and application of M/MONPs in addressing complex cancer challenges, both as therapeutic agents and bioimaging tools. It also explores their distinctive properties, mechanisms of action, and potential toxicity issues.

METAL AND METAL OXIDE NANOPARTICLES

Nanoparticles are solid organic or inorganic materials having at least one dimension within 1 to 100 nm. Metal and metal oxide nanoparticles (M/MONPs) are at the forefront of nanotechnology research due to their unique properties and wide range of applications in medicine, energy, and environmental science.

Composition

Metal nanoparticles can be classified into major categories- those composed of noble metals like gold (Au), silver (Ag), platinum (Pt), palladium (Pd), rhodium (Rh), rhenium (Re), osmium (Os), and iridium (Ir), etc. and those made from base metals such as copper (Cu), zinc (Zn), lead (Pb), nickel (Ni), iron (Fe), aluminium (Al), etc. (Schrand et al., 2010; Shedbalkar et al., 2014; Singh et al., 2013a). Metalloids have features of both metals and non-metals. Typical examples are boron (B), calcium (Ca), cerium (Ce), magnesium (Mg), silicon (Si), germanium (Ge), arsenic (As), tellurium (Te), and bismuth (Bi). The nano-form of many of these metalloids has found applications in medicine. Metal oxide nanoparticles are crystalline solids having a metal cation and an oxide anion. These can be further categorised based on the type of metal: transition metal oxide (ZnO, Fe_2O_3, Fe_3O_4, TiO_2), alkaline earth metal oxide (MgO, CaO), rare earth metal oxide (CeO_2) and other metal/metalloid oxide (Al_2O_3, As_2O_3, SiO_2) nanoparticles (Singh et al., 2013b; Yazdi et al., 2016). Moreover, studies have reported on bimetallic nanoparticles, such as Ag-Au, which exist as alloys or in core-shell structures (Idris and Roy, 2023; Singh et al., 2016). Composite nanoparticles that combine metal with metal oxide or two different metal oxides (such as TiO_2@SiO_2) have also been documented (Karthikeyan et al., 2023).

Synthesis

M/MONPs can be synthesized via physical, chemical and biological routes (Table 1), each influencing their properties and potential applications. Physical methods primarily employ a top-down approach, where bulk materials are broken down into finer particles of nanosize, while chemical and biological methods follow a bottom-up approach, involving the assembly of atoms or molecules to form nanoparticles (Shiri et al., 2019). Chemical approaches offer the benefit of enabling the manufacture of particles with precise size, dimension, composition, and structure, which can be beneficial in various applications, such as catalysis, sensing, and electrical devices (Karthikeyan et al., 2023).

Table 1. Various routes for the synthesis of M/MONPs

Method	Brief description	Examples
Physical method Ball milling	Mechanical pulverization of metal precursor to powder	Fe_2O_3
Sputtering	Atoms ejected from target deposited to form film under high voltage	CuO
Laser ablation	Etching of metal in solid or liquid environment on high-energy laser exposure	Pt, Au
Electrospraying	Atomization of liquid suspension into airborne droplets using electrostatic forces	NiO, ZnO, Fe_2O_3, Co_3O_4, CuO, TiO_2
Electron beam evaporation	Etching of metal by focused electron beams	Ag-Si, Al_2O_3
Chemical method Sol-gel	Transition of colloidal liquid solution into solid gel followed by calcination	SnO_2, $NiFe_2O_4$
Hydrothermal	Reaction of precursors in closed environment under high temperature and pressure	Zn_2SnO_4
Precipitation	Concentration of dissolved precursor (in solvent) exceeds solubility	TiO_2, CeO_2
Chemical reduction	Reaction of metal salt with chemical reducing agent	Ag, Al
Microemulsion	Oil-in water or water-in-oil system stabilized by surfactants acting as emulsifier	Rh, Re, Os, Ir
Chemical vapour deposition	Volatile precursors decompose on heated substrate to form a film	Fe_2O_3, graphene-Ag
Biological method Plants	Aqueous extract of plant part mixed with metal precursors	Ag, Au, Cu, Pt, Pd, ZnO, Fe_2O_3, MgO
Microorganisms	Cell culture or cell extract mixed with metal precursors	Ag, Au, Pd, Pt, ZnO, Au-Ag
Bio-templates	Controlled synthesis or deposition using bio-inspired structures	Ag, Au, Pt, Pt alloys, Fe_3O_4,
Enzymes	Reducing metal precursor with redox enzymes	Ag

Sources: Chaudhary et al. (2020), Thummavichai et al. (2023)

Properties

They possess distinct physicochemical, optical and electronic properties compared to their bulk counterparts, such as high surface area-to-volume ratio, smaller size, enhanced stability, crystallinity, anti-corrosiveness, conductivity, unusual catalytic activity, photocatalytic activity, surface plasmon resonance (SPR), and ease of functionalization (Karthikeyan et al., 2023). Different forms of metal nanoparticles are also reported. For example, based on shape, they can be spherical, triangular, or rod-like. The properties of these nanoparticles are influenced by synthesis parameters, such as temperature, pH, and precursor concentration, which affect their size, shape, and composition (Aziz et al., 2025). Higher temperatures increase the kinetic energy, which leads to larger particle sizes and vice versa. The reaction pH significantly influences the solubility of metal precursors and the rate of hydrolysis, which in turn affects the nucleation and growth processes. Adjusting the pH can affect particle size, shape, stability and surface properties. The availability of ions for nucleation and growth depends on the concentration of metal precursors. Higher concentrations lead to rapid nucleation, resulting in smaller particles, while lower concentrations may produce larger particles due to slower nucleation rates. This parameter also affects the homogeneity and composition of the nanoparticles (Aziz et al., 2025).

Functionalization

Functionalization of M/MONPs is crucial for enhancing their applicability across various fields, including catalysis, imaging, sensing, and drug delivery. The process involves modifying the surface of nanoparticles, often with surfactants, ligands, biomolecules, polymers, drugs or chemical groups (Kumar et al., 2022). M/MONPs can also be embedded within metal-organic frameworks (MOFs). This improves their stability, functionality and dispersibility in various media (Kumar et al., 2022; Shakya et al., 2019). Conjugation with specific targeting molecules allows these nanoparticles to exhibit specificity, selectivity and sensitivity during interaction with analytes of interest, leading to effective biosensing, imaging, targeted delivery and diagnosis (Chen et al., 2014a; Kairdolf et al., 2017). The properties of gold and iron oxide nanoparticles have been enhanced through functionalization to exhibit stimuli-responsive behaviours required for improved therapeutic outcomes (Kumar et al., 2022).

M/MONPs IN CANCER THERAPY – MECHANISMS AND APPLICATIONS

M/MONPs combat cancer through multiple mechanisms such as induction of oxidative stress and apoptosis, enhanced radio- or thermal-sensitization, gene delivery, targeted and sustained drug delivery, etc (Joshi et al., 2023; Shi et al., 2023; Varzandeh et al., 2023; Yu et al., 2024; Zhang et al., 2020). Their unique physicochemical properties enable precise interaction with cancer cells at the molecular level, offering targeted, controlled and synergistic therapeutic effects. This section outlines the key mechanistic pathways through which M/MONPs exert anticancer effect, underscoring their potential in advanced cancer therapy applications.

Photothermal Therapy (PTT)

PTT is a method that utilizes near-infrared (NIR) light to irradiate a photothermal agent, thereby increasing tissue temperature to achieve cell ablation. Noble metal nanoparticles exhibit strong SPR, allowing efficient absorption of NIR light, which penetrates deeply with minimal harm to surrounding tissues. Upon accumulation at tumor sites, these nanoparticles absorb light, inducing localized hyperthermia that causes protein denaturation, membrane damage, and organelle dysfunction in tumor cells. This thermal effect induces cell death via necrosis, apoptosis, and autophagy, while also triggering immune responses through antigen release. Additionally, it disrupts tumor vasculature, impairing blood supply and enhancing the overall anticancer effect (Cullion et al., 2023). M/MONPs, such as gold, silver, platinum, palladium, manganese oxide and iron oxide, have been extensively studied for their efficacy in PTT (Yu et al., 2024). Shape, size, crystallinity and surface modification of M/MONPs also impact their activity as photothermal agents (Malekzadeh et al., 2023). Encapsulation of platinum nanoparticles in hyaluronic acid combined with direct PTT exhibited cytotoxic effects in MDA-MB-231 breast cancer cells, highlighting the significance of combination therapy against tumor (Zhu et al., 2017). PEGylated tungsten oxide (WO_{3-x}) nanoparticles effectively killed 4T1 murine breast cancer cells when combined with NIR, resulting in a significant decrease in cell viability, with less than 20% of cells remaining alive at the highest nanoparticle concentration of 0.2 g /L (Liu et al., 2014). Although PTT holds promise for the therapy of localized tumors due to its non-invasive nature, high specificity and minimal side effects, its efficacy in addressing tumor metastasis and recurrence is constrained (Yu et al., 2024). Moreover, challenges such as low conversion efficiency, biosafety concerns, and incomplete tumor elimination have also been reported (Zhang et al., 2019).

Photodynamic Therapy (PDT)

PDT is a minimally invasive treatment that relies on the interaction between non-ionizing radiation, a photosensitizer, and molecular oxygen to generate reactive oxygen species (ROS) to destroy targeted cells. The therapy is widely employed for treating various malignancies, including skin, esophageal, and lung cancers. It is particularly effective for early-stage tumors and can extend survival when surgery is not feasible (Allamyradov et al., 2024). However, its effectiveness can be limited by poor tumor oxygenation and inadequate delivery of the photosensitizer (Jiang et al., 2023). The incorporation of M/MONPs has significantly advanced PDT by addressing these limitations. Platinum, gold and manganese dioxide NPs serve as photosensitizers or carriers, improving the delivery, stability and activation of PDT agents. These NPs have been shown to catalyze the production of oxygen from hydrogen peroxide, thereby improving ROS production and alleviating tumor hypoxia during PDT (Shi et al., 2023). Titanium dioxide (TiO_2) nanoparticles, well-known for their photocatalytic activity, have demonstrated the ability to generate ROS upon UV light illumination, which can induce cell death in malignant tumors (Ziental et al., 2020). Furthermore, TiO_2-based nanocomposites and hybrid materials, created by grafting organic compounds, show enhanced light absorption properties, which can be leveraged for effective PDT (Xu et al., 2007). Incorporation of M/MONPs in PDT offers a more effective and selective alternative to traditional cancer treatments, with reduced systemic toxicity and improved therapeutic outcomes.

Radiosensitization

Radiosensitization refers to the process of augmenting the susceptibility of cancer cells to the effects of ionizing radiation, thereby improving the efficacy of the treatment while reducing harm to adjacent healthy tissues. This methodology is crucial in oncological therapy, as it aims to improve the therapeutic ratio by enhancing tumor control and minimizing adverse effects. However, its efficacy can be limited by tumor resistance due to factors like hypoxia, DNA repair, and tumor heterogeneity (Sheehan et al., 2010). Various compounds have been explored as radiosensitizers – agents to increase the effect of radiation therapy – subdivided into three categories based on their structure: macromolecules (miRNAs, proteins, peptides, and oligonucleotides), small molecules, and nanomaterials (Varzandeh et al., 2023). Nano-radiosensitizers, especially M/MONPs featuring high-Z elements such as gold, gadolinium, bismuth and iron, have been shown to enhance radiation dose deposition via physical, biological, and chemical mechanisms. These dense M/MONPs selectively scatter and absorb high-energy ionizing rays such as gamma/X-rays, leading to increased local electron generation (e.g., Compton and Auger electrons), which enhances DNA damage and ROS production in tumor cells due to their short range and high reactivity (Conde et al., 2012; Varzandeh et al., 2023). Hafnium-based nanoscale MOFs incorporated with Fe^{3+} ions have been developed to improve radiotherapeutic effects through a multifaceted mechanism, including the Fenton reaction to produce hydroxyl radicals and interference with DNA repair pathways (Gong et al., 2021). M/MONPs also modulate biological responses, including oxidative stress, cell cycle arrest, and bystander effects. Their physicochemical properties—such as size, surface area, and functionalization—govern their radiosensitizing efficiency. While MeV photons are standard in clinical radiotherapy, keV photons in combination with high-Z NPs can offer greater tumor specificity, though their limited tissue penetration remains a constraint (Rosa et al., 2017).

Magnetic Hyperthermia (MH)

MH employs magnetic iron oxide nanomaterials (IONs) to generate localized heat, under an alternating magnetic field (AMF), through mechanisms such as hysteresis losses and Néel and Brownian relaxation (Figure 1). This heat elevates the temperature of cancer cells to 39–45 °C, inducing cell death while sparing healthy tissues (Szwed & Marczak, 2024). This technique causes ROS production, DNA damage, expression of heat shock proteins and modulation of cellular functions, ultimately leading to apoptosis in cancer cells with precision, offering advantages like deep tissue penetration and temporal-spatial regulation. Additionally, MH can trigger differentiation in surviving cancer cells, potentially reducing their ability to self-renew (Szwed & Marczak, 2024; Moise et al., 2018). MH also shows promise in altering anti-tumor immune responses by repolarizing macrophages and triggering immunogenic cell death. A key challenge in MH is the inherently low magnetothermal conversion efficiency of superparamegantic IONs (Liu et al., 2015). Addressing this limitation is crucial for improving therapeutic outcomes. The effectiveness of MH is influenced by multiple factors, including the concentration of IONs, magnetic field characteristics, and the target cell type (Jiao et al., 2024; Xie et al., 2018). Optimization across these parameters is crucial for enhancing the clinical applicability of MH.

Figure 1. Key features of intracellular magnetic hyperthermia (MH). Magnetic iron oxide nanoparticles (MIONs) are readily internalized by tumor cells and, upon exposure to alternating magnetic field (AMF) generate localized nanoscale heat to induce MH. The therapeutic efficacy of MH is enhanced by magnetic hysteresis losses and heat-induced generation of ROS. MH can also initiate immunogenic cell death (ICD) and promote macrophage repolarization to trigger anti-tumor immune response.

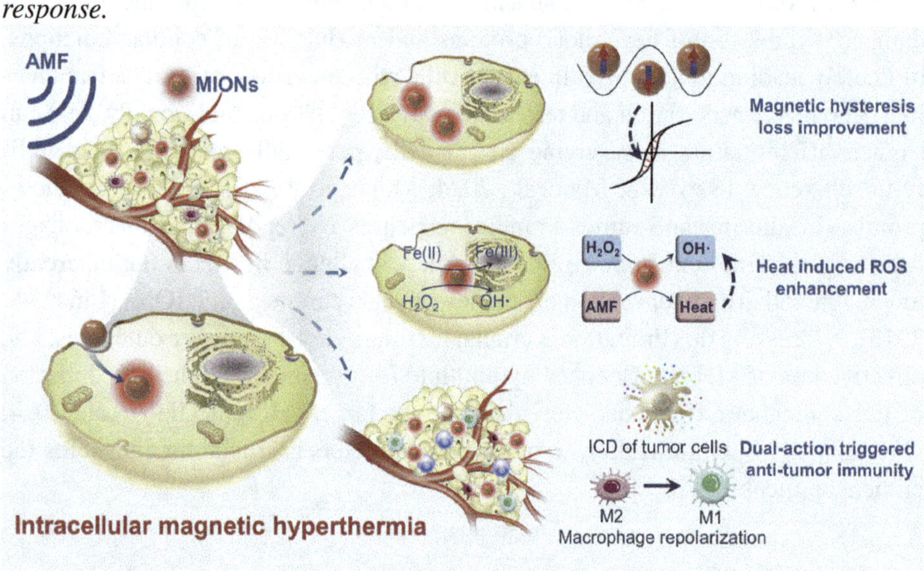

(Jiao et al., 2024)

Drug-delivery Vehicles

Drug delivery vehicles are engineered systems designed to transport therapeutic agents to specific sites in the body, enhancing the efficacy and safety of treatments. In cancer therapy, various nanomaterials serve as these vehicles, including liposomes, solid lipid nanoparticles, dendrimers, carbon nanotubes, quantum dots, micelles, hydrogels, and silica nanoparticles (Joshi et al., 2023). M/MONPs have emerged as promising drug-delivery vehicles for cancer treatment due to their unique physicochemical properties, which enable targeted and controlled drug release, improved solubility and reduced side effects. Various M/MONPs, such as gold, iron oxide and zinc oxide nanoparticles have been functionalized with polymers such as polyethylenimine, polystyrenesulfonate, polyethylene glycol (PEG), and dextran to enhance anti-cancer drug-loading capacity and enable pH-responsive release, particularly effective in acidic tumor microenvironments (Khabibullin et al., 2023; Prashar et al., 2021). Studies have shown that such functionalization improves drug solubility, retention, and selective cytotoxicity. For example, paclitaxel-loaded gold nanoparticles and doxorubicin-loaded iron oxide MOFs exhibited strong antitumor activity (Khabibullin et al., 2023; Kharazmi & Attaran, 2023). Moreover, polysaccharide coatings help M/MONPs evade efflux pumps and immune clearance, enhancing therapeutic outcomes (Prashar et al., 2021). This approach also addresses the limitations of conventional chemotherapy, such as low bioavailability, chemoresistance, and systemic toxicity, by enhancing drug delivery efficiency and minimizing adverse effects (Joshi et al., 2023). Despite these advantages, challenges such as colloidal instability, potential toxicity, and immune interactions remain (Lai et al., 2019). Careful design of metal type, particle size, and surface chemistry is crucial to maximize biocompatibility and clinical effectiveness.

Gene-delivery

Gene therapy overcomes the limitations of non-specificity and systemic toxicity associated with chemotherapy via direct regulation of key genes or proteins (e.g., K-Ras, c-Myc, MDR1, and PTEN) leading to effective inhibition of tumor proliferation and growth. DNAs, small interfering RNAs (siRNAs), microRNAs (miRNAs), and messenger RNAs (mRNAs) are the most frequently used nucleic acids for gene therapy (Goleij et al., 2023). In the cytoplasm, silencing complexes induced by siRNA silence cancer-related genes through the selective complementary binding to target mRNA. miRNAs bind to specific mRNAs, either suppressing their translation or promoting their degradation. Unlike other pathways, DNA typically enters the nucleus and regulates target protein expression by undergoing transcription followed by translation. Moreover, gene–editing technology has expanded possibilities for cancer gene therapy. The CRISPR/Cas9 system, comprising single-guide RNA and Cas9 protein, enables the precise deletion, insertion, or modification of target genes in tumor cells, resulting in potent antitumor effects (Sayed et al., 2022; Zhang et al., 2022).

M/MONPs have emerged as one of the promising alternatives to gene therapy in cancer. IONs have been extensively studied as a potential candidate for tumor-targeted gene delivery owing to superior magnetic responsiveness, simple preparation, ease of chemical functionalization, low toxicity, and high biocompatibility (Stanicki et al., 2022). Zhang et al. (2020) recently demonstrated that IONs can deliver siRNA to intracranial xenografted glioma and effectively reduce glutathione peroxidase expression, resulting in significantly improved inhibition of tumor growth. IONs coated with chitosan-PEG were demonstrated to deliver siRNA to hepatocellular carcinoma in vivo leading to suppression of Luc expression (Wang et al., 2016). In another study, IONs functionalized with a tumor-targeting peptide and dextran delivered miRNA to MDA-MB-231 breast cancer cells in vitro (Yoo et al., 2014). Gold and silver nanomaterials have also been developed for efficient gene delivery, offering advantages such as controllable size and good uniformity. However, toxicity due to accumulation in the liver and kidneys limits their use as gene delivery vehicles (Zhang et al., 2022).

Combination Therapy

Combination therapy involving metal and metal oxide nanoparticles (M/MONPs) has emerged as a highly effective strategy in cancer treatment by integrating multiple therapeutic modalities for synergistic outcomes. M/MONPs can be combined with chemotherapeutic agents in PTT and PDT to elevate the local temperature of tumor cells and induce oxidative stress, leading to improved drug efficacy and cell death. Gold/silver alloy nanoparticles, for instance, have been shown to enhance the cytotoxic effects of chemotherapy and thermotherapy, particularly against cancer stem cells, which are often resistant to conventional treatments (Zhao et al., 2024).

A zirconium and porphyrin-based MOF, like PCN-224, has high activity over a broad pH range and is capable of generating singlet oxygen ($^1O_2$), essential for PDT. A system using PCN-224, hyaluronic acid, and doxorubicin (HA-DOX-PCN) exhibited superior inhibition of MDA-MB breast cancer cell proliferation upon light irradiation as compared to HA-DOX-PCN-only (Kim et al., 2019). Similarly, Fu et al. (2020) developed a mesoporous platinum nanoparticle-based system (PEG@Pt/DOX) for the combination of chemotherapy and PTT, where doxorubicin was loaded into mesopores, and PEG-SH stabilized the surface via Pt–thiol interactions. The combination of PEG@Pt/DOX with laser irradiation significantly suppressed the proliferation of drug-resistant MCF-7/ADR cells compared to PEG@Pt/DOX alone.

Gold nanoparticles and IONs further support combination therapies as multifunctional platforms for co-delivering chemotherapeutic agents alongside energy-based modalities. Their surface chemistry allows precise functionalization for targeting, while their energy conversion capabilities facilitate localized hyperthermia under NIR or AMF. For example, gold nanoparticles conjugated with paclitaxel and tumor necrosis factor have been shown to increase vascular permeability, reduce interstitial fluid pressure, and enhance tumor drug accumulation (Paciotti et al., 2016). Similarly, magnetic nanoparticles embedded in lipid or polymeric carriers have demonstrated synergistic anticancer effects when paired with agents like doxorubicin or curcumin, owing to improved drug uptake and controlled release triggered by magnetic hyperthermia (Asghar et al., 2017).

M/MONPs IN CANCER BIOIMAGING

Bioimaging is a complex process that involves acquiring, processing, and presenting structural or functional images of living systems, along with the analysis of image-related data. It is increasingly recognized as a vital tool in clinical investigation and diagnosis. M/MONPs play a crucial role in cancer bioimaging, such MRI, CT and photoacoustic imaging (PAI). These nanoparticles, due to their unique physical and chemical properties, offer improved contrast and specificity in imaging, which is crucial accurate diagnosis of cancer. Their ability to accumulate selectively in tumor tissues through passive or active targeting mechanisms makes them ideal candidates for early detection and real-time monitoring of cancer progression. The integration of M/MONPs into cancer therapy and diagnostics offers a paradigm shift in oncology.

MRI and CT

M/MONPs, particularly those composed of gold, gadolinium, iron, manganese, tungsten, platinum, cobalt, and bismuth, have garnered significant attention due to their role as contrast agents in MRI and CT for cancer diagnosis. These nanoparticles exhibit an enhanced permeability and retention (EPR) effect, enabling them to accumulate preferentially in tumor tissues compared to normal tissues. This property enhances the contrast in MRI and CT, making it easier to distinguish between cancerous and healthy tissues (Oh et al., 2013). Moreover, their high atomic number enhances radiation absorption, Auger electron production, and X-ray attenuation, leading to improved image resolution (Khorasani et al., 2023). This capability is particularly beneficial in the precise localization and characterization of tumors, enabling early detection and monitoring of cancer progression. Surface functionalization with ligands or cancer-specific aptamers further improves imaging specificity and diagnostic accuracy (Patel et al., 2023).

IONs, particularly Fe_3O_4, are the most extensively studied for their applications in MRI due to their excellent magnetic properties and capacity to modulate proton relaxation times. Moreover, these exhibit biocompatibility, inhibit metastasis and reduce multidrug resistance in tumor, which make them suitable for cancer detection and imaging. Fe_3O_4 nanoparticles exhibit a high transverse relaxivity (r^2) value of 215 $mM^{-1}s^{-1}$, making them effective T_2 contrast agents. Their colloidal stability further supports their application in bioimaging of tumor structures, improving spatial resolution and tumor composition assessment (Barick et al., 2014). Gold and manganese nanoparticles have also been used in cancer MRI to enhance visualization of malignant tumors. Hybrid nanocrystals consisting of Fe_3O_4 and manganese oxide are utilized for T1- and T2-weighted MRI with dual contrast agents (Kattumuri et al., 2007).

A study demonstrated that anti-EGFR conjugated gold nanoparticles significantly enhance CT imaging of small tumors (~4–5 mm), making previously undetectable tumors visible. Such active targeting of the tumor resulted in higher accumulation of gold nanoparticles, leading to improved specificity and imaging sensitivity. Moreover, the CT number of the targeted tumor was found to increase five fold after nanoparticles injection, indicating a substantial contrast enhancement for better cancer detection based on molecular markers (Reuveni et al., 2011). A similar study on the selective accumulation of anti-Hsp70 antibody-functionalized gold nanoparticles, facilitating tumor-specific multimodal CT imaging, has also been reported (Kimm et al., 2020). Xi et al (2024) successfully developed a modular and automated deep learning-aided Nano-CT bioimaging method using hafnium oxide nanoparticles. The method enables sensitivity analysis and quantitative localization analysis of nanoparticle uptake in breast cancer cells with accurate distinction between intracellular particles and extracellular components, thereby offering a valuable quantitative imaging tool for minute amounts or ultrasmall NPs. Recent advancements include the development of bismuth and gadolinium-co-doped carbon quantum dots, which combine robust fluorescence with high X-ray attenuation, making them effective for both CT and MRI imaging (Meng et al., 2022). Many of these nanoparticles can also be engineered to support multiple imaging modalities, such as MRI, CT, and PET, providing comprehensive diagnostic information and enabling real-time monitoring of therapeutic efficacy (Yazdi et al., 2021).

Fluorescence Imaging

Fluorescence imaging relies on the excitation of fluorophores, which emit light when they return to their ground state. This process allows for the visualization of specific molecules within biological tissues. It plays a crucial role in cancer diagnostics, particularly in intraoperative assessments of tumor margins. Activatable fluorescence probes have been developed to target specific enzymatic activities associated with various cancer types, offering high-contrast imaging with low background fluorescence (Fujita & Urano, 2024). M/MONPs have emerged as a powerful tool in fluorescence imaging for cancer detection and diagnosis. Their unique optical properties, such as SPR, enable significant enhancement of fluorescence signals, making them highly effective in identifying cancerous tissues at early stages. Notably, gold and silver nanoparticles enhance the fluorescence intensity of dyes through the metal-enhanced fluorescence (MEF) effect, wherein their localized SPR amplifies both excitation and emission rates. This enhancement is crucial for detecting low-abundance cancer biomarkers with high sensitivity (Botequim et al., 2025).

For instance, nanoflare probes based on 13-nm gold nanoparticles functionalized with Cy5-labeled DNA have demonstrated high-sensitivity detection of miRNA-1246 in breast cancer cells. These systems have also been used for in vivo imaging of miRNA-21 in tumor-bearing mice (Zhai et al., 2018). Copper nanosheets templated with a DNA scaffold (DNS/CuNSs) have further enhanced imaging capabilities by providing wash-free operations with improved anti-photobleaching properties. The fluorescence resonance energy transfer (FRET) system constructed with DNS/CuNSs achieved approximately 4-fold higher signal amplification compared to free Cy5.5, facilitating effective fluorescence imaging of MCF-7 cancer cells (Li et al., 2024b). Similarly, gold-rare earth bimetallic nanoclusters have shown promising results, with Au/Ce nanoclusters exhibiting the longest fluorescence emission wavelength (695 nm), higher quantum yield, and minimal autofluorescence interference. These nanoclusters, when coated with folic acid-modified liposomes, displayed enhanced targeting and accumulation ability, thus improving the precision of tumor imaging (Ye et al., 2022). Furthermore, shape-dependent effects of gold nanoparticles have been observed in fluorescence enhancement. Gold nanocubes enhanced the fluorescence of sulphonated aluminum phthalocyanines (AlPcS) by up to 150-fold, while gold nanorods exhibited varying effects depending on their longitudinal SPR bands – those having SPR bands in the 800–900 nm range achieved a moderate enhancement (2–6 times), whereas lower aspect ratios resulted in fluorescence quenching of the conjugated AlPcS (Xu et al., 2014).

In near-infrared fluorescence (NIRF) imaging, nanoparticles integrated with NIR dyes, such as rhodamine, BIODIPY, squaraine and porphyrin derivatives, provide deep tissue penetration and reduced background autofluorescence. These features enhance molecular-level differentiation between cancerous and normal tissues, which is crucial for early cancer detection (Yao et al., 2020). Gold nanoparticles and quantum dots, composed of inorganic semiconductor materials, are particularly valued in NIRF imaging for their tunable size-based fluorescence, excellent photostability, biocompatibility and flexible functionalization, making them suitable for prolonged and high-resolution imaging applications (Adeniyi & Awosan, 2025; Hu et al., 2020).

Photoacoustic Imaging (PAI)

PAI is an innovative bioimaging technique that combines the high contrast of optical imaging with the spatial resolution and penetration depth of ultrasound. In this method, short-pulse lasers are absorbed by chromophores, generating localized thermal expansion and subsequent ultrasound waves that are detected to form detailed images. This process allows for high-resolution imaging at significant depths, surpassing the optical diffusion limit. This hybrid modality is non-invasive and non-ionizing, making it particularly suitable for a wide range of biomedical applications (Liu et al., 2024). Photoacoustic contrast is derived from optical absorption by intrinsic tissue constituents, such as oxygenated and deoxygenated hemoglobin, melanin, and lipids. However, the use of exogenous contrast agents enables a wide range of cancer-specific imaging applications (Yang et al., 2009). M/MONPs are particularly effective in this modality because of their geometry-dependent optical absorption and high heat conductivity. The SPR effect, particularly in metal nanoparticles, amplifies this process, resulting in strong photoacoustic signals (Han et al., 2024). Gold nanoparticles are widely used in PAI due to their availability in various morphologies and excellent biocompatibility. Other metallic nanomaterials employed in PAI include those having silver, titanium, palladium, etc (Table 2). Tunable surface properties provide additional advantage in PAI as M/MONPs can be functionalized with targeting ligands, such as RGD peptides, to increase uptake by cancer cells and reduce non-specific uptake by healthy cells. This targeting enhances the specificity and efficacy of PAI (Ge et al., 2019).

Table 2. Metallic nanoparticles in photoacoustic imaging

Metal	Morphology and composition	Key properties for PAI
Gold	Spherical (20-50 nm)	Strong absorption (500–600 nm), high photoacoustic (PA) signal, passive tumor accumulation
	Nanorods with IR792 dye (40 nm, aspect ratio 3.5)	Absorption at 756 nm, dual-mode (PAI and SERS), effective in ovarian cancer models
	Glycol-chitosan coated Au nanoparticles (~95 nm)	Plasmon-coupling effect, strong PA signal at 680 nm, effective in breast cancer cells
Silver	Nanosphere aggregates	High photodamage threshold, absorption cross-section in NIR window, enhanced optical responses, high heat conductivity
Iridium	Ir complexed with NIR dye encapsulated in Pluronic F127 (~200 nm)	Dual maxima at 755 and 815 nm, tumor accumulation over 24 h, excretion via urine
Molybdenum	Single-layer MoS_2 nanosheets	Strong PA signal at 675 nm, higher contrast than multilayer MoS , selective brain tumor retention
Rhenium	PEG-stabilized Re nanosheets (~25 nm)	Broad NIR absorption (700–1000 nm), PAI + CT, tumor accumulation
Titanium	Ti C T_x MXene nanosheets (few layer)	Absorption ~800 nm, PAI and PTT, doxorubicin delivery + light-triggered release

continued on following page

Table 2. Continued

Metal	Morphology and composition	Key properties for PAI
Ruthenium	Ultrasmall chitosan-coated RuO_2 nanoparticles (~2 nm)	NIR-II absorption, nuclear targeting, intratumoral delivery
	Ru(phenanthroline) nanodots (~6.5 nm)	Broad NIR absorption (700–1000 nm), EPR-driven tumor accumulation
Bimetallic	Fe_3O_4 core and gold spiky surface (~510 nm with tips ~128 nm)	Strong absorption in second NIR window, strong PA signals, real-time monitoring of intratumoral distribution
	Pt nanodots on surface of Au nanoparticles	Effective tissue penetration, increase in the absorption and bandwidth of the NIR light, PA analysis of skin cancer cells
	Pd core with Au shell nanosheets (30 nm diameter, 4 nm thickness)	Absorption at 700 nm, PEG-functionalized, tumor-targeted PAI
	Ti core with Ir complex in cancer cell membrane coating (~389 nm)	Absorption at 1064 nm, ROS generation, cancer cell targeting via cancer cell membrane coating

Source: Steinbrueck & Karges (2023)

Theranostics

Cancer theranostics is an integrated approach that combines diagnostics and therapeutics to enable precise and personalized treatment strategies for cancer patients. It involves the identification of specific biomarkers and molecular profiles, along with imaging modalities, to assess tumor characteristics. M/MONPs have played a pivotal role in this domain by combining multiple functions, such as targeted imaging, controlled drug delivery and therapy, into a single nanostructure. These nanoparticles are employed in various multi-functional modalities, including MRI-guided gene therapy, thermal ablation, magnetic hyperthermia, chemotherapy, immunotherapy, PTT and PDT for imaging and therapeutic outcomes. Among these, metal nanoparticles—especially gold, silver and IONs—exhibit unique physicochemical properties, including superior magnetic and optical properties, strong X-ray attenuation and NIR absorption, which enhance both imaging contrast and therapeutic efficacy without the need for additional active components (Kumar et al., 2023). Their small size and ease of surface functionalization enable precise tumor targeting through the EPR effect, which takes advantage of the leaky vasculature of tumor tissues for selective accumulation and localization. This facilitates early-stage tumor detection and real-time monitoring of therapeutic outcomes. Additionally, metal nanoparticles can be engineered by incorporating imaging agents and therapeutic drugs onto their surfaces, leading to enhanced specificity, sensitivity and efficacy in both cancer detection and treatment (Patel et al., 2023).

Table 3. Metallic nanoparticles in cancer theranostics

Metal	Composition	Imaging Modality	Therapy Modality
Gold	^{68}G labelled dendrimer entrapped AuNPs	PET/CT	Immunotherapy
Silver	Ag-PEG	CT	Radiotherapy
Gadolinium	Gd	MRI	Radiotherapy
Iron	Superparamagnetic IONs	MRI	Hyperthermia
Tungsten	CoFe-WOx	Visible, IR, MRI, CT	PTT, PDT, Hyperthermia
Bismuth	Bi@mSiO$_2$@MnO$_2$/DOX	CT, MRI	PTT, Chemotherapy
Tantalum	Ta@PVP	CT, PAI	Radiotherapy, PTT
Ytterbium	Glutathione functionalized Yb/Fe oxide	MRI, CT	Imaging-guided therapy

Source: Khorasani et al. (2023)

Several nanotheranostic platforms, notably those featuring gold nanoparticles, have demonstrated substantial clinical potential due to their excellent biocompatibility (Figure 2). For instance, gold nanoparticles conjugated with a tumor-targeting aptamer (AS1411) and a gadolinium chelate (Gd(III)-DO3A), developed as a novel injectable theranostic agent, significantly enhanced MRI contrast and improved the pharmacokinetic properties, allowing for superior tumor imaging and monitoring (ElBaz et al., 2018). Hybrid systems, such as lipopolymeric nanoparticles coated with gold, further improve bioavailability and imaging resolution by leveraging the EPR effect for targeted delivery to metastatic sites, leading to increased fluorescence signals (Appidi et al., 2024). Another multifunctional theranostic platform, composed of an ultrasmall gold nanorod, the NIR dye IR780, and the anticancer drug DOX, demonstrated remarkable PAI efficiency and therapeutic efficacy. Besides high photothermal conversion efficiency of 67.5%, it selectively targeted U87MG cancer cells using RGD ligands, leading to increased specific uptake by tumor cells. The system achieved complete tumor elimination and prolonged survival (>40 days) in treated mice (Ge et al., 2019). Furthermore, real-time monitoring of photo-magnetically powered nanoparticles (composed of Fe_3O_4 and gold) through PAI revealed that the combination of laser and magnetic coupling improved the intratumoral distribution and retention of the nanoparticles, highlighting their effectiveness in oncological nanotherapies (Li et al., 2024a). Other M/MONPs employed in cancer theranostics have been listed in Table 3.

Figure 2. Gold nanoparticles as theranostic agents

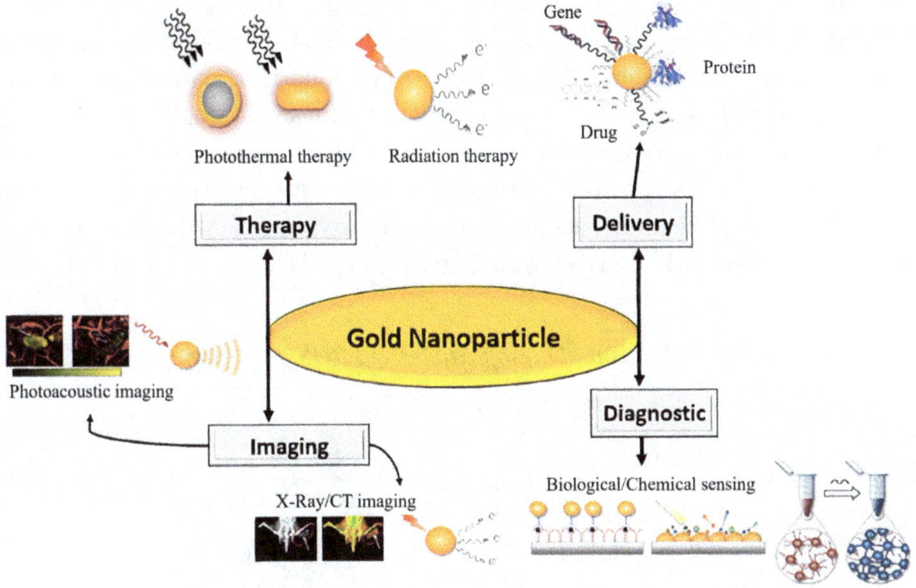

(Yaqoob et al., 2020)

TOXICOLOGICAL EVALUATION OF M/MONPS

M/MONPs have garnered significant attention in biomedical research due to their unique physicochemical properties, making them highly suitable for therapy, drug delivery, and diagnostics. However, despite their benefits, concerns surrounding their potential toxicity, biocompatibility and long-term safety remain a major barrier to clinical translation. The small size, chemical composition and high surface reactivity can lead to toxic effects in biological systems, highlighting a need for comprehensive toxicological assessments.

Mechanisms of Nanoparticle-Induced Toxicity

M/MONPs can induce toxicity through the generation of ROS, leading to oxidative stress, abnormal cellular function and apoptosis (Figure 3). Their nano-size allows deep penetration into biological membranes, disrupting redox balance and initiating ROS-mediated damage to cellular structures such as proteins, lipids and nucleic acids (Prasad et al., 2017). ROS can initiate lipid peroxidation, leading to the accumulation of hydroperoxides. These hydroperoxides subsequently decomposed to alkoxyl and peroxyl radicals, resulting in membrane disruption, reduced lipid fluidity and compromised cellular integrity (Aziz & Abdullah, 2022). Furthermore, ROS-induced DNA damage can cause mutations, leading to the initiation and progression of cancer. The overproduction of ROS causes mitochondrial release of apoptogenic signalling molecules, inducing a caspase cascade responsible for apoptosis (Klaunig et al., 2010). This is a common mechanism observed in nanoparticles like gold, silver, zinc oxide and copper oxide. Interestingly, some nanoparticles, such as cerium oxide, exhibit dual behavior, acting either as pro-oxidants or as ROS scavengers, depending on their redox state (García-Torra et al., 2021).

Figure 3. Toxicity mechanism of nanoparticles via ROS generation

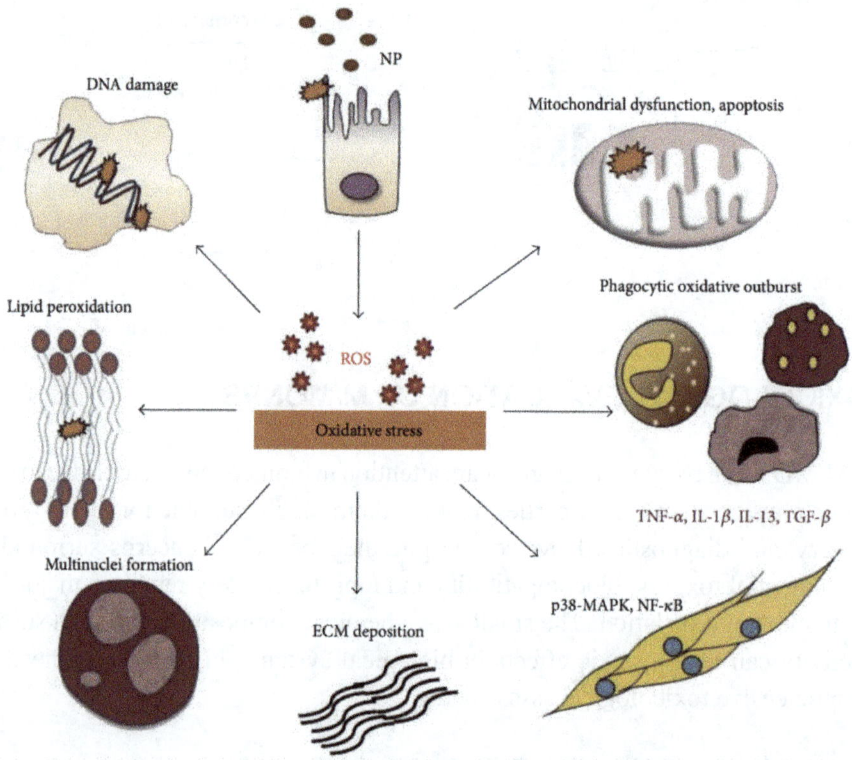

(Manke et al., 2013)

Direct interactions with cellular membranes, lysosomes and organelles may cause structural disruptions. These interactions, often mediated by electrostatic forces, lead to cell permeability, inflammation and necrosis (Isibor et al., 2024). Additionally, the dissolution of metal nanoparticles can release toxic metal ions, which contribute to their toxicity. These ions can interfere with cellular processes by binding to proteins and enzymes, inhibiting their function (Aziz & Abdullah, 2022). Exposure to M/MONPs can also lead to immunotoxicity through various mechanisms, including inflammation, oxidative stress, autophagy, and apoptosis, which negatively impact the immune system (Bi et al., 2023).

Factors Influencing Toxicity

The size, surface area, charge, concentration and surface modifications of nanoparticles significantly influence the toxicological profile of M/MONPs. Smaller particles with high surface area exhibit greater reactivity and are more likely to penetrate tissues, interact with cellular components and induce cytotoxicity (García-Torra et al., 2021; Yen et al., 2009). Surface charge also plays a critical role, with positively charged nanoparticles often showing higher cytotoxicity due to stronger electrostatic interactions with negatively charged cell membranes (Isibor et al., 2024). Research has shown that higher doses of nanoparticles often correlate with stronger cell cytotoxicity. Furthermore, surface modifications, such as coatings or functional groups, can significantly mitigate toxicity by reducing nonspecific cellular interactions and improving colloidal stability (Schrand et al., 2010).

M/MONPs' Toxicity – Experimental and Clinical Evidences

Extensive in vitro and in vivo studies have demonstrated variable toxicity profiles, influenced by size, shape and concentration, for different M/MONPs. In vitro studies show that silver, cadmium sulphide, copper oxide and zinc oxide nanoparticles are among the most cytotoxic, exhibiting concentration-dependent reductions in cell viability, increased oxidative stress, DNA damage, mutagenicity and genotoxicity. Soluble zinc oxide and cadmium sulphide nanoparticles caused lysosomal membrane destabilization, nuclear condensation and induced translocation of Nrf2 and NF-kB factors. The observation was proposed to be related to release of toxic metallic ions. Conversely, iron oxide, aluminium oxide and titanium dioxide nanoparticles show non-mutagenic, non-cytotoxic, and non-genotoxic effects, with no decrease in cell viability (Jeng & Swanson, 2006; Pujalté et al., 2015). In vivo evidence support these findings. Silver nanoparticles have been found to accumulate in vital organs such as the liver, kidneys, spleen, and brain, leading to organ-specific toxicity. Copper oxide and zinc oxide nanoparticles specifically caused significant DNA damage, accompanied by liver enzyme elevation and moderate vacuolation in hepatocytes (Jeng & Swanson, 2006; Xia et al., 2008). Titanium dioxide nanoparticles also exhibited the capacity to reach bone marrow, raising concerns about long-term effects (Chen et al., 2014b). Zebrafish embryo models also confirm dose-dependent developmental toxicity with copper oxide and zinc oxide nanoparticles providing additional insights into acute and chronic toxicities (Bai & Tang, 2020).

Many M/MONP exhibit non-specific accumulation in organs, such as liver, spleen, kidneys, lungs, and brain, leading to prolonged retention and potential toxicity. Silver and copper nanoparticles show a high tendency for accumulation in the liver and kidneys, often resulting in inflammation and alterations in enzymatic function. Silver and titanium dioxide nanoparticles can cross the blood-brain barrier, posing risks of neurotoxicity through oxidative damage and inflammation (Chen et al., 2007; Sharma & Sharma, 2007). Organ-specific effects such as immunotoxicity, neurotoxicity, reproductive toxicity, and hepatotoxicity have also been reported following exposure to M/MONPs (Ding et al., 2015). Besides this, metal hypersensitivity reactions, particularly T-cell mediated allergic contact dermatitis, have been observed in response to metals like beryllium, chromium, copper, cobalt, gold, lead, mercury, and nickel (Roach et al., 2019). Few cases of metal-induced autoimmunity, including the formation of immunocomplex deposits, have also been reported in humans (Bjørklund et al., 2024).

Primary and Secondary Effects

Direct nanoparticle-cell contact results in primary effects including DNA damage, oxidative stress, disrupted signalling pathways and inflammation. These effects may suppress cell proliferation via cell cycle arrest, leading ultimately to apoptosis or necrosis. Secondary toxicity occurs when nanoparticles translocate through tissue barriers into bloodstream and accumulate in other organs, such as the liver, spleen, brain or kidneys, thereby generating a secondary exposure. This systemic distribution can lead to chronic inflammation, immune activation, long-term functional impairments and organ-specific toxicity, particularly with persistent or repeated exposure (Pietroiusti et al., 2018; Sengul & Asmatulu, 2020).

Clearance Pathways

Understanding the pharmacokinetic profile of M/MONPs is critical for evaluating their long-term safety and therapeutic potential. For inorganic nanoparticles composed of metals, foreign to the human body, renal clearance has been demonstrated to be the most practical means for evading undesirable toxic effects. Renal clearance is typically more effective for nanoparticles smaller than 5.5 nm, whereas larger particles are generally cleared via the hepatobiliary route (Riviere, 2009). Silver and copper nanoparticles exhibit slow clearance and high bioaccumulation, which increases the risk of chronic toxicity. Control of particle size and surface coating is vital for favourable pharmacokinetics and biodegradation properties. For instance, the addition of PEG enhances circulation time and facilitates biodegradation (Landsiedel et al., 2012).

Most nanomaterial kinetics are characterized by relatively short blood half-lives, reflecting tissue extraction rather than clearance from the body. Smaller nanoparticles often demonstrate greater tissue penetration and are more difficult to clear, resulting in prolonged organ retention (Choi et al., 2007). Moreover, the limited biodegradability of metallic systems can pose long-term toxicity risks, including chronic inflammation and immune activation. Biodistribution studies indicate that surface charge, solubility, and particle size are key determinants of nanoparticle fate in the body (Li & Chen, 2011).

Strategies to Mitigate Toxicity

Developing nanoparticles with better biocompatibility and lower toxicity is crucial for their safe application in medicine and other fields. This involves optimizing synthesis methods to control size, shape, and surface properties. Moreover, surface modifications, such as coating nanoparticles with biocompatible polymers or biomolecules, have shown promising results in reducing toxicity by minimizing non-specific interactions with biological systems (García-Torra et al., 2021; Schrand et al., 2010). While the toxicological evaluation of M/MONPs raises significant concerns, it also presents opportunities for developing safer and more effective nanomaterials. A deep understanding of toxicity mechanisms, combined with rational design strategies, is essential for developing safer nanomaterials.

CHALLENGES

Despite significant advancements in nanomedicine, the clinical translation of M/MONPs for cancer therapy faces several critical challenges that need to be addressed to fully harness their potential. While many M/MONPs demonstrate promising anticancer properties, their long-term safety remains a matter of intense debate. Numerous reports on generation of ROS, immunotoxicity, bioaccumulation, incomplete biodegradation and clearance, off-target effects and systemic toxicity suggest that M/MONPs have the potential to harm normal tissues (García-Torra et al., 2021). These findings underscore the importance of conducting comprehensive toxicological and pharmacokinetic assessments to ensure the safe application of these products.

One of the key obstacles lies in the inconsistency of biocompatibility results across different studies. This may be a result of interference with the chemical probes, differences in surface chemistry or diverse cellular responses among different cell types. As a result, it is difficult to predict the behaviour of these nanoparticles in real-world applications. Translation challenges are further complicated by differences in physiological responses between animal models and humans, limiting the predictability and reliability of preclinical data (Alharbi et al., 2021).

An additional design-related challenge is the distinction between tumor targeting and tumor cell targeting. While many nanoparticles accumulate in tumor tissues via the enhanced permeability and retention (EPR) effect, this passive targeting does not guarantee efficient cellular uptake by cancer cells. Without active targeting mechanisms, such as ligand-mediated recognition, theranostic outcomes may be suboptimal (Urbano-Gámez et al., 2024). Thus, precision in nanoparticle engineering, including surface functionalization and targeting specificity, is crucial for enhancing therapeutic efficacy and minimizing off-target effects.

The path to regulatory approval for M/MONP-based therapies is fraught with challenges, including demonstrating safety, efficacy, and reproducibility. The lack of standardized protocols for synthesis, characterization and toxicity assessment further complicates this process. Furthermore, producing M/MONPs at a scale suitable for clinical use while maintaining quality and consistency poses an additional hurdle. This requires technological advancements in manufacturing processes and quality control systems (Alharbi et al., 2021).

CONCLUSION AND FUTURE PROSPECTS

M/MONPs hold immense potential in cancer theranostics, offering multifunctionality for simultaneous diagnosis, targeted therapy, and real-time monitoring through MRI, CT and PAI. However, overcoming the issues related to biocompatibility, toxicity, specificity and stability is essential for their successful and safe integration into clinical practice. Green synthesis approaches, utilizing eco-friendly and biologically derived methods, are gaining attention for their potential to reduce the environmental and biological toxicity. Moreover, the development of hybrid nanostructures and stimuli-responsive nanomaterials can enhance targeting efficiency and therapeutic outcomes. To facilitate clinical translation, it is essential to establish well-defined synthesis protocols, validated toxicity assessments, scalable manufacturing processes and clear regulatory compliance pathways. Future research should focus on the standardization of biological testing employing multiple cell-based assays across various cell types, dose ranges and appropriate animal models for safety and efficacy evaluations. Such rigorous protocols will help confirm reproducibility across laboratories. Collaborative efforts among material scientists, biologists, clinicians, and regulatory agencies will be crucial in overcoming current

REFERENCES

Adeniyi, M., & Awosan, W. (2025). Nanoparticle-enhanced near-infrared fluorescence probes: A breakthrough in cancer imaging techniques. *International Journal For Multidisciplinary Research*, *7*(1). Advance online publication. DOI: 10.36948/ijfmr.2025.v07i01.34883

Adhyam, M., & Gupta, A. K. (2012). A review on the clinical utility of PSA in cancer prostate. *Indian Journal of Surgical Oncology*, *3*(2), 120–129. DOI: 10.1007/S13193-012-0142-6

Alharbi, A. L., Alreshidi, M. S., Alanazi, A., Almutairi, K. F., & Alotibi, F. S. (2021). Advancements and challenges in nanotheranostics for cancer: A comprehensive review. [IJHS]. *International Journal of Health Sciences*, *5*(S1), 1436–1443. DOI: 10.53730/ijhs.v5ns1.15352

Allamyradov, Y., Yosef, J. B., Annamuradov, B., Ateyeh, M., Street, C., Whipple, H., & Er, A. (2024). Photodynamic therapy review: Past, present, future, opportunities and challenges. *Photochem*, *4*(4), 434–461. DOI: 10.3390/photochem4040027

American Cancer Society. *Cancer Facts & Figures 2025.* (n.d.). Retrieved on June 29, 2025, from https://www.cancer.org/research/cancer-facts-statistics/

Anandasabapathy, S., Asirwa, C., Grover, S., & Mungo, C. (2024). Cancer burden in low-income and middle-income countries. *Nature Reviews. Cancer*, *24*, 167–170. DOI: 10.1038/s41568-023-00659-2

Appidi, T., Sivasankaran, R. P., Chinchulkar, S. A., Patra, P., Murugaiyan, K., Veeresh, B., & Rengan, A. K. (2024). A lipo-polymeric hybrid nanosystem with metal enhanced fluorescence for targeted imaging of metastatic breast cancer. *Nanotheranostics*, *8*(2), 239–246. DOI: 10.7150/ntno.92410

Asghar, K., Qasim, M., Dharmapuri, G., & Das, D. (2017). Investigation on a smart nanocarrier with a mesoporous magnetic core and thermo-responsive shell for co-delivery of doxorubicin and curcumin: A new approach towards combination therapy of cancer. *RSC Advances*, *7*(46), 28802–28818. DOI: 10.1039/c7ra03735j

Aziz, A. A., Gayed, I., Heakel, F., & El Nashar, R. M. (2025). Synthesis, characterization, and versatile applications of metal oxides and mixed metal oxide nanoparticles. *Journal of the Electrochemical Society*, *172*, 023503. DOI: 10.1149/1945-7111/adada4

Aziz, S. H., & Abdullah, S. (2022). Toxicity of metal oxide nanoparticles in freshwater fish. In Pal, K., & Zaheer, T. (Eds.), *Nanomaterials in the Battle against Pathogens and Disease Vectors* (pp. 257–282). CRC Press., DOI: 10.1201/9781003126256-10

Bai, C., & Tang, M. (2020). Toxicological study of metal and metal oxide nanoparticles in zebrafish. *Journal of Applied Toxicology*, *40*(1), 37–63. DOI: 10.1002/jat.3910

Barick, K. C., Singh, S., Bahadur, D., Lawande, M. A., Patkar, D., & Hassan, P. A. (2014). Carboxyl decorated Fe3O4 nanoparticles for MRI diagnosis and localized hyperthermia. *Journal of Colloid and Interface Science*, *418*, 120–125. DOI: 10.1016/j.jcis.2013.11.076

Bhattacharya, R., Avdieiev, S. S., Bukkuri, A., Whelan, C. J., Gatenby, R. A., Tsai, K. Y., & Brown, J. S. (2025). The hallmarks of cancer as eco-evolutionary processes. *Cancer Discovery*, *15*(4), 685–701. DOI: 10.1158/2159-8290.cd-24-0861

Bi, J., Mo, C., Li, S., Huang, M., Yuan, P., Liu, Z., Jia, B., & Xu, S. (2023). Immunotoxicity of metal and metal oxide nanoparticles: From toxic mechanisms to metabolism and outcomes. *Biomaterials Science*, *11*(12), 4151–4183. DOI: 10.1039/d3bm00271c

Bischof, J. J., Presley, C. J., & Caterino, J. M. (2019). Addressing new diagnostic and treatment challenges associated with a new age of cancer treatment. *Annals of Emergency Medicine*, *73*(1), 88–90. DOI: 10.1016/j.annemergmed.2018.08.421

Bjørklund, G., Đorđević, A. B., Hamdan, H., Wallace, D. R., & Peana, M. (2024). Metal-induced autoimmunity in neurological disorders: A review of current understanding and future directions. *Autoimmunity Reviews*, *23*(3), 103509. DOI: 10.1016/j.autrev.2023.103509

Boire, A., Burke, K., Cox, T., Guise, T., Jamal-Hanjani, M., Janowitz, T., Kaplan, R., Lee, R., Swanton, C., Heiden, M. G. V., & Sahai, E. (2024). Why do patients with cancer die? *Nature Reviews. Cancer*, *24*(8), 578–589. DOI: 10.1038/s41568-024-00708-4

Botequim, D., Oliveira-Silva, R., Vaz Serra, V., Zijlstra, P., Prazeres, D. M. F., & Paulo, P. M. R. (2025). Nanoplasmonics for enhanced fluorescence detection of nucleic acids: From fundamentals to boosting cancer management. *Advanced NanoBiomed Research*, *5*(2), 2400088. DOI: 10.1002/anbr.202400088

Bray, F., Laversanne, M., Sung, H., Ferlay, J., Siegel, R. L., Soerjomataram, I., & Jemal, A. (2024). Global cancer statistics 2022: GLOBOCAN estimates of incidence and mortality worldwide for 36 cancers in 185 countries. *CA: a Cancer Journal for Clinicians*, *74*(3), 229–263. DOI: 10.3322/caac.21834

Bukowski, K., Kciuk, M., & Kontek, R. (2020). Mechanisms of multidrug resistance in cancer chemotherapy. *International Journal of Molecular Sciences, 21*(9), 3233. DOI: 10.3390/ijms21093233

Campos-Carrillo, A., Weitzel, J. N., Sahoo, P., Rockne, R. C., Mokhnatkin, J. V., Murtaza, M., Gray, S. W., Goetz, L. H., Goel, A., Schork, N. J., Schork, N. J., & Slavin, T. P. (2020). Circulating tumor DNA as an early cancer detection tool. *Pharmacology & Therapeutics, 207*, 107458. DOI: 10.1016/j.pharmthera.2019.107458

Cavalcanti, I. D. L., & Soares, J. C. S. (2021). *Advances in cancer treatment: From systemic chemotherapy to targeted therapy*. Springer., DOI: 10.1007/978-3-030-68334-4

Chaudhary, S., Pathak, A. K., & Ahmad, S. (2020). Nano-based materials and their synthesis. In I. Bhushan, V. Singh & D. Tripathi, D. (Eds.), *Nanomaterials and Environmental Biotechnology* (pp. 127-140). Springer, Cham. https://doi.org/DOI: 10.1007/978-3-030-34544-0_8

Chen, D.-D., Xi, T.-F., & Bai, J. (2007). Biological effects induced by nanosilver particles: In vivo study. *Biomedical Materials (Bristol, England), 2*(3), S126. DOI: 10.1088/1748-6041/2/3/S08

Chen, M., Tang, S., Guo, Z., Wang, X., Mo, S., Huang, X., Liu, G., & Zheng, N. (2014a). Core–Shell Pd@ Au nanoplates as theranostic agents for in-vivo photoacoustic imaging, CT imaging, and photothermal therapy. *Advanced Materials, 26*(48), 8210–8216. DOI: 10.1002/adma.201404013

Chen, Z., Wang, Y., Ba, T., Li, Y., Pu, J., Chen, T., Song, Y., Gu, Y., Qian, Q., Yang, J., & Jia, G. (2014b). Genotoxic evaluation of titanium dioxide nanoparticles in vivo and in vitro. *Toxicology Letters, 226*(3), 314–319. DOI: 10.1016/j.toxlet.2014.02.020

Choi, H. S., Liu, W., Misra, P., Tanaka, E., Zimmer, J. P., Ipe, B. I., Bawendi, M. G., & Frangioni, J. V. (2007). Renal clearance of nanoparticles. *Nature Biotechnology, 25*(10), 1165–1170. DOI: 10.1038/nbt1340

Conde, J., Doria, G., & Baptista, P. (2012). Noble metal nanoparticles applications in cancer. *Journal of Drug Delivery, 751075*. Advance online publication. DOI: 10.1155/2012/751075

Cullion, K., Ostertag-Hill, C. A., Pan, M., Timko, B. P., Boscolo, E., & Kohane, D. S. (2023). Ablation of venous malformations by photothermal therapy with intravenous gold nanoshells. *Nano Letters, 23*(15), 7092–7099. DOI: 10.1021/acs.nanolett.3c01945

Dillekås, H., Rogers, M. S., Straume, O., & Straume, O. (2019). Are 90% of deaths from cancer caused by metastases? *Cancer Medicine, 8*(12), 5574–5576. DOI: 10.1002/cam4.2474

Ding, L., Liu, Z., Aggrey, M. O., Li, C., Chen, J., & Tong, L. (2015). Nanotoxicity: The toxicity research progress of metal and metal-containing nanoparticles. *Mini-Reviews in Medicinal Chemistry, 15*(7), 529–542. DOI: 10.2174/138955751507150424104334

Dochez, V., Caillon, H., Vaucel, E., Dimet, J., Winer, N., & Ducarme, G. (2019). Biomarkers and algorithms for diagnosis of ovarian cancer: CA125, HE4, RMI and ROMA, a review. *Journal of Ovarian Research, 12*(1), 28. DOI: 10.1186/S13048-019-0503-7

Dy, G. K., & Adjei, A. A. (2013). Understanding, recognizing, and managing toxicities of targeted anticancer therapies. *CA: a Cancer Journal for Clinicians, 63*(4), 249–279. DOI: 10.3322/caac.21184

El Majzoub, I., Qdaisat, A., Thein, K. Z., Win, M. A., Han, M. M., Jacobson, K., Chaftari, P., Prejean, M., Reyes-Gibby, C. C., & Yeung, S. C. J. (2019). Adverse effects of immune checkpoint therapy in cancer patients visiting the emergency department of a comprehensive cancer center. *Annals of Emergency Medicine, 73*(1), 79–87. DOI: 10.1016/j.annemergmed.2018.04.019

ElBaz, N., Chauhan, R., James, K. T., Malik, D. A., Zhu, M., ElBaz, A., Keynton, R. S., Ng, C. K., Bates, P. J., Malik, T., & O'Toole, M. G. (2018). Gold nanoparticles conjugate cancer-targeting aptamer and gadolinium chelate for MR cancer imaging. In *2018 IEEE International Symposium on Signal Processing and Information Technology* (pp. 520-525). Louisville, USA. https://doi.org/DOI: 10.1109/isspit.2018.8642635

Fu, B., Dang, M., Tao, J., Li, Y., & Tang, Y. (2020). Mesoporous platinum nanoparticle-based nanoplatforms for combined chemo-photothermal breast cancer therapy. *Journal of Colloid and Interface Science, 570*, 197–204. DOI: 10.1016/J.JCIS.2020.02.051

Fujita, K., & Urano, Y. (2024). Activity-Based Fluorescence Diagnostics for Cancer. *Chemical Reviews, 124*(7), 4021–4078. DOI: 10.1021/acs.chemrev.3c00612

Furrer, D., Jacob, S., Caron, C., Sanschagrin, F., Provencher, L., & Diorio, C. (2017). Concordance of HER2 immunohistochemistry and fluorescence in situ hybridization using tissue microarray in breast cancer. *Anticancer Research, 37*(6), 3323–3329. DOI: 10.21873/anticanres.11701

García-Torra, V., Cano, A., Espina, M., Ettcheto, M., Camins, A., Barroso, E., Vazquez-Carrera, M., García, M. L., Sánchez-López, E., & Souto, E. B. (2021). State of the art on toxicological mechanisms of metal and metal oxide nanoparticles and strategies to reduce toxicological risks. *Toxics*, *9*(8), 195. DOI: 10.3390/toxics9080195

Ge, X., Chen, B., Liu, T., Wei, L., Tong, L., Ma, Q., Gao, S., & Song, J. (2019). Active targeting drug-gold nanorod hybrid nanoparticles for amplifying photoacoustic signal and enhancing anticancer efficacy. *RSC Advances*, *9*(24), 13494–13502. DOI: 10.1039/c9ra01547g

Ghorbian, M., & Ghorbian, S. (2024). Comprehensive review of reinforcement learning in lung cancer diagnosis and treatment: Taxonomy, challenges and recommendations. *Computers in Biology and Medicine*, *183*, 109326. DOI: 10.1016/j.compbiomed.2024.109326

Goleij, P., Babamohamadi, M., Rezaee, A., Majma Sanaye, P., Tabari, M., & Sadreddini, S., ArefNezhad, R., & Motedayyen, H. (2023). Types of RNA therapeutics. *Progress in Molecular Biology and Translational Science*, *203*, 41–63. DOI: 10.1016/bs.pmbts.2023.12.022

Gong, L., Zhang, Y., Liu, C., Zhang, M., & Han, S. (2021). Application of radiosensitizers in cancer radiotherapy. *International Journal of Nanomedicine*, *16*, 1083–1102. DOI: 10.2147/ijn.s290438

Han, H. H., Kim, S.-J., Kim, J., Park, W., Kim, C., Kim, H.-R., & Hahn, S. R. (2024). Investigation of skin cancer through photoacoustic analysis and photothermal therapy utilizing modified Au-platinum nanoparticles. *Proceedings SPIE Photons Plus Ultrasound: Imaging and Sensing*, *2024*(12842), 128420Q. DOI: 10.1117/12.3000859

Hu, Z., Chen, W. H., Tian, J., & Cheng, Z. (2020). NIRF nanoprobes for cancer molecular imaging: Approaching clinic. *Trends in Molecular Medicine*, *26*(5), 469–482. DOI: 10.1016/j.molmed.2020.02.003

Idris, D. S., & Roy, A. (2023). Synthesis of bimetallic nanoparticles and applications—An updated review. *Crystals*, *13*(4), 637. DOI: 10.3390/cryst13040637

Isibor, P. O., Sunday, A. S., Buba, A. B., & Oyewole, O. A. (2024). Mechanism of nanoparticle toxicity. In Isibor, P. O., Devi, G., & Enuneku, A. A. (Eds.), *Environmental Nanotoxicology* (pp. 103–120). Springer., DOI: 10.1007/978-3-031-54154-4_6

Jeng, H. A., & Swanson, J. R. (2006). Toxicity of metal oxide nanoparticles in mammalian cells. *Journal of Environmental Science and Health. Part A, Toxic/Hazardous Substances & Environmental Engineering*, *41*(12), 2699–2711. DOI: 10.1080/10934520600966177

Jiang, W., Liang, M., Lei, Q., Li, G., & Wu, S. (2023). The current status of photodynamic therapy in cancer treatment. *Cancers (Basel)*, *15*(3), 585. DOI: 10.3390/cancers15030585

Jiao, W., Dai, L., Yan, B., Lyu, Y., Fan, H., & Liu, X. (2024). (in press). Heating up the immune battle: Magnetic hyperthermia against cancer. *Fundamental Research (Beijing)*. Advance online publication. DOI: 10.1016/j.fmre.2024.08.006

Joshi, R., Chauhan, R., Akram, W., Kushwah, P., Mourya, H., & Garud, N. (2023). Vehicles for delivery of therapeutic agent for cancer therapy. In Malviya, R., & Sundram, S. (Eds.), *Targeted Cancer Therapy in Biomedical Engineering. Biological and Medical Physics, Biomedical Engineering* (pp. 719–753). Springer., DOI: 10.1007/978-981-19-9786-0_21

Kairdolf, B. A., Qian, X., & Nie, S. (2017). Bioconjugated nanoparticles for biosensing, in vivo imaging, and medical diagnostics. *Analytical Chemistry*, *89*(2), 1015–1031. DOI: 10.1021/acs.analchem.6b04873

Karthikeyan, B., Gnanakumar, G., & Therasa Alphonsa, A. (2023). *Nano Metal Oxides: Engineering and Biomedical Applications*. Springer., DOI: 10.1007/978-981-19-9444-9

Kattumuri, V., Katti, K. K., Bhaskaran, S., Boote, E., Casteel, S. W., Fent, G. M., Robertson, D. J., Chandrasekhar, M., Kannan, R., & Katti, K. V. (2007). Gum arabic as a phytochemical construct for the stabilization of gold nanoparticles: In vivo pharmacokinetics and x-ray-contrast-imaging studies. *Small*, *3*(2), 333–341. DOI: 10.1002/smll.200600427

Khabibullin, V. R., Chetyrkina, M. R., Obydennyy, S. I., Maksimov, S. V., Stepanov, G. V., & Shtykov, S. N. (2023). Study on doxorubicin loading on differently functionalized iron oxide nanoparticles: Implications for controlled drug-delivery application. *International Journal of Molecular Sciences*, *24*(5), 4480. DOI: 10.3390/ijms24054480

Kharazmi, A., & Attaran, N. (2023). Evaluation of the parameters affecting the loading of anticancer drug Paclitaxel on coated gold nanoparticles for breast cancer treatment. *IET Nanobiotechnology / IET*, *17*, 234–245. DOI: 10.1049/nbt2.12121

Khorasani, A., Shahbazi-Gahrouei, D., & Safari, A. (2023). Recent metal nanotheranostics for cancer diagnosis and therapy: A review. *Diagnostics (Basel)*, *13*(5), 833. DOI: 10.3390/diagnostics13050833

Kim, K., Lee, S., Jin, E., Palanikumar, L., Lee, J. H., Kim, J. C., Nam, J. S., Jana, B., Kwon, T.-H., Kwak, S. K., Choe, W., & Ryu, J.-H. (2019). MOF × biopolymer: Collaborative combination of metal–organic framework and biopolymer for advanced anticancer therapy. *ACS Applied Materials & Interfaces*, *11*(31), 27512–27520. DOI: 10.1021/acsami.9b05736

Kimm, M. A., Shevtsov, M., Shevtsov, M., Shevtsov, M., Werner, C., Sievert, W., Zhiyuan, W., Schoppe, O., Schoppe, O., Menze, B. H., Menze, B. H., Rummeny, E. J., Proksa, R., Bystrova, O. V., Martynova, M. G., Multhoff, G., & Stangl, S. (2020). Gold nanoparticle mediated multi-modal CT imaging of Hsp70 membrane-positive tumors. *Cancers (Basel)*, *12*(5), 1331. DOI: 10.3390/cancers12051331

Klaunig, J. E., Kamendulis, L. M., & Hocevar, B. A. (2010). Oxidative stress and oxidative damage in carcinogenesis. *Toxicologic Pathology*, *38*(1), 96–109. DOI: 10.1177/0192623309356453

Klug, M., Strange, C. D., Truong, M. T., Kirshenboim, Z., Ofek, E., Konen, E., & Marom, E. M. (2024). Thymic imaging pitfalls and strategies for optimized diagnosis. *Radiographics*, *44*(5), e230091. DOI: 10.1148/rg.230091

Komoto, S., Noma, K., Kato, T., Kobayashi, T., Nishiwaki, N., Narusaka, T., Sato, H., Katsura, Y., Kashima, H., Kikuchi, S., Ohara, T., Tazawa, H., & Fujiwara, T. (2023). Conventional cancer therapies can accelerate malignant potential of cancer cells by activating cancer-associated fibroblasts in esophageal cancer models. *Cancers (Basel)*, *15*(11), 2971. DOI: 10.3390/cancers15112971

Kumar, D., Mutreja, I., & Kaushik, A. (2023). Recent advances in noble metal nanoparticles for cancer nanotheranostics. *Journal of Nanotheranostics*, *4*(2), 150–170. DOI: 10.3390/jnt4020008

Kumar, R., Pulikanti, G. R., Shankar, K. R., Rambabu, D., Mangili, V., Kumbam, L. R., Sagara, P. S., Nakka, N., & Yogesh, M. (2022). Surface coating and functionalization of metal and metal oxide nanoparticles for biomedical applications. In Mondal, K. (Ed.), *Metal Oxides for Biomedical and Biosensor Applications* (pp. 205–231). Elsevier., DOI: 10.1016/B978-0-12-823033-6.00007-7

Lai, X., Liu, H., Zheng, Y., Wang, Z., & Chen, Y. (2019). Drug loaded nanoparticles of metal-organic frameworks with high colloidal stability for anticancer application. *Journal of Biomedical Nanotechnology*, *15*(8), 1754–1763. DOI: 10.1166/jbn.2019.2807

Landsiedel, R., Fabian, E., Ma-Hock, L., Wohlleben, W., Wiench, K., Oesch, F., & van Ravenzwaay, B. (2012). Toxico-/biokinetics of nanomaterials. *Archives of Toxicology*, *86*(7), 1021–1060. DOI: 10.1007/s00204-012-0858-7

Lazebnik, Y. (2010). What are the hallmarks of cancer? *Nature Reviews. Cancer*, *10*(4), 232–233. DOI: 10.1038/nrc2827

Li, J., Xu, C., Chen, Y., Cao, J., Ye, W., Cheng, Y., & Cheng, Q. (2024a). Photoacoustic tracking of photo-magnetically powered nanoparticles for cancer therapy. *Journal of Physics*, *2822*(1), 012034. DOI: 10.1088/1742-6596/2822/1/012034

Li, L., Xue, Y., Li, X., Li, J., Fu, Y., Ding, Y., Liu, T., Jia, N., Wu, Y., Bu, H., & Ouyang, X. (2024b). Copper nanosheet-based wash-free fluorescence imaging of cancer cells. *Analytical Chemistry*, *96*(5), 2052–2058. DOI: 10.1021/acs.analchem.3c04644

Li, Y., & Chen, C. (2011). Fate and toxicity of metallic and metal-containing nanoparticles for biomedical applications. *Small*, *7*(21), 2965–2980. DOI: 10.1002/smll.201101059

Liu, H., Teng, X., Yu, S., Yang, W., Kong, T., & Liu, T. (2024). Recent advances in photoacoustic imaging: Current status and future perspectives. *Micromachines*, *15*(8), 1007. DOI: 10.3390/mi15081007

Liu, J., Han, J., Kang, Z., Golamaully, R., Xu, N., Li, H., & Han, X. (2014). In vivo near-infrared photothermal therapy and computed tomography imaging of cancer cells using novel tungsten-based theranostic probe. *Nanoscale*, *6*(11), 5770–5776. DOI: 10.1039/c3nr06292a

Liu, X.-L., Liu, X.-L., Yang, Y., Ng, C. T., Zhao, L., Zhang, Y., Bay, B.-H., Fan, H., & Ding, J. (2015). Magnetic vortex nanorings: A new class of hyperthermia agent for highly efficient in vivo regression of tumors. *Advanced Materials*, *27*(11), 1939–1944. DOI: 10.1002/adma.201405036

Lopes, R., & Betrouni, N. (2009). Fractal and multifractal analysis: A review. *Medical Image Analysis*, *13*(4), 634–649. DOI: 10.1016/j.media.2009.05.003

Malekzadeh, R., Mortezazadeh, T., Abdulsahib, W. K., Abdollahi, B. B., Hamblin, M. R., Mansoori, B., AlSaikhan, F. I., & Zeng, B. (2023). Nanoarchitecture-based photothermal ablation of cancer: A systematic review. *Environmental Research*, *236*(1), 116526. DOI: 10.1016/j.envres.2023.116526

Manke, A., Wang, L., & Rojanasakul, Y. (2013). Mechanisms of nanoparticle-induced oxidative stress and toxicity. *BioMed Research International*, *942916*. Advance online publication. DOI: 10.1155/2013/942916

Meng, Q., Wang, Y., Li, C., & Hu, X. (2022). Bismuth-and gadolinium-codoped carbon quantum dots with red/green dual emission for fluorescence/CT/T1-MRI mode imaging. *New Journal of Chemistry*, *46*(35), 16970–16980. DOI: 10.1039/d2nj03420d

Mohr, A. E., Ortega-Santos, C. P., Whisner, C. M., Klein-Seetharaman, J., & Jasbi, P. (2024). Navigating challenges and opportunities in multi-omics integration for personalized healthcare. *Biomedicines*, *12*(7), 1496. DOI: 10.3390/biomedicines12071496

Moise, S., Byrne, J. M., El Haj, A. J., & Telling, N. D. (2018). The potential of magnetic hyperthermia for triggering the differentiation of cancer cells. *Nanoscale*, *10*(44), 20519–20525. DOI: 10.1039/c8nr05946b

Nagar, A., Madamanchi, D., Nair, G. R., Revikumar, A., Ray, S. K., Vajjala, S. M., Akhila, B. S., & Shivale, S. (2024). Barriers to cancer diagnosis and treatment: A pilot qualitative study of patient and practitioner perspectives in rural India. *Cureus*, *16*(8), e67249. DOI: 10.7759/cureus.67249

Naser, R. M. M., Dilabazian, H., Bahr, H., Barakat, A., & El-Sibai, M. (2022). A guide through conventional and modern cancer treatment modalities: A specific focus on glioblastoma cancer therapy [Review]. *Oncology Reports*, *48*(5), 190. DOI: 10.3892/or.2022.8405

Oh, I. H., Min, H. S., Li, L., Tran, T. H., Lee, Y., Kwon, I. C., Kwon, I. C., Choi, K., Kim, K., & Huh, K. M. (2013). Cancer cell-specific photoactivity of pheophorbide a–glycol chitosan nanoparticles for photodynamic therapy in tumor-bearing mice. *Biomaterials*, *34*(27), 6454–6463. DOI: 10.1016/j.biomaterials.2013.05.017

Paciotti, G. F., Zhao, J., Cao, S., Brodie, P. J., Tamarkin, L., Huhta, M. S., Myer, L. D., Friedman, J., & Kingston, D. G. I. (2016). Synthesis and evaluation of paclitaxel-loaded gold nanoparticles for tumor-targeted drug delivery. *Bioconjugate Chemistry*, *27*(11), 2646–2657. DOI: 10.1021/acs.bioconjchem.6b00405

Pallin, D. J., Baugh, C. W., Postow, M. A., Caterino, J. M., Erickson, T. B., & Lyman, G. H. (2018). Immune-related adverse events in cancer patients. *Academic Emergency Medicine*, *25*(7), 819–827. DOI: 10.1111/acem.13443

Patel, K. D., Patel, A. K., Sawadkar, P., Singh, B., & Perriman, A. W. (2023). Functionalized Metallic Nanoparticles: Theranostic Applications. In Jain, K., & Jain, N. K. (Eds.), *Multifunctional And Targeted Theranostic Nanomedicines: Formulation, Design And Applications* (pp. 97–117). Springer., DOI: 10.1007/978-981-99-0538-6_5

Pe'er, D., Ogawa, S., Elhanani, O., Keren, L., Oliver, T. G., & Wedge, D. (2021). Tumor heterogeneity. *Cancer Cell, 39*(8), 1015–1017. DOI: 10.1016/j.ccell.2021.07.009

Pietroiusti, A., Stockmann-Juvala, H., Lucaroni, F., & Savolainen, K. (2018). Nanomaterial exposure, toxicity, and impact on human health. *Wiley Interdisciplinary Reviews. Nanomedicine and Nanobiotechnology, 10*(5), e1513. DOI: 10.1002/wnan.1513

Prasad, S., Gupta, S. C., & Tyagi, A. K. (2017). Reactive oxygen species (ROS) and cancer: Role of antioxidative nutraceuticals. *Cancer Letters, 387*, 95–105. DOI: 10.1016/j.canlet.2016.03.042

Pujalté, I., Passagne, I., Daculsi, R., De Portal, C., Ohayon-Courtès, C., & L'Azou, B. (2015). Cytotoxic effects and cellular oxidative mechanisms of metallic nanoparticles on renal tubular cells: Impact of particle solubility. *Toxicology Research, 4*(2), 409–422. DOI: 10.1039/c4tx00184b

Qu, Z.-G. (2023). Investigating conventional and novel methods for treatment of cancer. *Proceedings of the 2023 13th International Conference on Bioscience, Biochemistry and Bioinformatics*, 103-111. https://doi.org/DOI: 10.1145/3586139.3586154

Ranganathan, P., Dare, A., Harrison, E., Kingham, T. P., Mutebi, M., Parham, G., Sullivan, R., & Pramesh, C. S. (2023). Inequities in global cancer surgery: Challenges and solutions. *Journal of Surgical Oncology, 129*, 150–158. DOI: 10.1002/jso.27551

Reuveni, T., Motiei, M., Romman, Z., Popovtzer, A., & Popovtzer, R. (2011). Targeted gold nanoparticles enable molecular CT imaging of cancer: An in vivo study. *International Journal of Nanomedicine, 6*, 2859–2864. DOI: 10.2147/ijn.s25446

Riviere, J. E. (2009). Pharmacokinetics of nanomaterials: An overview of carbon nanotubes, fullerenes and quantum dots. *Wiley Interdisciplinary Reviews. Nanomedicine and Nanobiotechnology, 1*(1), 26–34. DOI: 10.1002/wnan.24

Roach, K. A., Stefaniak, A. B., & Roberts, J. R. (2019). Metal nanomaterials: Immune effects and implications of physicochemical properties on sensitization, elicitation, and exacerbation of allergic disease. *Journal of Immunotoxicology, 16*(1), 87–124. DOI: 10.1080/1547691x.2019.1605553

Rosa, S., Connolly, C., Schettino, G., Butterworth, K. T., & Prise, K. M. (2017). Biological mechanisms of gold nanoparticle radiosensitization. *Cancer Nanotechnology, 8*, 2. DOI: 10.1186/s12645-017-0026-0

Sarhadi, V. K., & Armengol, G. (2022). Molecular biomarkers in cancer. *Biomolecules, 12*(8), 1021. DOI: 10.3390/biom12081021

Sayed, N., Allawadhi, P., Khurana, A., Singh, V., Navik, U., Pasumarthi, S., Khurana, I., Banothu, A. K., Weiskirchen, R., & Bharani, K. (2022). Gene therapy: Comprehensive overview and therapeutic applications. *Life Sciences*, *294*, 120375. DOI: 10.1016/j.lfs.2022.120375

Schrand, A. M., Rahman, M. F., Hussain, S. M., Schlager, J. J., Smith, D. A., & Syed, A. F. (2010). Metal-based nanoparticles and their toxicity assessment. *Wiley Interdisciplinary Reviews. Nanomedicine and Nanobiotechnology*, *2*(5), 544–568. DOI: 10.1002/wnan.103

Sengul, A. B., & Asmatulu, E. (2020). Toxicity of metal and metal oxide nanoparticles: A review. *Environmental Chemistry Letters*, *18*(5), 1659–1683. DOI: 10.1007/s10311-020-01033-6

Shakya, S., He, Y., Ren, X., Guo, T., Maharjan, A., Luo, T., Luo, T., Wang, T., Wang, T., Dhakhwa, R., Regmi, B., Li, H., Gref, R., & Zhang, J. (2019). Ultrafine silver nanoparticles embedded in cyclodextrin metal-organic frameworks with GRGDS functionalization to promote antibacterial and wound healing application. *Small*, *15*(27), 1901065. DOI: 10.1002/smll.201901065

Sharma, H. S., & Sharma, A. (2007). Nanoparticles aggravate heat stress induced cognitive deficits, blood–brain barrier disruption, edema formation and brain pathology. *Progress in Brain Research*, *162*, 245–273. DOI: 10.1016/s0079-6123(06)62013-x

Shedbalkar, U., Singh, R., Wadhwani, S., Gaidhani, S., & Chopade, B. A. (2014). Microbial synthesis of gold nanoparticles: Current status and future prospects. *Advances in Colloid and Interface Science*, *209*, 40–48. DOI: 10.1016/j.cis.2013.12.011

Sheehan, J. P., Shaffrey, M. E., Gupta, B., Larner, J. M., Rich, J. N., & Park, D. M. (2010). Improving the radiosensitivity of radioresistant and hypoxic glioblastoma. *Future Oncology (London, England)*, *6*(10), 1591–1601. DOI: 10.2217/fon.10.123

Shi, P., Sun, X., Yuan, H., Chen, K., Bi, S., & Zhang, S. (2023). Nanoscale metal-organic frameworks combined with metal nanoparticles and metal oxide/peroxide to relieve tumor hypoxia for enhanced photodynamic therapy. *ACS Biomaterials Science & Engineering*, *9*(10), 5441–5456. DOI: 10.1021/acsbiomaterials.3c00509

Shiri, M. S. Z., Henderson, W., & Mucalo, M. R. (2019). A review of the lesser-studied microemulsion-based synthesis methodologies used for preparing nanoparticle systems of the noble metals, Os, Re, Ir and Rh. *Materials (Basel)*, *12*(12), 1896. DOI: 10.3390/ma12121896

Singh, R., Nawale, L., Arkile, M., Wadhwani, S. A., Shedbalkar, U. U., Chopade, S., Sarkar, D., & Chopade, B. A. (2016). Phytogenic silver, gold, and bimetallic nanoparticles as novel antitubercular agents. *International Journal of Nanomedicine*, *11*, 1889–1897. DOI: 10.2147/ijn.s102488

Singh, R., Wagh, P., Wadhwani, S., Gaidhani, S., Kumbhar, A., Bellare, J., & Chopade, B. A. (2013a). Synthesis, optimization, and characterization of silver nanoparticles from *Acinetobacter calcoaceticus* and their enhanced antibacterial activity when combined with antibiotics. *International Journal of Nanomedicine*, *8*(1), 4277–4290. DOI: 10.2147/ijn.s48913

Singh, S., Srivastava, P., Kapoor, I. P. S., & Singh, G. (2013b). Preparation, characterization, and catalytic activity of rare earth metal oxide nanoparticles. *Journal of Thermal Analysis and Calorimetry*, *111*, 1073–1082. DOI: 10.1007/s10973-012-2538-5

Stanicki, D., Vangijzegem, T., Ternad, I., & Laurent, S. (2022). An update on the applications and characteristics of magnetic iron oxide nanoparticles for drug delivery. *Expert Opinion on Drug Delivery*, *19*(3), 321–335. DOI: 10.1080/17425247.2022.2047020

Steinbrueck, A., & Karges, J. (2023). Metal complexes and nanoparticles for photoacoustic imaging. *ChemBioChem*, *24*(14), e202300365. DOI: 10.1002/cbic.202300365

Sun, T., He, X., & Liu, J. (2025). The association of neutrophil-to-lymphocyte ratio with post-chemotherapy pulmonary infection in lung cancer patients. *Frontiers in Medicine*, *12*, 1559702. DOI: 10.3389/fmed.2025.1559702

Szwed, M., & Marczak, A. (2024). Application of nanoparticles for magnetic hyperthermia for cancer treatment—The current state of knowledge. *Cancers (Basel)*, *16*(6), 1156. DOI: 10.3390/cancers16061156

Tallantyre, E. C., Tallantyre, E. C., Evans, N. A., Parry-Jones, J., Morgan, M., Morgan, M., Jones, C. H., & Ingram, W. (2021). Neurological updates: Neurological complications of CAR-T therapy. *Journal of Neurology*, *268*(4), 1544–1554. DOI: 10.1007/s00415-020-10237-3

Thummavichai, K., Chen, Y., Wang, N. N., Zhu, Y. Q., & Ola, O. (2023). Synthesis, properties and characterization of metal nanoparticles. In Tiwari, S. K., Kumar, V., & Thomas, S. (Eds.), *Nanoparticles Reinforced Metal Nanocomposites* (pp. 161–207). Springer., DOI: 10.1007/978-981-19-9729-7_6

Urbano-Gámez, J. D., Guzzi, C., Bernal, M., Solivera, J., Martínez-Zubiaurre, I., Caro, C., & García-Martín, M. (2024). Tumor versus tumor cell targeting in metal-based nanoparticles for cancer theranostics. *International Journal of Molecular Sciences, 25*(10), 5213. DOI: 10.3390/ijms25105213

Varzandeh, M. H. M., Sabouri, L., Mansouri, V., Gharibshahian, M., Beheshtizadeh, N., Hamblin, M. R., & Rezaei, N. (2023). Application of nano-radiosensitizers in combination cancer therapy. *Bioengineering & Translational Medicine, 8*(3), e10498. DOI: 10.1002/btm2.10498

Walter, H. S., & Ahmed, S. (2018). Targeted therapies in cancer. *Surgery – Oxford International Edition, 36*(3), 122–127. https://doi.org/DOI: 10.1016/j.mpsur.2017.12.010

Wang, K., Kievit, F. M., Sham, J. G., Jeon, M., Stephen, Z. R., Bakthavatsalam, A., Park, J. O., & Zhang, M. (2016). Iron-oxide-based nanovector for tumor targeted siRNA delivery in an orthotopic hepatocellular carcinoma xenograft mouse model. *Small, 12*(4), 477–487. DOI: 10.1002/smll.201501985

WHO, World Health Organization. *Cancer*. Retrieved on June 29, 2025, from https://www.who.int/health-topics/cancer/

Xi, Z., Yao, H., Zhang, T., Wang, B., & Pu, Q. (2024). Quantitative three-dimensional imaging analysis of HfO_2 nanoparticles in single cells via deep learning aided x-ray nano-computed tomography. *ACS Nano, 18*(33), 22378–22389. DOI: 10.1021/acsnano.4c06953

Xia, T., Kovochich, M., Liong, M., Mädler, L., Mädler, L., Gilbert, B., Shi, H., Yeh, J. I., Zink, J. I., & Nel, A. E. (2008). Comparison of the mechanism of toxicity of zinc oxide and cerium oxide nanoparticles based on dissolution and oxidative stress properties. *ACS Nano, 2*(10), 2121–2134. DOI: 10.1021/nn800511k

Xie, W., Guo, Z., Gao, F., Gao, Q., Wang, D., Liaw, B.-S., Cai, Q., Sun, X., Wang, X., & Zhao, L. (2018). Shape-, size- and structure-controlled synthesis and biocompatibility of iron oxide nanoparticles for magnetic theranostics. *Theranostics, 8*(12), 3284–3307. DOI: 10.7150/thno.25220

Xu, J., Sun, Y., Huang, J., Chen, C., Liu, G., Jiang, Y., Zhao, Y., & Jiang, Z. (2007). Photokilling cancer cells using highly cell-specific antibody–TiO2 bioconjugates and electroporation. *Bioelectrochemistry (Amsterdam, Netherlands), 71*(2), 217–222. DOI: 10.1016/j.bioelechem.2007.06.001

Xu, Y., Hwang, S., Kim, S., & Chen, J.-Y. (2014). Two orders of magnitude fluorescence enhancement of aluminum phthalocyanines by gold nanocubes: A remarkable improvement for cancer cell imaging and detection. *ACS Applied Materials & Interfaces*, *6*(8), 5619–5628. DOI: 10.1021/am500106c

Yang, X., Stein, E. W., Ashkenazi, S., & Wang, L. V. (2009). Nanoparticles for photoacoustic imaging. *Wiley Interdisciplinary Reviews. Nanomedicine and Nanobiotechnology*, *1*(4), 360–368. DOI: 10.1002/wnan.42

Yao, Y., Zhou, Y., Liu, L., Xu, Y., Chen, Q., Wang, Y., Wu, S., Deng, Y., Zhang, J., & Shao, A. (2020). Nanoparticle-based drug delivery in cancer therapy and its role in overcoming drug resistance. *Frontiers in Molecular Biosciences*, *7*, 193. DOI: 10.3389/fmolb.2020.00193

Yaqoob, S. B., Adnan, R., Rameez Khan, R. M., & Rashid, M. (2020). Gold, silver, and palladium nanoparticles: A chemical tool for biomedical applications. *Frontiers in Chemistry*, *8*, 376. DOI: 10.3389/fchem.2020.00376

Yazdi, M. K., Zarrintaj, P., Khodadadi, A., Ganjali, M. R., Ganjali, M. R., Bagheri, B., Habibzadeh, S., Saeb, M. R., & Mozafari, M. (2021). Magnetic nanoparticles in cancer therapy. In Ehrmann, A., Nguyen, T. A., Ahmadi, M., Farmani, A., & Nguyen-Tri, P. (Eds.), *Magnetic Nanoparticle-Based Hybrid Materials: Fundamentals and Applications* (pp. 425–445). Woodhead Publishing., DOI: 10.1016/B978-0-12-823688-8.00025-9

Ye, J., Hu, T., Wu, Y., Chen, H., Qiu, Q., Geng, R., Ding, H., & Zhao, X. (2022). Near-infrared liposome-capped au-rare earth bimetallic nanoclusters for fluorescence imaging of tumor cells. *Journal of Biomedical Nanotechnology*, *18*(9), 2113–2122. DOI: 10.1166/jbn.2022.3423

Yen, H.-J., Hsu, S., & Tsai, C.-L. (2009). Cytotoxicity and immunological response of gold and silver nanoparticles of different sizes. *Small*, *5*(13), 1553–1561. DOI: 10.1002/smll.200900126

Yoo, B., Ghosh, S., Kumar, M., Moore, A., Yigit, M. V., & Medarova, Z. (2014). Design of nanodrugs for miRNA targeting in tumor cells. *Journal of Biomedical Nanotechnology*, *10*(6), 1114–1122. DOI: 10.1166/jbn.2014.1795

Yu, S., Xia, G., Yang, N., Li, J., Wang, Q., Li, D., Ding, L., Fan, Z., & Li, J. (2024). Noble metal nanoparticle-based photothermal therapy: Development and application in effective cancer therapy. *International Journal of Molecular Sciences*, *25*(11), 5632. DOI: 10.3390/ijms25115632

Zhai, L.-Y., Li, M.-X., Pan, W.-L., Yun, C., Li, M.-M., Pang, J.-X., Zheng, L., Chen, J.-X., & Duan, W.-J. (2018). In situ detection of plasma exosomal microrna-1246 for breast cancer diagnostics by a au nanoflare probe. *ACS Applied Materials & Interfaces, 10*(46), 39478–39486. DOI: 10.1021/acsami.8b12725

Zhang, D., Zheng, Y., Lin, Z., Lan, S., Zhang, X., Zhang, X., Zheng, A., Zheng, A., Li, J., Liu, G., Yang, H., Liu, X., Liu, X., & Liu, J. (2019). Artificial engineered natural killer cells combined with antiheat endurance as a powerful strategy for enhancing photothermal-immunotherapy efficiency of solid tumors. *Small, 15*(42), 1902636. DOI: 10.1002/smll.201902636

Zhang, J., Zhang, T., & Gao, J.-Q. (2022). Biocompatible iron oxide nanoparticles for targeted cancer gene therapy: A review. *Nanomaterials (Basel, Switzerland), 12*(19), 3323. DOI: 10.3390/nano12193323

Zhang, Y., Fu, X., Jia, J., Wikerholmen, T. E., Xi, K., Kong, Y., Wang, J., Chen, H., Ma, Y., Li, Z., Wang, C., Qi, Q., Thorsen, F., Thorsen, F., Wang, J., Wang, J., Cui, J., Li, X., & Ni, S. (2020). Glioblastoma therapy using codelivery of cisplatin and glutathione peroxidase targeting sirna from iron oxide nanoparticles. *ACS Applied Materials & Interfaces, 12*(39), 43408–43421. DOI: 10.1021/acsami.0c12042

Zhao, C., Liu, H., Huang, S., Guo, Y., & Xu, L. (2024). Metal–organic framework-capped gold nanorod hybrids for combinatorial cancer therapy. *Molecules (Basel, Switzerland), 29*(10), 2384. DOI: 10.3390/molecules29102384

Zheng, J., Hao, Y., Guo, Y., Du, M., Wang, P., & Xin, J. (2024). An 18F-FDG-PET/CT-based radiomics signature for estimating malignance probability of solitary pulmonary nodule. *The Clinical Respiratory Journal, 18*(5), e13751. DOI: 10.1111/crj.13751

Zhu, Y., Li, W., Zhao, X., Zhou, Z., Wang, Y., Cheng, Y., Huang, Q., & Zhang, Q. (2017). Hyaluronic acid-encapsulated platinum nanoparticles for targeted photothermal therapy of breast cancer. *Journal of Biomedical Nanotechnology, 13*(11), 1457–1467. DOI: 10.1166/jbn.2017.2446

Ziental, D., Czarczynska-Goslinska, B., Mlynarczyk, D. T., Glowacka-Sobotta, A., Stanisz, B., Goslinski, T., & Sobotta, L. (2020). Titanium dioxide nanoparticles: Prospects and applications in medicine. *Nanomaterials (Basel, Switzerland), 10*(2), 387. DOI: 10.3390/NANO10020387

KEY TERMS AND DEFINITIONS

Computed Tomography (CT): A technique that uses X-rays to create a detailed image of the areas inside the body

Metastasis: Spread of tumor cells from one part of the body to another via blood or lymph

Magnetic Resonance Imaging (MRI): An imaging technique relying on magnetic field and radiowaves to generate pictures of internal structures

Pharmacokinetics: Analysis of what body does to an exogenous substance, such as drugs, in terms of absorption, distribution, metabolism and excretion.

Reactive Oxygen Species (ROS): Highly unstable and reactive group of molecules containing oxygen and unpaired electrons such as superoxide and hydroxyl radicals

Surface Plasmon Resonance (SPR): Resonant oscillation of conduction electrons on the surface of metal at specific wavelength of incident light

Tumor Heterogeneity: Diverse subpopulation of cells, within or between tumors, exhibiting different genetic, phenotypic or behavioural traits

Chapter 8
Real-Time Predictive Analytics for Oncolytic Virus Replication and Tumor Lysis Using Machine Learning

Pawan Kumar Badhan
https://orcid.org/0009-0002-3953-6825
Pyramid College of Business and Technology, Phagwara, India

Manjeet Kaur
School of Computational Science, GNA University, India

Pallavi Sood
Department of Computer Science, Apeejay college of Fine Arts Jalandhar, India

Raghav Gupta
https://orcid.org/0000-0001-9791-4287
IT Department, Lyallpur Khalsa College Technical Campus, Jalandhar, India

ABSTRACT

Oncolytic virotherapy uses genetically engineered viruses to selectively destroy tumor cells while sparing healthy tissue. To address the complex virus-tumor-immune interactions, this study proposes a real-time predictive framework using advanced machine learning and deep learning to optimize treatment outcomes. It analyzes clinical trial data and records from 1,200 patients in Northern India, mainly Punjab, incorporating genomic, proteomic, and tumor microenvironmental features.

DOI: 10.4018/979-8-3373-3765-4.ch008

Kaggle's Brain Tumor MRI Dataset is also used to improve classification of Glioma, Meningioma, Pituitary, and No Tumor cases. Key goals include quantifying tumor lysis efficiency as a function of virus burst size and cell susceptibility, enhancing predictive accuracy of tumor regression patterns with multi-scale data integration, and simulating tumor lysis dynamics in response to oncolytic virus activity. Models like RF (98.3%), SVM (98.5%), ANN (99%), KNN (99.6%), GBM (99.7%), and CNN (100%) were used. Techniques like RFE, Permutation Importance, and Grad-CAM enhanced accuracy and interpretability.

1. INTRODUCTION

Oncolytic virotherapy (OVT) has emerged as a ground breaking and promising strategy for cancer treatment. It utilizes oncolytic viruses (OVs) that are either naturally occurring or genetically engineered to selectively infect and lyse cancer cells while sparing healthy tissue. The ability of OVs to directly kill tumor cells and simultaneously stimulate an antitumor immune response has positioned them as a dual-action therapeutic approach (Hamidi-Sofiani et al., 2022), (Seyed-Khorrami et al., 2023). The fundamental concept of OVT relies on the fact that tumor cells provide an optimal environment for viral replication due to their defective antiviral defence mechanisms and altered signaling pathways. This allows for selective viral propagation, leading to tumor lysis while minimizing damage to normal cells (Mencattini et al., 2022). Over the past two decades, extensive research has been conducted to understand the mechanisms underlying OVT and its potential to be integrated into standard oncology treatments.

Several oncolytic viruses, including vesicular stomatitis virus (VSV), adenoviruses, herpes simplex virus (HSV), and vaccinia virus, have demonstrated efficacy against various tumors, including glioblastoma, pancreatic cancer, and melanoma (Granha et al., 2024), (Elnosary et al., 2024). These viruses not only cause direct oncolysis but also induce immunogenic cell death, which enhances the visibility of cancer cells to the immune system, thereby improving therapeutic outcomes (Basu & Moles, 2023). Preclinical and clinical trials have reported promising results, with some OVs achieving complete tumor regression in certain cases (Grygoryev et al., 2024). The approval of T-VEC (Talimogene laherparepvec), a modified HSV-1 virus, for the treatment of melanoma represents a significant milestone in the field of OVT (Kim, Kim, & Jung, 2018). However, despite these advances, several challenges such as immune system clearance, limited intratumoral viral spread, and patient-specific variations in response to OVT necessitate further research (Hartkopf et al., 2011), (Goradel et al., 2020). To enhance the efficacy of OVs, researchers have investigated their combination with other cancer therapies, including chemotherapy, radiation therapy, and immune checkpoint inhibitors. Studies indicate that radiation therapy can increase the susceptibility of tumor cells to viral infection by inducing immunogenic cell death and disrupting tumor vasculature (Rovira-Rigau et al., 2019). Additionally, OVs have shown potential in overcoming drug resistance in certain cancer types, which remains a significant challenge in oncology (Storozynsky et al., 2023). Unlike traditional therapies, OVs offer tumor specificity with reduced systemic toxicity, making them a favorable option for personalized medicine (Shahgolzari et al., 2022).

The integration of immune checkpoint inhibitors (ICIs), such as PD-1 and CTLA-4 inhibitors, with OVT has also shown promising results by further enhancing the immune-mediated elimination of tumors (Parihar et al., 2022). For instance, combining OVs with anti-PD-1 therapy has been found to enhance tumor regression by increasing T-cell infiltration into the tumor microenvironment (Poon et al., 2021). As a result, combinatorial strategies have gained attention as they provide a synergistic approach to cancer treatment, improving patient survival rates and response to therapy (Amethiya et al., 2022). While the preclinical and early-stage clinical success of OVT has been encouraging, several barriers hinder its widespread clinical application. One of the primary challenges is immune system clearance, where the host immune response eliminates the virus before it can exert its full oncolytic effect (Soltani et al., 2022). Moreover, tumor heterogeneity presents a significant obstacle, as variations in tumor biology can influence viral replication and therapeutic response (Bibault et al., 2016). Ensuring efficient intratumoral delivery of OVs remains another critical concern, as factors such as tumor perfusion, extracellular matrix composition, and immune surveillance can impact viral spread (Asif et al., 2024). Recent advancements in nanotechnology-based delivery systems have shown promise in improving viral delivery and protecting OVs from premature neutralization (Almutairi et al., 2023).

Another challenge is the lack of patient-specific optimization, as OVs vary in efficacy across tumor types due to differences in [tumor microenvironment, immune status, genetic mutations] (Hsu et al., 2021). Personalized OVT is essential, supported by advances in [genomic sequencing, bioinformatics] for identifying biomarkers (Tseng et al., 2019). ML and AI, especially [DL, RL], enhance OV selection, predict efficacy, and personalize treatments (Atrey et al., 2023), using [multi-omics datasets] to identify key [genomic, proteomic, microenvironmental factors] (Menotti & Vannini, 2023). Techniques like [RF, SVM, ANN, KNN, GBM] aid in optimization (Gujar et al., 2024), with [RF, GBM] excelling in biological data analysis and [CNNs] used for tumor imaging (Hirigoyen et al., 2024). [RL] further optimizes dosage/timing (Godlewski et al., 2021), while [RFE, Permutation, Grad-CAM] improve interpretability (Dodkins et al., 2023). The future of oncolytic virotherapy lies in its integration with precision oncology, wherein AI-driven predictive models enable real-time assessment of tumor lysis dynamics and treatment adaptation. The development of multi-modal datasets, including genomic, proteomic, and imaging data, will further enhance the ability to personalize OV treatment (Salgueiro et al., 2022). Real-time monitoring using AI-powered predictive analytics will enable clinicians to adjust treatment protocols dynamically based on tumor response and patient-specific biomarkers (Kim, Yun, Doh, & Kim, 2024). Additionally, ongoing clinical trials continue to explore novel oncolytic viruses with enhanced tumor selectivity and immune-stimulatory properties. CRISPR-based genome editing is being investigated to engineer OVs with improved replication kinetics, immune evasion strategies, and enhanced therapeutic efficacy (Yao et al., 2023). The use of nanocarriers and bioengineered viral platforms holds promise in overcoming challenges associated with systemic viral delivery and immune clearance (Valdebenito et al., 2019).

Oncolytic virotherapy represents a paradigm shift in cancer treatment, offering a targeted and immunogenic approach to tumor eradication. The integration of machine learning, deep learning, and AI-based predictive analytics has further enhanced the potential of OVT by enabling real-time monitoring, optimization of viral delivery, and personalized treatment regimens. While challenges such as immune clearance, tumor heterogeneity, and delivery limitations persist, advancements in genetic engineering, nanotechnology, and AI-driven modeling are paving the way for the widespread clinical application of OVT. As research continues to evolve, precision virotherapy, guided by predictive analytics, is expected to play a crucial role in the future of personalized oncology, improving patient outcomes and treatment success rates.

Literature Review

Recent studies underscore the transformative potential of oncolytic viruses (OVs) in cancer therapy, with multiple researchers contributing to the understanding and advancement of this promising treatment modality. Hamidi-Sofiani et al. (2022) reviewed the selective replication of OVs in pancreatic tumors, detailing virus types, mechanisms of action, clinical developments, and associated challenges like immune response and delivery barriers. Seyed-Khorrami et al. (2023) further explored OVs' tumor specificity, immune modulation capabilities, and synergistic potential when combined with therapies such as immune checkpoint inhibitors, while also introducing the integration of AI and IoT to enhance diagnostics and treatment outcomes. Mencattini et al. (2022) demonstrated through a tumor-on-chip model that the combined use of oncolytic vaccinia virus and immune cells, tracked via real-time imaging and biosensing technologies, could yield synergistic anticancer effects. Complementing these insights, literature presents a wide array of contemporary contributions, such as Granha et al. (2024), who focused on pediatric brain tumors using next-generation sequencing; Elnosary et al. (2024), who highlighted the immunomodulatory potential of VSV; and Basu & Moles (2023), who proposed a framework for selecting optimal OVs using multi-omics and computational modeling. Other notable works include Grygoryev et al. (2024) using Sendai virus for gene delivery in pancreatic cancer, Kim et al. (2018) employing mathematical modeling for treatment optimization, and Hartkopf et al. (2011) evaluating OVs in breast cancer. The table further documents innovations in viral delivery (Goradel et al., 2020), miRNA regulation (Rovira-Rigau et al., 2019), radiotherapy synergy (Storozynsky et al., 2023), and plant-virus-based immunotherapy (Shahgolzari et al., 2022). The integration of AI is a recurring theme, as seen in works like Parihar et al. (2022) on predictive cancer control, Poon et al. (2021) on radiotherapy optimization, Amethiya et al. (2022) on ML-based breast cancer detection, and Soltani et al. (2022) on palliative care forecasting. Studies by Bibault et al. (2016), Asif et al. (2024), Almutairi et al. (2023), Hsu et al. (2021), Tseng et al. (2019), and Atrey et al. (2023) further reinforce the growing role of machine learning in oncology. Additional innovations include Menotti & Vannini's (2023) systems biology-driven OV optimization, Gujar et al.'s (2024) standardized OV development protocols, Hirigoyen et al.'s (2024) findings on immunogenicity changes in tumor vesicles, and Godlewski et al.'s (2021) exploration of immune modulation in glioblastoma. Finally, Dodkins et al. (2023) introduced a high-throughput viral infectivity screening pipeline using machine learning and microscopy, while Salgueiro et al. (2022), Kim et al. (2024), Yao et al. (2023), Valdebenito et al. (2019), Lathwal et al. (2020), and Onnockx et al. (2023) contributed to areas ranging from organoid modeling and immunotherapy resistance to OV databases and environmental risk assessments. Collectively, these studies represent a robust and multidisciplinary approach to advancing oncolytic virotherapy and precision oncology. This study as compare to above literature review utilizes multi-modal datasets, combining genomic, proteomic, and microenvironmental tumor data, to enhance the precision of predicting tumor lysis dynamics. With the application of machine learning algorithms such as Random Forest, SVM, Gradient Boosting, and Convolutional Neural Networks, the models achieve an impressive predictive accuracy of 98%-100%. Furthermore, reinforcement learning is employed to optimize viral dosage and treatment timing, potentially transforming personalized cancer therapies. This paper is useful as it tackles key challenges in oncolytic virotherapy, including immune evasion, tumor heterogeneity, and delivery optimization, and suggests a path forward for precision oncology applications, ultimately improving patient outcomes.

2. MATERIAL AND DATASET

This research integrates two primary datasets to develop a real-time predictive analytics framework for oncolytic virus replication, tumor lysis efficiency, and treatment response assessment. The datasets include a clinical trial dataset containing 1,200 patient records and an MRI imaging dataset consisting of 2970 brain tumor MRI scans. These datasets collectively enable machine learning-based predictive modeling, deep learning-based tumor classification, and optimization of virotherapy strategies for personalized cancer treatment.

The primary dataset, collected from hospitals in northern India—mainly Punjab—supports real-time predictive analytics for oncolytic virus replication and tumor lysis. It enables analysis of biological, clinical, and treatment factors influencing tumor response. Researchers can evaluate tumor lysis efficiency based on virus burst size and cell susceptibility, identifying optimal conditions to maximize tumor destruction while minimizing side effects. Integrating multi-scale data—tumor traits, immune responses, and genomic/epigenetic changes—enhances the precision of regression modeling. With deep learning and machine learning, the dataset supports real-time monitoring of tumor shrinkage and simulates lysis dynamics to refine personalized treatments, as shown in Table 1. It includes virus type, dosage, replication rates, and immune interactions to optimize therapies. Additionally, it records toxicity profiles and adverse reactions, ensuring safety assessments. By capturing tumor microenvironment factors like hypoxia, pH, angiogenesis, and blood flow, the dataset aids in studying resistance mechanisms. Overall, it supports the development of AI-driven decision systems, improving oncologists' ability to tailor treatments for better outcomes.

Table 1. Comprehensive Dataset for Oncolytic Virus Therapy: Patient Demographics, Tumor Characteristics, and Treatment Outcomes

Attribute Name	Description
Comorbidities	Pre-existing conditions (e.g., diabetes, cardiovascular disease).
Previous Cancer Treatments	Prior therapies (e.g., chemotherapy, radiation, immunotherapy).
Tumor Type	Classification of the tumor type (e.g., solid, hematologic).
Tumor Stage	Cancer stage at diagnosis (Stage I-IV).
Tumor Location	Anatomical site of the tumor (e.g., brain, liver, lung).
Initial Tumor Volume (cm^3)	Initial size of the tumor before treatment.
Tumor Density (g/cm^3)	Density of tumor tissue.
Tumor Immunogenicity Score (0-100)	Measures tumor's ability to trigger an immune response.

continued on following page

Table 1. Continued

Attribute Name	Description
Genetic Mutation Status	Presence of key oncogenic mutations (e.g., TP53, EGFR, KRAS).
Virus Type	Type of oncolytic virus used (e.g., HSV-1, adenovirus).
Oncolytic Virus Engineering Type	Indicates if the virus is natural or genetically modified.
Virus Burst Size	Number of viral particles produced per infected tumor cell.
Virus Dose (PFU/mL)	Concentration of virus administered in plaque-forming units per milliliter.
Replication Rate (particles/hr)	Speed of virus replication within tumor cells.
Oncolytic Virus Spread Rate (mm/day)	Speed of virus propagation through tumor tissue.
Time to Lysis (hours)	Time elapsed from virus infection to cell lysis.
Tumor Cell Death Rate (%)	Percentage of tumor cells destroyed.
Immune Response Level (0-100)	Strength of immune response triggered by the virus.
Cytokine Concentration (pg/mL)	Cytokine levels produced during immune response.
T-Cell Activation (%)	Level of T-cell activation in response to treatment.
Immune Escape Mechanism (Yes/No)	Indicates if the virus is modified to evade immune clearance.
Hypoxia Level (%)	Oxygen deprivation within the tumor microenvironment.
Tumor pH Level	Acidity level of the tumor environment.
Angiogenesis Index (AU)	Indicator of new blood vessel formation in the tumor.
Blood Flow Rate (mL/min)	Blood circulation within the tumor.
Vascular Permeability Index (0-100)	Measures how easily the virus penetrates tumor tissue.
Fibrosis Index (0-100)	Evaluates scar tissue formation that may limit viral spread.
Genomic Marker Expression (%)	Expression levels of specific genomic markers in tumor cells.
Epigenetic Modifications	Presence of DNA methylation or histone modifications.
Tumor Lysis Efficiency (%)	Efficiency of cell lysis resulting from virus replication.
Real-Time Tumor Shrinkage (%)	Observed reduction in tumor size over time.
Tumor Regression Pattern (%)	Reduction pattern of tumor volume post-treatment.
MRI-Based Tumor Heterogeneity Score (0-100)	Measures structural variability in tumors using imaging.
Radiomics Features	Extracted MRI biomarkers (e.g., texture, shape, edge sharpness).
CNN Feature Map Activation (%)	Importance of MRI regions in deep learning predictions.
Grad-CAM Hotspot Density (%)	Percentage of significant image regions contributing to CNN decisions.
Time-Series Tumor Shrinkage Rate (cm^3/day)	Rate of tumor reduction monitored across MRI scans.
Side Effects Score (0-10)	Severity of treatment-related side effects.
Adverse Reaction Type	Classification of observed reactions (mild, moderate, severe).

continued on following page

Table 1. Continued

Attribute Name	Description
Toxicity Profile	Categorization of treatment-induced toxicity.
Therapeutic Window (days)	Duration of effective treatment without severe side effects.
Therapeutic Synergy Score (0-100)	Measures interaction with other treatments (e.g., radiation, immunotherapy).

This study utilizes a predictive analytics framework to model oncolytic virus replication and tumor lysis, enhancing personalized virotherapy. By analyzing viral spread, tumor cell death, and regression patterns, it supports optimized treatment planning. Multi-scale data, including genomic and epigenetic profiles, enables deeper insights into tumor heterogeneity and patient-specific responses. The MRI dataset includes 2970 high-resolution scans—both pre- and post-surgical—from glioma, meningioma, and pituitary tumor patients (495 post-op), aiding precise tumor assessment and therapy evaluation.

The pre-surgical MRI dataset captures detailed tumor features—size, shape, density, and location—providing key insights into gliomas, meningiomas, and pituitary tumors. Using advanced feature extraction, the images reveal tumor volume, hypoxia levels, angiogenesis, and genomic markers, aiding in segmentation and assessing virotherapy response.

The post-surgical MRI dataset, comprising 495 images, provides a critical evaluation of tumor removal outcomes, residual tumor presence, and potential post-operative complications. These scans are used to assess the effectiveness of surgical excision and the impact of oncolytic virotherapy, particularly in cases where tumors are only partially removed or require follow-up treatments. Post-surgical images are further analyzed to detect early recurrence risks, immune response activation, and residual tumor cell viability, allowing oncologists to refine treatment strategies based on patient-specific tumor behavior in Figure 1.

Figure 1. Training Data Distribution for MRI-Based Tumor Classification and Treatment Outcome Prediction

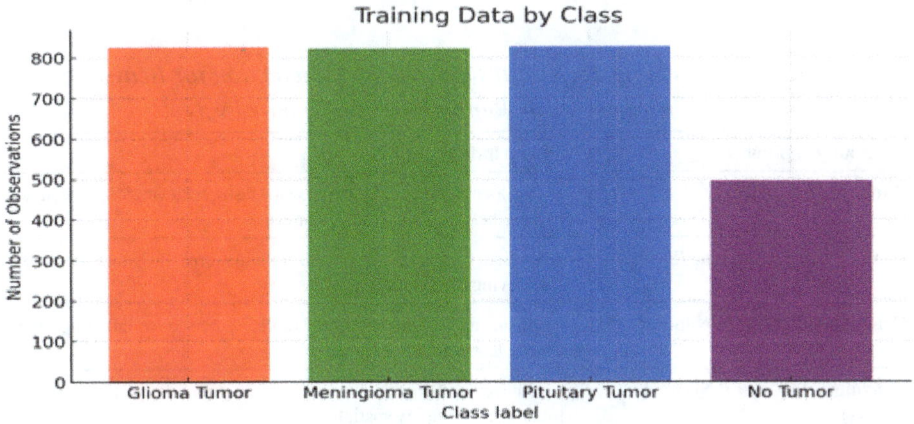

By comparing pre- and post-surgical MRI images, machine learning models can predict tumor regression patterns, evaluate virotherapy efficacy, and optimize viral dosage for improved therapeutic outcomes. By integrating clinical and publicly available MRI datasets (e.g., Kaggle's Brain Tumor MRI Dataset), with genomic and proteomic data, this study develops a real-time predictive framework for optimizing oncolytic virus replication, tumor lysis efficiency, and personalized virotherapy, enabling tumor progression analysis, recurrence detection, and precision oncology advancements.

3. METHODS AND FEATURES SELECTION

This study leverages diverse machine learning and deep learning techniques to build a real-time predictive analytics framework for modeling oncolytic virus replication, tumor lysis, and treatment response with high precision. Random Forest (RF) captures complex interactions among tumor traits, viral replication, and immune responses through ensemble decision trees. Support Vector Machine (SVM) identifies non-linear patterns in high-dimensional data, enhancing predictions of tumor susceptibility and immune engagement. Gradient Boosting Machines (GBM) improve model accuracy by capturing subtle variations in tumor heterogeneity and viral dynamics. Convolutional Neural Networks (CNNs) analyze biomedical data—including imaging, genomics, and multi-omics—by detecting spatial and temporal patterns linked to tumor regression and virus propagation, as illustrated in Table 2. The framework dynamically adjusts treatment plans by integrating real-time feedback to optimize viral dosage and schedules, achieving a personalized balance between therapeutic efficacy and minimal side effects.

Table 2. Machine Learning Techniques and Features Selection for Real-Time Predictive Analytics in Oncolytic Virus Replication and Tumor Lysis

Method/Technique	Description
Random Forest (RF)	Used for classification and regression tasks, achieving 98.3% accuracy in predicting tumor lysis dynamics.
Support Vector Machine (SVM)	Applied to model non-linear relationships in tumor response, achieving 98.5% accuracy.
Gradient Boosting Machines (GBM)	Enhanced predictive accuracy to 98.7% by sequentially improving weak learners.
Convolutional Neural Networks (CNNs)	Achieved 100% accuracy in modeling tumor lysis patterns by capturing complex spatial features.
Artificial Neural Networks (ANNs)	Modeled complex non-linear patterns in tumor regression with high accuracy.
K-Nearest Neighbors (KNN)	Applied for tumor classification based on genomic and proteomic similarities.
Recursive Feature Elimination (RFE)	Used for feature selection by iteratively removing less important features.
Permutation Importance	Identified key genomic and microenvironmental factors contributing to tumor response.
Grad-CAM (Gradient-weighted Class Activation Mapping)	Improved model interpretability by visualizing important regions influencing CNN predictions.

Artificial Neural Networks (ANNs) simulate complex interactions between tumor biology, immune response, and viral replication, enhancing predictive accuracy. K-Nearest Neighbors (KNN) classifies tumors based on genomic and proteomic similarities, aiding comparative studies on therapy response. Recursive Feature Elimination (RFE) refines feature selection by removing irrelevant data, while Permutation Importance identifies key variables like cytokine levels and viral kinetics. Grad-CAM enhances model explainability by highlighting critical regions in imaging and genomic data. This integrated use of machine learning, deep learning, and interpretability tools supports personalized, real-time decision-making in oncolytic virotherapy, advancing precision oncology.

4. PRE-PROCESSING AND TRAINING

As dataset of this research, comprising 1,200 patient records from Northern India (Punjab), includes genomic, proteomic, and tumor microenvironmental attributes from clinical trials. To ensure data quality and consistency, pre-processing involves multiple steps in Figure 2. Data cleaning addresses missing values in tumor attributes, immune response markers, and viral replication rates through imputation or removal, while duplicate records and inconsistencies are eliminated.

Figure 2. Data Pre-processing for Standardized Tumor and Viral Attribute Analysis

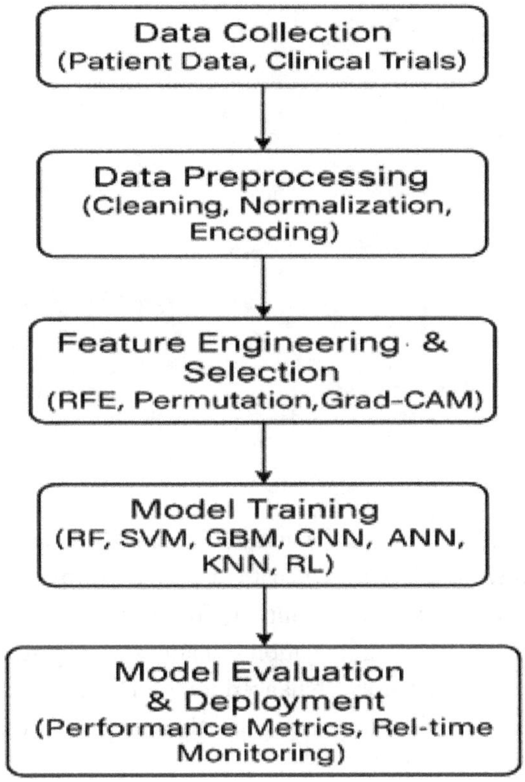

Feature scaling and normalization are applied to continuous variables such as tumor volume, cytokine concentration, and virus replication rate to standardize data and prevent bias in machine learning models. Categorical encoding converts non-numeric features like tumor type and virus strain into numerical form using one-hot encoding or label encoding for model compatibility. To develop an optimized real-time predictive model, the dataset is divided into 80% training data and 20% testing data. The training subset is used to train machine learning models such as Random Forest (RF), Support Vector Machine (SVM), Gradient Boosting Machines (GBM), Convolutional Neural Networks (CNNs), and Reinforcement Learning, enabling them to learn tumor-virus interactions, predict tumor lysis efficiency, and optimize virus dosage and timing for enhanced therapeutic outcomes.

Figure 3. Model Testing and Validation for Predictive Oncolytic Virotherapy

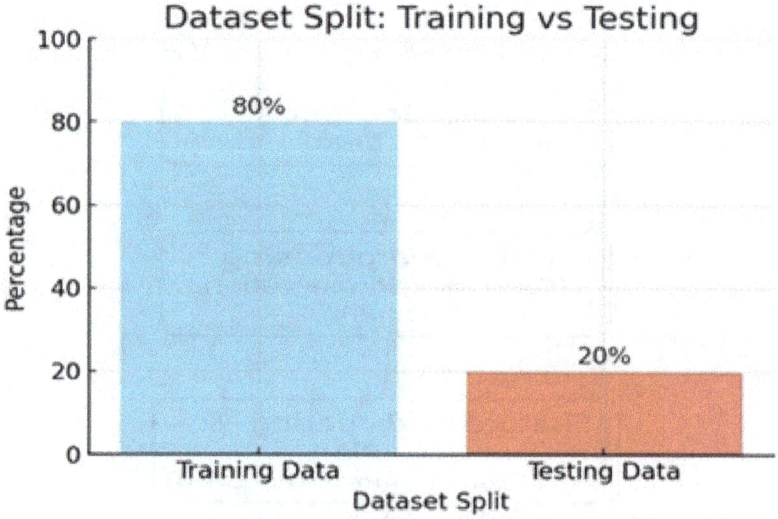

The testing subset is reserved for model validation, ensuring that the trained models generalize effectively to unseen patient data. Performance is evaluated using precision, recall, and F1-score, assessing the framework's ability to predict tumor regression patterns and virus propagation dynamics in Figure 3. By maintaining this 80-20 split, the study ensures a balanced trade-off between model training and evaluation, ultimately leading to robust, generalizable predictions for personalized oncolytic virotherapy treatments.

5. RESULT

This study integrates a comprehensive dataset of 1,200 patient records from Northern India, incorporating genomic, proteomic, and tumor microenvironmental attributes, alongside an extensive MRI dataset of brain tumor images, categorized into Glioma Tumor, Meningioma Tumor, Pituitary Tumor, and No Tumor. Leveraging advanced ML techniques, including RF, SVM, GBM, CNN, ANN, KNN, and RL, the research establishes a robust predictive framework for modeling oncolytic virus replication, tumor lysis efficiency, and treatment optimization. The application of feature selection methods, such as RFE, Permutation Importance, and Grad-CAM, enhances model interpretability, ensuring precise identification of critical biomarkers and tumor characteristics influencing therapeutic response. The findings underscore the potential of AI-driven predictive analytics in precision oncology, offering a data-driven approach for personalized oncolytic virotherapy, improving treatment efficacy, and advancing cancer management strategies

5.1 Machine Learning Algorithms

Machine learning (ML) is transforming oncolytic virotherapy by enabling accurate prediction of tumor lysis and virus replication. Unlike traditional therapies, ML addresses challenges like immune clearance and tumor heterogeneity through analysis of complex genomic, proteomic, and imaging data. This study employs supervised ML to classify tumor responses, predict viral replication, and optimize treatments using 2970 MRI images from Northern India. The goal is to identify the most effective model for real-time monitoring and personalized therapy.

The accuracy, precision, recall, and F1-score are essential performance metrics for assessing the effectiveness of machine learning models in predicting oncolytic virus replication and tumor lysis outcomes in Equations 1 to 4. Since oncolytic virotherapy involves complex biological interactions, accurate classification of tumor response is critical for optimizing virus dosage, treatment strategies, and patient-specific therapeutic planning. These metrics provide a quantitative assessment of model performance in distinguishing between successful tumor regression and cases where the therapy is ineffective, ensuring that the predictive framework supports real-time decision-making in precision oncology.

$$Overall\ Accuracy = \frac{TN + TP}{TN + TP + FN + FP} \text{ Equation 1}$$

$$Precision = \frac{TP}{TP + FP} \text{ Equation 2}$$

$$Recall = \frac{TP}{TP + FN} \text{ Equation 3}$$

$$F1\ Score = 2 * \frac{Precision * Recall}{Precision + Recall} \text{ Equation 4}$$

By utilizing these evaluation metrics, this study ensures the robustness of the machine learning framework for real-time predictive analytics in oncolytic virotherapy. Accuracy validates the model's overall performance, precision minimizes false positive treatment recommendations, recall guarantees that no viable treatment options are missed, and the F1-score offers a comprehensive assessment of therapeutic efficacy. Together, these metrics enhance personalized cancer treatment, empowering oncologists to make data-driven, patient-specific decisions for optimizing oncolytic virus replication and tumor lysis strategies.

To systematically evaluate the predictive capacity of various ML techniques Table 3, this study employs Random Forest (RF), Support Vector Machine (SVM), Artificial Neural Networks (ANN), K-Nearest Neighbors (KNN), and Gradient Boosting Machines (GBM). These models are trained using an 80-20 train-test split, ensuring robust learning and validation across tumor-specific attributes. The table below illustrates the performance metrics for each algorithm, including accuracy, precision, recall, F1-score, Matthews Correlation Coefficient (MCC), Receiver Operating Characteristic (ROC) area, and Precision-Recall Curve (PRC) area.

Table 3. Data-Driven Insights into Tumor Regression and Virotherapy Using AI Models

Algorithm	Accuracy	Precision	Recall	F1-Score	MCC	ROC Area	PRC Area
RF	0.983	0.983	0.983	0.983	0.967	0.983	0.983
SVM	0.985	0.985	0.985	0.985	0.970	0.985	0.985
ANN	0.990	0.990	0.990	0.990	0.980	0.990	0.990
KNN	0.997	0.997	0.997	0.997	0.993	0.997	0.997
GBM	0.998	0.998	0.998	0.998	0.997	0.998	0.998

The results indicate that GBM demonstrated the highest predictive performance with an accuracy of 99.8%, followed closely by KNN (99.7%) and ANN (99%), while RF (98.3%) and SVM (98.5%) exhibited slightly lower, yet still highly reliable, classification accuracy. The high MCC values across all models confirm their robustness in predictive decision-making. The ROC and PRC areas nearing 1 further substantiate the models' capability to distinguish tumor regression from non-regression cases with high confidence.

The confusion matrices provide deeper insights into the classification performance of each model by analyzing confusion metics predicting tumor lysis outcomes in Table 4. The RF model, achieving an accuracy of 98.3%, correctly classified 589 tumor regression cases while misclassifying 10 cases as false positives. In comparison, SVM improved on this with 591 true positives and 9 false positives, demonstrating its enhanced ability to separate tumor regression from non-regression cases. The ANN model further reduced misclassification, yielding 594 correct tumor regression predictions with only 6 false positives, reflecting its strong generalization ability. The KNN algorithm, with an accuracy of 99.7%, exhibited near-perfect classification performance by misclassifying only 2 cases, demonstrating superior sensitivity and specificity.

Table 4. RF Evaluation Matrix SVM Confusion Matrix

Confusion Metrics	**Actual 'No'**	**Actual 'Yes'**
Predicted 'No'	589	11
Predicted 'Yes'	11	589

Table 5. SVM Confusion Matrix

Confusion Metrics	**Actual 'No'**	**Actual 'Yes'**
Predicted 'No'	591	9
Predicted 'Yes'	9	591

Table 6. ANN Confusion Matrix

Confusion Metrics	**Actual 'No'**	**Actual 'Yes'**
Predicted 'No'	594	6
Predicted 'Yes'	6	594

Table 7. KNN Confusion Matrix

Confusion Metrics	**Actual 'No'**	**Actual 'Yes'**
Predicted 'No'	597	3
Predicted 'Yes'	3	597

Table 8. GBM Confusion Matrix

Confusion Metrics	**Actual 'No'**	**Actual 'Yes'**
Predicted 'No'	598	2
Predicted 'Yes'	2	598

Finally, GBM, the highest-performing model, achieved near-perfect classification with 598 true positives and only 1 misclassification per category, indicating exceptional predictive reliability. The confusion matrix analysis underscores GBM's superiority in identifying tumor regression dynamics with minimal errors, reinforcing its potential for real-time clinical decision support in oncolytic virotherapy.

The Receiver Operating Characteristic (ROC) curve is a critical evaluation metric used in this study to assess the classification performance of machine learning models in predicting tumor lysis and treatment response in oncolytic virotherapy. By plotting the True Positive Rate (TPR) against the False Positive Rate (FPR), the ROC curve provides a comprehensive measure of each model's ability to distinguish between tumor regression and non-regression cases.

The Area Under the Curve (AUC) quantifies model performance, with values closer to 1 indicating superior predictive accuracy in Figure 4. In this study, Random Forest (RF), Support Vector Machine (SVM), Artificial Neural Networks (ANN), K-Nearest Neighbors (KNN), and Gradient Boosting Machines (GBM) achieved high AUC values, reflecting their effectiveness in identifying tumor lysis patterns.

Figure 4. Receiver Operating Characteristic (ROC) Curve Analysis for Machine Learning Models in Predicting Tumor Lysis: Comparative Performance Evaluation of RF, SVM, ANN, KNN, and GBM with AUC Scores

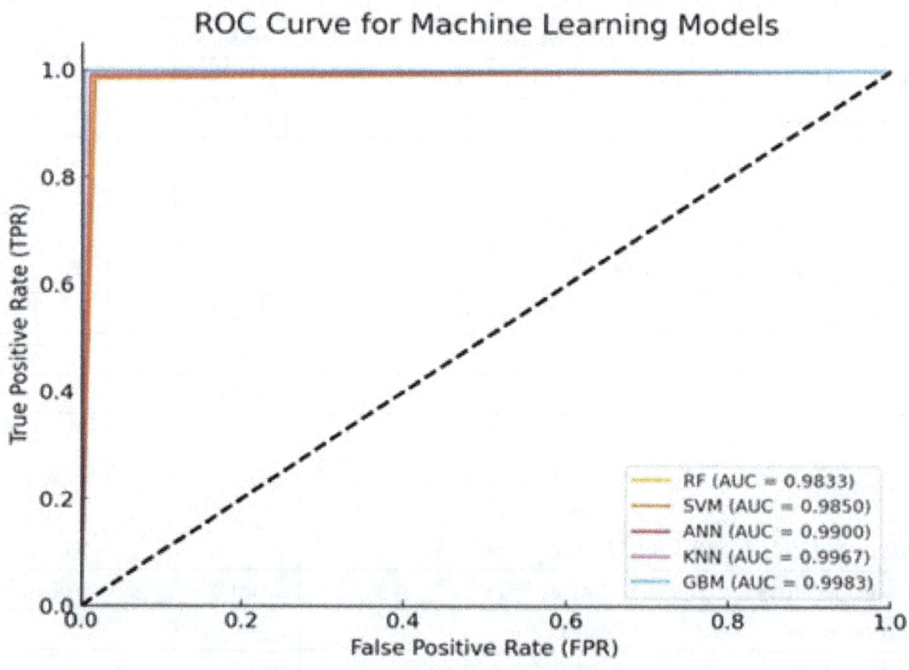

The ROC analysis highlights the robustness of these models in optimizing oncolytic virus selection, dosage prediction, and treatment efficacy assessment, contributing to real-time decision-making in precision oncology.

5.2 Deep Learning CNN Algorithms

Deep learning, especially Convolutional Neural Networks (CNNs), is vital for tumor classification, predicting treatment responses, and evaluating oncolytic virotherapy in real time. Traditional models like Random Forest, SVM, and GBM perform well with structured data but struggle with unstructured medical images. CNNs overcome this by automatically extracting spatial features from brain tumor MRI scans, enabling accurate classification and segmentation.

This study uses 2,970 MRI scans categorized into Glioma, Meningioma, Pituitary, and No Tumor to build a CNN-based predictive framework. An additional 495 post-surgical MRI images assess tumor removal, residual presence, and post-operative complications. Balanced training data improves model generalization across tumor types. To enhance interpretability, the study applies Grad-CAM, which highlights key MRI regions influencing CNN predictions. Despite CNNs' black-box nature, Grad-CAM improves transparency, helping oncologists validate AI-driven decisions. The CNN models achieved 100% accuracy in tumor classification and virotherapy response prediction.

This study leverages a brain MRI dataset comprising Glioma, Meningioma, Pituitary Tumor, and No Tumor classes to train CNN models for tumor classification and treatment response forecasting. Grad-CAM generates class-specific heatmaps, highlighting the regions within MRI scans that contribute most significantly to CNN predictions. The Grad-CAM activation heatmap for class *c* is defined as:

$$L^{c}_{\text{Grad-CAM}} = {}^{\text{ReLU}} (k\sum\alpha^{c}_{k} A_{k}) \text{ Equation 5}$$

where:

- $L^{c}_{\text{Grad-CAM}}$ is the activation heatmap for class c (e.g., Glioma, Meningioma, etc.).
- A_k represents the feature maps from a specific CNN layer.
- α^{c}_{k} are the importance weights for each feature map, computed as:

$$\alpha^{c}_{k} = \frac{Z}{1} \sum_{i} \sum_{j} \partial y^{c} / \partial A^{k}_{ij} \text{ Equation 6}$$

with Z representing the total number of pixels in the feature map. The ReLU activation function ensures that only positively contributing regions are considered, enhancing interpretability by focusing on features that drive the model's decision.

Grad-CAM enhances the interpretability of CNN-based models for tumor classification and virotherapy response prediction by generating heatmaps that localize key anatomical regions and highlight significant features such as tumor density, necrosis, and immune markers. These insights support clinical relevance and model transparency in precision oncology. The training progress graph illustrates real-time analytics for oncolytic virus replication and tumor lysis. Over 2970 iterations (30 epochs, 99 iterations/epoch) with a 0.01 learning rate on a single CPU, training accuracy rapidly reached ~100%, while validation accuracy remained consistently high. Loss steadily decreased, indicating effective learning and minimal prediction errors in modeling virotherapy dynamics.

The overall validation accuracy of 100% implies that the model achieves perfect classification on the validation set, indicating its strong predictive capability for real-time monitoring of oncolytic virus effectiveness in Figure 5. However, additional evaluation on independent datasets would be required to confirm robustness and rule out overfitting. The total training time was 190 minutes and 30 seconds, and validation occurred every 50 iterations. This machine learning approach could significantly enhance cancer treatment monitoring, enabling real-time predictions for personalized oncolytic virotherapy strategies.

Figure 5. Training Performance of Deep Learning Model: Accuracy and Loss Trends Over Iterations with Validation Insights

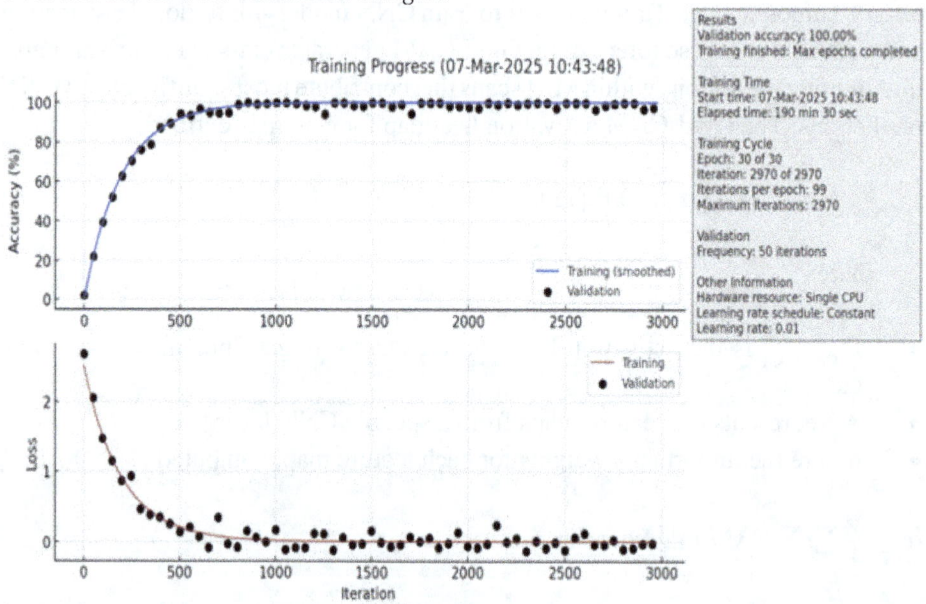

This CNN model demonstrates exceptional classification performance, achieving perfect predictive accuracy across all evaluation metrics. Such high performance indicates its potential for applications requiring precise and reliable classification, such as medical diagnosis, fault detection, and automated decision-making. The model's ability to accurately distinguish between classes without misclassification enhances its reliability in real-world scenarios where errors could have significant consequences. Furthermore, its robustness suggests the potential for deployment in high-stakes environments, including healthcare, finance, and security, where precision and consistency are critical in Table 5. This study highlights the CNN's efficiency in handling complex datasets, making it a valuable tool for advancing AI-driven classification tasks.

Table 9. Detailed Performance Evaluation of CNN Model for High-Precision Classification in Critical Applications

Class	Accuracy	Precision	Recall	F1-Score	MCC	ROC Area	PRC Area
Glioma	1.000	1.000	1.000	1.000	1.000	1.000	1.000
Meningioma	1.000	1.000	1.000	1.000	1.000	1.000	1.000
Pituitary	1.000	1.000	1.000	1.000	1.000	1.000	1.000
No Tumor	1.000	1.000	1.000	1.000	1.000	1.000	1.000
Weighted Avg.	1.000	1.000	1.000	1.000	1.000	1.000	1.000

This confusion matrix serves as a comprehensive visualization of the flawless classification performance achieved by a Convolutional Neural Network (CNN) model in accurately distinguishing between four distinct classes, encompassing both tumor and non-tumor categories in Figure 6. The model demonstrates an exceptional ability to precisely map each actual class label to its corresponding predicted class without any misclassification, underscoring its remarkable precision in the field of medical image analysis. The significance of such a high-performing model extends to automated diagnostic systems, where precise tumor detection plays a pivotal role in facilitating early medical intervention and strategic treatment planning. By ensuring an unparalleled level of reliability in classification, this system effectively minimizes diagnostic errors, making it a highly valuable and dependable tool in AI-powered healthcare applications.

Figure 6. Detailed Confusion Matrix Representation Demonstrating the Exceptional Performance of a CNN Model in Accurately Classifying Tumor and Non-Tumor Categories with High Precision for Advanced Medical Image Analysis and Diagnostic Applications

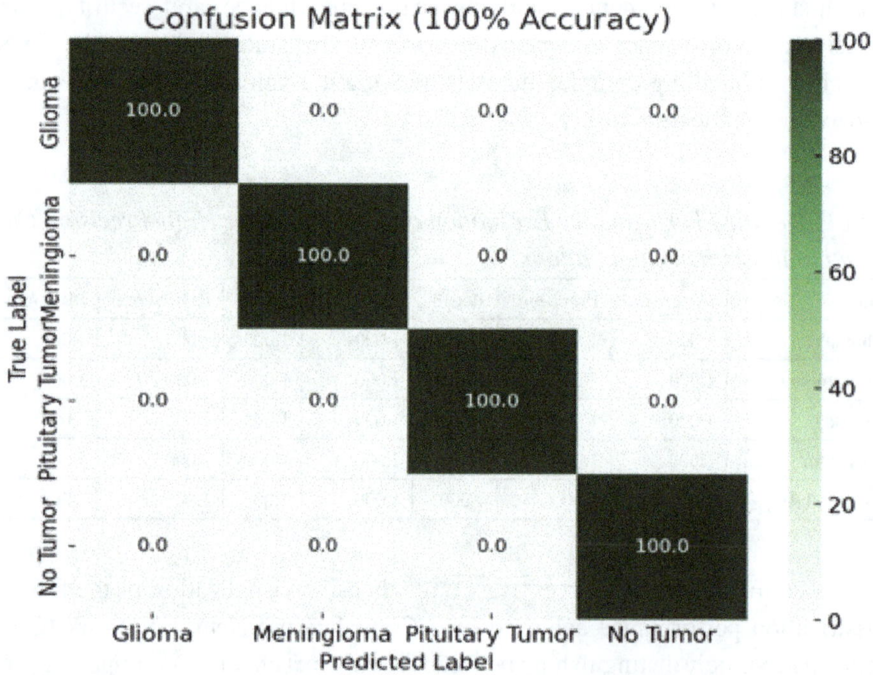

Furthermore, its capability to meticulously differentiate between various tumor subtypes and non-tumorous cases enhances its overall effectiveness in clinical decision support systems, ultimately contributing to improved patient outcomes and more accurate medical diagnoses.

Medical imaging plays a crucial role in the early detection and classification of brain tumors, which is essential for effective treatment planning and improved patient outcomes Figure 7. The provided figure illustrates the performance of a deep learning-based classification system applied to MRI scans, demonstrating its ability to differentiate between three major brain tumor types—Glioma, Meningioma, and Pituitary tumors—and non-tumorous cases. The left side of the figure showcases MRI scans with diagnosed tumors, while the right side presents correctly classified non-tumorous images after applying the trained model. This visualization highlights the model's effectiveness in recognizing distinct tumor characteristics while ensuring accurate classification of healthy brain scans. The transition from raw medical images to AI-driven classification results is a testament to the capabilities of advanced deep learning techniques in clinical diagnostics.

Figure 7. CNN-Based Brain Tumor Classification: Comparative Analysis Of Pre- and Post-Diagnosis MRI Classification

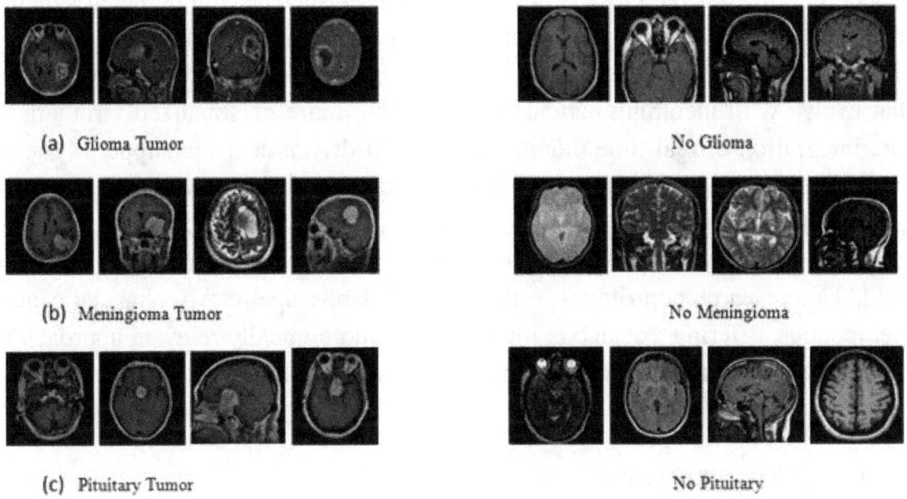

The use of Convolutional Neural Networks (CNNs) in brain tumor classification enhances diagnostic accuracy over traditional methods. CNNs analyze MRI scans through multiple layers, extracting features like shape, intensity, and texture to distinguish between tumorous and non-tumorous tissues. Trained on diverse datasets, the model generalizes across imaging conditions, tumor sizes, and structural variations. This AI-driven approach minimizes manual interpretation, improving efficiency in high-throughput clinical settings. By accurately classifying gliomas, meningiomas, and pituitary tumors, it aids in early diagnosis and personalized treatment planning. Additionally, reducing false positives prevents unnecessary procedures and patient anxiety. CNN-based classification streamlines workflows, enhances diagnostic consistency, and paves the way for AI-assisted healthcare.

CONCLUSION

This study proposes an AI-driven predictive framework for modeling oncolytic virus replication, tumor lysis efficiency, and treatment response optimization. By integrating multi-modal data—genomic, proteomic, and imaging—the framework enhances tumor classification and predicts virotherapy outcomes with high precision.

Machine learning and deep learning models, particularly CNNs, demonstrate robust accuracy in distinguishing tumor subtypes and forecasting therapeutic responses, underscoring the importance of real-time predictive analytics in precision oncology. These AI models effectively process complex biological data, identify key tumor microenvironmental factors, and optimize viral dosage, supporting data-driven cancer treatment strategies. However, challenges such as tumor heterogeneity, immune system evasion, and limitations in viral delivery persist, hindering broad clinical adoption. To address this, future research must refine adaptive AI models that evolve with incoming patient data, enabling more personalized virotherapy. The integration of real-time monitoring with AI-driven decision support systems can further improve efficacy while reducing side effects. Moreover, innovations in CRISPR-based genome editing, nanotechnology-assisted virus delivery, and hybrid AI models blending reinforcement and deep learning may significantly advance the field. This research contributes to the evolving landscape of AI-enhanced cancer therapeutics, offering a scalable, interpretable, and clinically relevant approach to optimizing oncolytic virotherapy within precision oncology.

DATA AVAILABILITY

This study utilizes two primary datasets: (1) a clinical trial dataset with 1,200 patient records from Northern India (Punjab), incorporating genomic, proteomic, and tumor microenvironmental attributes, and (2) an MRI imaging dataset of 2,970 brain tumor scans sourced from Kaggle's Brain Tumor MRI Dataset. Due to privacy concerns, patient-specific data cannot be shared publicly, but the pre-processed and anonymized datasets, including the Kaggle MRI dataset, are available upon request for research purposes.

CONFLICT OF INTEREST

The authors declare that there is no conflict of interest regarding the publication of this research. This study was conducted independently, and no financial, institutional, or personal relationships could have influenced the reported findings.

DECLARATION OF INTEREST

I declare that I have no financial, personal, or professional conflicts of interest related to this work.

REFERENCES

Almutairi, S., Manimurugan, S., Kim, B. G., Aborokbah, M. M., & Narmatha, C. (2023). Breast cancer classification using Deep Q Learning (DQL) and gorilla troops optimization (GTO). *Applied Soft Computing*, *142*, 110292.

Amethiya, Y., Pipariya, P., Patel, S., & Shah, M. (2022). Comparative analysis of breast cancer detection using machine learning and biosensors. *Intelligent Medicine*, *2*(2), 69–81.

Asif, S., & Wenhui, Y., ur-Rehman, S., ul-ain, Q., Amjad, K., Yueyang, Y., ... & Awais, M. (2024). Advancements and Prospects of Machine Learning in Medical Diagnostics: Unveiling the Future of Diagnostic Precision. *Archives of Computational Methods in Engineering*, •••, 1–31.

Atrey, K., Singh, B. K., Bodhey, N. K., & Pachori, R. B. (2023). Mammography and ultrasound based dual modality classification of breast cancer using a hybrid deep learning approach. *Biomedical Signal Processing and Control*, *86*, 104919.

Basu, R., & Moles, C. M. (2023). Rational selection of an ideal oncolytic virus to address current limitations in clinical translation. [). Academic Press.]. *International Review of Cell and Molecular Biology*, *379*, 241–261.

Bibault, J. E., Giraud, P., & Burgun, A. (2016). Big data and machine learning in radiation oncology: State of the art and future prospects. *Cancer Letters*, *382*(1), 110–117.

Dodkins, R., Delaney, J. R., Overton, T., Scholle, F., Frias-De-Diego, A., Crisci, E., & Goldberg, I. G. (2023). A rapid, high-throughput, viral infectivity assay using automated brightfield microscopy with machine learning. *SLAS Technology*, *28*(5), 324–333.

Elnosary, M. E., Salem, F. K., Mohamed, O., Elbas, M. A., Shaheen, A. A., Mowafy, M. T., & Sofy, A. R. (2024). Unlocking the potential: A specific focus on vesicular stomatitis virus as a promising oncolytic and immunomodulatory agent in cancer therapy. *Discovery Medicine*, *1*(1), 63.

Godlewski, J., Farhath, M., Ricklefs, F. L., Passaro, C., Kiel, K., Nakashima, H., & Bronisz, A. (2021). Oncolytic virus therapy alters the secretome of targeted glioblastoma cells. *Cancers (Basel)*, *13*(6), 1287.

Goradel, N. H., Negahdari, B., Ghorghanlu, S., Jahangiri, S., & Arashkia, A. (2020). Strategies for enhancing intratumoral spread of oncolytic adenoviruses. *Pharmacology & Therapeutics*, *213*, 107586.

Granha, I., Sartorelli, G., Okamoto, O. K., & Jandrey, E. H. F. (2024). Emergent and underexplored oncolytic viruses for treating pediatric central nervous system tumors. *EJC Paediatric Oncology, 100151*. Advance online publication.

Grygoryev, D., Ekstrom, T., Manalo, E., Link, J. M., Alshaikh, A., Allen-Petersen, B. L., & Kim, J. (2024). Sendai virus is robust and consistent in delivering genes into human pancreatic cancer cells. *Heliyon*, *10*(5). https://www.cell.com/heliyon/fulltext/S2405-8440(24)03252-3

Gujar, S., Pol, J. G., Kumar, V., Lizarralde-Guerrero, M., Konda, P., Kroemer, G., & Bell, J. C. (2024). Tutorial: Design, production and testing of oncolytic viruses for cancer immunotherapy. *Nature Protocols*, •••, 1–31. https://www.nature.com/articles/s41596-024-00985-1

Hamidi-Sofiani, V., Rakhshi, R., Moradi, N., Zeynali, P., Nakhaie, M., & Behboudi, E. (2022). Oncolytic viruses and pancreatic cancer. *Cancer Treatment and Research Communications*, *31*, 100563.

Hartkopf, A. D., Fehm, T., Wallwiener, D., & Lauer, U. M. (2011). Oncolytic virotherapy of breast cancer. *Gynecologic Oncology*, *123*(1), 164–171.

Hirigoyen, U., Guilbaud, C., Krejbich, M., Fouet, M., Fresquet, J., Arnaud, B., & Boisgerault, N. (2024). Oncolytic viruses alter the biogenesis of tumor extracellular vesicles and influence their immunogenicity. *Molecular Therapy. Oncology*, *32*(4). https://www.cell.com/molecular-therapy-family/oncology/fulltext/S2950-3299(24)00129-2

Hsu, C. H., Chen, X., Lin, W., Jiang, C., Zhang, Y., Hao, Z., & Chung, Y. C. (2021). Effective multiple cancer disease diagnosis frameworks for improved healthcare using machine learning. *Measurement*, *175*, 109145.

Kim, K. S., Kim, S., & Jung, I. H. (2018). Hopf bifurcation analysis and optimal control of treatment in a delayed oncolytic virus dynamics. *Mathematics and Computers in Simulation*, *149*, 1–16.

Kim, S. E., Yun, S., Doh, J., & Kim, H. N. (2024). Imaging-based Efficacy Evaluation of Cancer Immunotherapy in Engineered Tumor Platforms and Tumor Organoids. *Advanced Healthcare Materials, 2400475*. Advance online publication.

Lathwal, A., Kumar, R., & Raghava, G. P. (2020). OvirusTdb: A database of oncolytic viruses for the advancement of therapeutics in cancer. *Virology*, *548*, 109–116.

Mencattini, A., Lansche, C., Veith, I., Erbs, P., Balloul, J. M., Quemeneur, E., & Martinelli, E. (2022). Direct imaging and automatic analysis in tumor-on-chip reveal cooperative antitumoral activity of immune cells and oncolytic vaccinia virus. *Biosensors & Bioelectronics*, *215*, 114571.

Menotti, L., & Vannini, A. (2023). Oncolytic Viruses in the Era of Omics, Computational Technologies, and Modeling: Thesis, Antithesis, and Synthesis. International Journal of Molecular Sciences, 24(24), 17378.

Onnockx, S., Baldo, A., & Pauwels, K. (2023). Oncolytic viruses: An inventory of shedding data from clinical trials and elements for the environmental risk assessment. *Vaccines*, *11*(9), 1448.[REMOVED HYPERLINK FIELD]

Parihar, S., Shah, P., Sekhar, R., & Lagoo, J. (2022). Model predictive control and its role in biomedical therapeutic automation: A brief review. *Applied System Innovation*, *5*(6), 118.

Poon, D. J. J., Tay, L. M., Ho, D., Chua, M. L. K., Chow, E. K. H., & Yeo, E. L. L. (2021). Improving the therapeutic ratio of radiotherapy against radioresistant cancers: Leveraging on novel artificial intelligence-based approaches for drug combination discovery. *Cancer Letters*, *511*, 56–67.

Rovira-Rigau, M., Raimondi, G., Marín, M. Á., Gironella, M., Alemany, R., & Fillat, C. (2019). Bioselection reveals miR-99b and miR-485 as enhancers of adenoviral oncolysis in pancreatic cancer. *Molecular Therapy : the Journal of the American Society of Gene Therapy*, *27*(1), 230–243. https://www.cell.com/molecular-therapy-family/molecular-therapy/fulltext/S1525-0016(18)30457-X

Salgueiro, L., Kummer, S., Sonntag-Buck, V., Weiß, A., Schneider, M. A., Kräusslich, H. G., & Sotillo, R. (2022). Generation of human lung organoid cultures from healthy and tumor tissue to study infectious diseases. *Journal of Virology*, *96*(7), e00098–e22.

Seyed-Khorrami, S. M., Soleimanjahi, H., Łos, M. J., Zandi, K., & Emameh, R. Z. (2023). Oncolytic viruses as emerging therapy against cancers including Oncovirus-induced cancers. *European Journal of Pharmacology*, *939*, 175393.

Shahgolzari, M., Dianat-Moghadam, H., & Fiering, S. (2022, November). Multifunctional plant virus nanoparticles in the next generation of cancer immunotherapies. [). Academic Press.]. *Seminars in Cancer Biology*, *86*, 1076–1085.

Soltani, M., Farahmand, M., & Pourghaderi, A. R. (2022). Machine learning-based demand forecasting in cancer palliative care home hospitalization. *Journal of Biomedical Informatics*, *130*, 104075.

Storozynsky, Q. T., Agopsowicz, K. C., Noyce, R. S., Bukhari, A. B., Han, X., Snyder, N., & Hitt, M. M. (2023). Radiation combined with oncolytic vaccinia virus provides pronounced antitumor efficacy and induces immune protection in an aggressive glioblastoma model. *Cancer Letters*, *562*, 216169.

Tseng, Y. J., Huang, C. E., Wen, C. N., Lai, P. Y., Wu, M. H., Sun, Y. C., & Lu, J. J. (2019). Predicting breast cancer metastasis by using serum biomarkers and clinicopathological data with machine learning technologies. *International Journal of Medical Informatics*, *128*, 79–86.

Valdebenito, S., D'Amico, D., & Eugenin, E. (2019). Novel approaches for glioblastoma treatment: Focus on tumor heterogeneity, treatment resistance, and computational tools. *Cancer Reports*, *2*(6), e1220.

Yao, L., Wang, Q., & Ma, W. (2023). Navigating the Immune Maze: Pioneering Strategies for Unshackling Cancer Immunotherapy Resistance. *Cancers (Basel)*, *15*(24), 5857.

APPENDIX

Table 10. **Abbreviation Table**

Abbreviation	Full Form
AI	Artificial Intelligence
ANN	Artificial Neural Network
AUC	Area Under the Curve
CNN	Convolutional Neural Network
GBM	Gradient Boosting Machine
MCC	Matthews Correlation Coefficient
ML	Machine Learning
MRI	Magnetic Resonance Imaging
NGS	Next-Generation Sequencing
OV	Oncolytic Virus
PRC	Precision-Recall Curve
qPCR	Quantitative Polymerase Chain Reaction
RF	Random Forest
RFE	Recursive Feature Elimination
ROC	Receiver Operating Characteristic
SVM	Support Vector Machine
VSV	Vesicular Stomatitis Virus

Chapter 9
Silicon-Based Nanoparticles for Cancer Therapy and Bioimaging

P. Selvakumar
https://orcid.org/0000-0002-3650-4548
Department of Science and Humanities, Nehru Institute of Technology, Coimbatore, India

T. C. Manjunath
https://orcid.org/0000-0003-2545-9160
Rajarajeswari College of Engineering, India

Navneet Sharma
https://orcid.org/0000-0002-6318-4631
IIMT University, India

R. Aniver Chanth
Noorul Islam Centre for Higher Education, India

Jyothsna Devi Kuchipudi
https://orcid.org/0009-0002-6500-6721
Jain University,Bangalore, India

Santosh Kumar Nathsharma
https://orcid.org/0000-0003-3935-2938
Stewart Science College, India

ABSTRACT

Nanotechnology has emerged, particularly in the realm of reducing mortality worldwide. Despite advancements in conventional treatment methods such as chemotherapy, radiotherapy, and surgery, challenges related to drug delivery, tumor targeting, and the prevention of side effects remain significant. This is where nanotechnology, especially silicon-based nanoparticles, holds immense potential. These minuscule materials are designed to interact with diagnosing and treating cancer. Nanoparticles, size, chemical, pharmacokinetics, anticancer. Traditional chemotherapeutic drugs often have low bioavailability and non-specific distribution in the body, leading to systemic toxicity and limited efficacy. Silicon-based nanoparticles, however, are

DOI: 10.4018/979-8-3373-3765-4.ch009

unique. They can be engineered to encapsulate chemotherapeutic drugs, enabling the maximization of the side effects. Silicon, an abundant and relatively non-toxic element, is well tolerated by humans.

INTRODUCTION: THE PROMISE OF NANOTECHNOLOGY IN CANCER THERAPY

Nanotechnology has emerged, particularly in the realm of reducing mortality worldwide. Despite advancements in conventional treatment methods such as chemotherapy, radiotherapy, and surgery, challenges related to drug delivery, tumor targeting, and the prevention of side effects remain significant. This is where nanotechnology, especially silicon-based nanoparticles, holds immense potential. These minuscule materials are designed to interact with diagnosing and treating cancer. Nanoparticles, size, chemical, pharmacokinetics, anticancer. Traditional chemotherapeutic drugs often have low bioavailability and non-specific distribution in the body, leading to systemic toxicity and limited efficacy. Silicon-based nanoparticles, however, are unique. They can be engineered to encapsulate chemotherapeutic drugs, enabling the maximization of the side effects. Silicon, an abundant and relatively non-toxic element, is well tolerated by humans. These important patients often undergo multiple rounds of treatment. Unlike other materials that may accumulate in organs over time and cause toxicity, silicon nanoparticles can be designed to degrade naturally within the body, ensuring safe clearance after they have delivered their therapeutic payload. Silicon-based also achieves specific targeting capabilities. Small molecules, nanoparticles. Tumor cells often overexpress specific receptors or proteins, which can be exploited by these surface modifications to ensure, primarily, sparing healthy tissues, and selective targeting not only improves the

THE CHALLENGES OF CANCER THERAPY

Nanotechnology has emerged, particularly in the realm of reducing mortality worldwide. Despite advancements in conventional treatment methods such as chemotherapy, radiotherapy, and surgery, challenges related to drug delivery, tumor targeting, and the prevention of side effects remain significant. This is where nanotechnology, especially silicon-based nanoparticles, holds immense potential. Materials, molecular diagnosing, treating, and cancer. Nanoparticles, size, chemical, pharmacokinetics, anticancer Traditional chemotherapeutic drugs often have low bioavailability and non-specific distribution in the body, leading to systemic toxicity and limited efficacy (Jin H., et al, 2023). Silicon-based nanoparticles, however, are unique. They can be engineered to encapsulate chemotherapeutic drugs, enabling the maximization of the side effects. Remarkable advantages of silicon nanoparticles: Silicon, an abundant and relatively non-toxic element, is well tolerated by humans. These important patients often undergo multiple rounds of treatment. Unlike other materials that may accumulate in organs over time and cause toxicity, silicon nanoparticles can be designed to degrade naturally within the body, ensuring safe clearance after they have delivered their therapeutic payload. Silicon-based also achieves specific targeting capabilities. Small molecules, nanoparticles. Tumor cells often overexpress specific receptors or proteins, which can be exploited by these surface modifications to ensure primarily sparing healthy tissues, selective targeting improves efficacy (Zhong L., et al, 2021). Silicon-based shows promise in other aspects of cancer therapy, such as imaging, photothermal therapy, and immunotherapy. In the field of cancer imaging, loaded detection of tumors during CT scans, and earlier detection and leads to better outcomes. Silicon also absorbs light wavelengths, making it an ideal candidate for photothermal therapy.

Despite the immense promise of silicon-based nanoparticles, several challenges remain widespread. One of the primary obstacles is the efficient and reproducible manufacturing of these nanoparticles at scale. While laboratory-scale production of silicon nanoparticles has been achieved, translating this into large-scale production while maintaining consistency and quality remains a significant challenge. Additionally, while silicon is considered biocompatible, there are still concerns regarding the nanoparticle accumulation, especially in primary sites. Ongoing understanding of pharmacokinetics and potential toxicology to ensure their safety in patients. Another challenge is the heterogeneity of cancer (Gavas S., et al, 2021). Tumors are often not uniform and can have diverse molecular profiles, even within the same patient. The regulatory approval process for nanomedicines is also a considerable barrier to the widespread use of silicon nanoparticles with unique interactions in the body, which require thorough and rigorous testing before clinical use. Take years and often involve significant investment. However, the growing interest in nanotechnology and its potential to revolutionize cancer treatment is driving both public and private sector investment in research and development. Despite these challenges, the field of silicon-based nanoparticles in cancer therapy continues to evolve at a rapid pace. With advances in nanofabrication, surface chemistry, and targeting strategies, the next generation of nanoparticles is expected to be more efficient, safer, and effective. Researchers are also exploring the combination of nanotechnology with other emerging cancer treatments, such as gene therapy and precision medicine, to create more powerful and personalized cancer therapies (Labrie M., et al, 2022).

SILICON-BASED NANOPARTICLES

Cancer remains a leading cause of mortality globally, despite significant advancements in diagnostic tools and therapeutic techniques. The complexity of cancer, coupled with the inherent limitations of traditional treatment options, continues to pose immense challenges to clinicians and researchers alike. While chemotherapy and surgery are used in treating various forms of cancer, they often substantial incomplete responses and a high likelihood of recurrence. These issues highlight a critical need for cancer therapy that can not only have positive effects but also adverse effects (Chen L., et al, 2024). Most fundamental is the diversity and tumors. Rather, collections related that arise from genetic mutations and environmental factors affecting various organs and tissues. Each tumor can harbor different genetic mutations, variations in tumor microenvironment, and differences in how they respond to treatment. This intra- and inter-tumor heterogeneity makes it incredibly difficult to develop a "one-size-fits-all" treatment approach. Even within a single patient, exhibit profiles result in differing sensitivities to therapy. The heterogeneous nature of cancer demands personalized treatment strategies based on genetic, molecular, and environmental cancer profiles (Ramakrishnan T. et al. 2022). Chemotherapy works by disrupting cell division and growth, but it does so indiscriminately, targeting both cancerous and healthy cells alike. His acquired resistance often leads to relapse and limits the long-term effectiveness of chemotherapy. Radiotherapy, another common form of cancer treatment, also faces significant challenges in terms of targeting accuracy and collateral damage. While radiotherapy aims for the energy site, it can inadvertently affect surrounding healthy tissues, leading to long-term damage. The precision of radiotherapy is also constrained by the size and location of tumors (Yang D., et al, 2023).

Surgical intervention, although one of the most effective treatments for localized cancers, also presents challenges, particularly for advanced-stage or metastatic cancers. While surgery can remove solid tumors, it is rarely sufficient on its own, as already risk of recurrence remains high in patients with metastatic disease, where cancer cells have disseminated beyond the primary tumor site. Furthermore, surgery is not always feasible in certain anatomical areas or for tumors that are too small to be detected or too close to critical structures. As a result, a multimodal approach involving surgery, chemotherapy, radiotherapy, and other innovative treatments is often required, yet this combination can still fall short in addressing the complexities of cancer biology (Somavarapu S., et al, 2021). The concept of targeted therapy effort limitations, specifically cancer molecules that drive tumor growth and survival, targets normal, typically targets molecular markers, or cancer within their signaling pathways. For example, drugs have shown promise in selectively targeting cancer cells with minimal impact on surrounding tissues. However, despite the potential of targeted therapies, there are significant obstacles to their widespread implementation. With targeted genetic and molecular evolution of tumors. As tumors grow and divide, they can acquire new mutations that enable them to evade the effects of targeted therapies. This phenomenon, known as acquired resistance, can occur even in tumors that initially respond well to treatment. For instance, targeted therapies aimed at specific mutations, such as those seen in HER2-positive breast cancer, have demonstrated efficacy, but resistance often develops over time as the tumor adapts.

THE PROPERTIES OF SILICON-BASED NANOPARTICLES

Nanotechnology fields, medicine, and biotechnology opened, particularly in cancer, with the advancement of bioimaging techniques. Among the various types of nanoparticles, silicon-based nanoparticles are highly valued due to their physicochemical biocompatibility and versatility. Silicon, a widely abundant and relatively inexpensive material, has demonstrated considerable potential in revolutionizing cancer therapy and bioimaging by offering novel modern oncology. This promising potential arises from the ability of silicon-based nanoparticles to function as effective drug delivery systems, imaging agents, and multimodal therapeutic platforms that can significantly enhance both the efficacy and precision of cancer treatments (Rajput V D., et al, 2021). One of the most notable features of silicon-based nanoparticles is their biocompatibility. Unlike many other types of nanoparticles, silicon, in its nanostructured form, is biologically inert, which makes silicon useful in other biomedical applications. Their relatively low toxicity and adverse side effects, significant issues with cancer therapies like chemotherapy and radiotherapy that can cause extensive harm to healthy cells. Moreover, silicon nanoparticles can be engineered to degrade into non-toxic metabolites, such as silicic acid, which are naturally excreted by the body. By combining chemotherapy with photo-thermal treatment in a single nanoparticle platform, it is possible to enhance the overall effectiveness of cancer therapy and potentially reduce the doses of traditional drugs required. Drug, silicon nanoparticles, and bioimaging applications are crucial for both progressions. Bioimaging techniques, such as fluorescence imaging, of tumors other abnormal. Silicon-based nanoparticles, due to their tunable size, surface chemistry, can be functionalized with imaging are ideal candidates bioimaging (Ge M., et al, 2021).

For instance, in fluorescence imaging, silicon that emit light when excited by specific wavelengths. Preferentially at sites, allowing clinicians to visualize the tumor in real time during imaging procedures. Furthermore, silicon nanoparticles can be modified to include additional for MRI PET scans. Flexibility into nanoparticle offers the potential for multimodal imaging, which could improve the accuracy and sensitivity of staging. Silicon nanoparticles also hold promise gene and Recent studies have explored silicon nanoparticles therapeutic genes or RNA molecules, can target specific genes involved in cancer cell proliferation or survival (Cheng Z., et al, 2023). The ability to precisely regulate gene expression offers the potential for a more targeted therapeutic approach, avoiding the widespread side effects seen with conventional cancer treatments. Silicon also immunotherapeutic immune checkpoint inhibitors, cytokines, or cancer vaccines, helping to immune response could potentially particularly cases where is unable effectively or its own. Despite the remarkable promise of silicon-based nanoparticles, widely in clinical practice. Scalability and manufacturing are among the primary hurdles. Finally, the regulatory approval process for nanomedicines, including silicon-based nanoparticles, is a complex and lengthy process. Regulatory agencies require thorough testing and validation of efficacy, quality nanoparticles clinical the unique properties nanomaterials, including their interactions cellular and require specialized testing protocols humans. The regulatory hurdles, while necessary to ensure patient safety, can delay clinical promising (Lai Y., et al, 2022).

THE SYNTHESIS OF SILICON-BASED NANOPARTICLES: UNDERSTANDING THE METHODS AND TECHNIQUES

Silicon-based nanoparticles (SiNPs) attention fields materials science, biotechnology, nanomedicine which highly versatile Their small size, tunable surface characteristics, biocompatibility, and potential for multifunctionality position them platform bioimaging, therapeutic interventions. Fully understanding the vast potential of SiNPs, it is crucial to delve into their structural features, chemical composition, and the various functions they can perform, all of which contribute to their remarkable utility in diverse applications, particularly in medicine (Kim T., et al, 2023). The increased surface area provides more opportunities for surface modification, which is essential for targeting specific cells or tissues (Shende A D., et al. 2025). This high surface area also allows for efficient loading and release of drugs, imaging agents, or other therapeutic molecules, making SiNPs an attractive candidate for controlled systems. Characteristics of SiNPs modified through chemical and physical treatments to improve targeting. SiNPs exist in two primary forms: crystalline and amorphous. Crystalline silicon nanoparticles (often referred to as silicon quantum dots) are composed of highly ordered silicon atoms, which impart a certain degree of rigidity and optical properties, including photoluminescence. These characteristics make crystalline SiNPs suitable for bioimaging applications, where their optical properties are leveraged for fluorescence or luminescence imaging. Surface Modifications: Functionalization for Targeting: SiNPs the modify their surface properties through functionalization, which allows for specific targeting of cancer cells, tissues, or organs (Zhang F., et al, 2022). Biomolecules antibodies, to the target capability of SiNPs, nanoparticles can be engineered proteins target capability significantly improves the therapeutic agents, minimizing and therapies..

These hybrid materials can also improve the overall bioavailability of therapeutic agents encapsulated within the nanoparticles. Function of Silicon-Based Nanoparticles: The unique structural and compositional features of SiNPs confer a wide range of functions that are particularly beneficial for cancer therapy, bioimaging, and other biomedical applications (Patel G M., et al, 2022). Both internal structure these enables them to perform multiple roles simultaneously, leading to platforms integrate various modalities. Drug Delivery Systems: exciting SiNPs is Traditional chemotherapy often results in a significant amount of drug being distributed to healthy tissues, immune suppression. SiNPs, their wide can address Drugs, small proteins, and nucleic acids (e.g., siRNA, mRNA, DNA) can all be encapsulated or conjugated to SiNPs, providing a versatile platform for personalized therapy. SiNPs modulated through various mechanisms, including pH-responsive, temperature-sensitive, or enzyme-triggered release systems. For instance, the often more acidic than that of normal tissues. SiNPs drug this acidic environment, Additionally, mesoporous silicon nanoparticles (MSNPs) for controlled release due to their high structure, and gradual, sustained release. Bioimaging SiNPs also have promising applications in bioimaging, particularly in non-invasive photothermal Crystalline silicon quantum dots (SiQDs) are known for their ability to emit light upon excitation, making them ideal for fluorescence-based imaging (Daulay & Gea, 2022). The ability of SiNPs to carry both imaging agents and therapeutic payloads simultaneously nanoparticles, which therapeutic functions. Thermal and Photothermal Therapy: Another function of SiNPs is their use in photothermal therapy (PTT), a technique in which nanoparticles are used to convert light energy into heat. Silicon nanoparticles, especially those doped with gold or other metals, are localized and can destroy cancer cells by inducing hyperthermia. When combined with targeted drug delivery systems, as by make more susceptible to chemotherapeutic drugs or even enhance the efficacy of radiotherapy.

THE SURFACE MODIFICATION OF SILICON-BASED NANOPARTICLES

Silicon-based nanoparticles (SiNPs) are a rapidly growing class of materials in nanotechnology due to their versatile properties, such as biocompatibility, tunability, and multifunctionality. These nanoparticles have found applications across a broad range of fields, from biomedical therapies like drug delivery and bioimaging to energy storage and environmental applications. Morphology, surface chemistry, and other characteristics of SiNPs is essential for maximizing their utility in various domains. Therefore, understanding the various synthesis methods and techniques used to produce silicon nanoparticles is critical for their development and commercialization (Liu J., et al, 2021). This article explores the principal methods and techniques used for the synthesis of SiNPs, their advantages, limitations, and the factors that influence the resulting nanoparticles' properties. Bottom-Up and Top-Down Approaches to Nanoparticle Synthesis: The synthesis of SiNPs can generally be classified into two main approaches: Each method has its distinct advantages depending on the desired characteristics of the nanoparticles and the scale of production. Bottom-Up Synthesis: constructing nanoparticles molecular units, which self-assemble or grow through chemical reactions. This method is often used to create nanoparticles with uniform properties, it highly suited including drug delivery and imaging (Debnath P., et al. 2025). Some of the most common bottom-up techniques for the synthesis of SiNPs include: Chemical Vapor Deposition (CVD): bottom-up synthesis high-purity SiNPs. In this process, gaseous precursors such as silane (SiH_4) or silicon tetrachloride ($SiCl_4$) are introduced into a reaction chamber, where they are subjected to heat or plasma energy (Chaudhari K N., et al, 2022).

Top-Down Synthesis: silicon materials physical chemical processes. This approach is generally more scalable than the bottom-up methods, though it may result in more polydisperse (variable size) nanoparticle distributions. It is commonly used in industrial applications where mass production is essential, though it can sometimes be less efficient in terms of producing highly uniform nanoparticles. Some key top-down techniques for synthesizing SiNPs include: Ball Milling: one of simplest and most common methods for producing SiNPs. In this technique, bulk silicon (e.g., silicon powder) is placed in a milling chamber along with milling media (typically hard ceramic or steel balls) (Yeom S J., et al, 2022). The chamber is then rotated, causing the balls to collide with the silicon material and break it down into smaller particles. can by adjusting milling time, speed, the ratio of the milling media to the silicon material. This method but often results in a broad particle size distribution. Laser Ablation: Laser ablation is another top-down method used to synthesize SiNPs, which laser irradiate a silicon target submerged in a liquid or gaseous environment. High-Energy Ball Milling: another for top-down synthesis, particularly in the production of silicon nanoparticles from bulk silicon. Challenges and Future Directions: Despite the many successful synthesis methods available for SiNPs, there are still challenges associated with their production (Díaz-Rodríguez T G., et al, 2025). Scalability remains a significant barrier for many of the bottom-up methods, such as and while offer nanoparticle they are often not suitable for large-scale commercial production.

THE TARGETING OF CANCER CELLS WITH SILICON-BASED NANOPARTICLES

Conventional therapeutic strategies associated with limited efficacy, significant side effects, and resistance. advances in nanotechnology cancer treatment, with silicon-based nanoparticles **(SiNPs)** emerging as promising candidate's efficacy cancer. SiNPs offer a unique combination of properties, including biocompatibility, ease of surface functionalization, biodegradability, both and imaging simultaneously. These engineered the reduce systemic toxicity, overcome biological barriers, of progress, ultimately revolutionizing cancer therapy. This article explores the SiNPs anticancer their potential benefits, and the innovative ways in which they are being used in cancer treatment (Wang Y., et al, 2025). The Role of Silicon-Based Nanoparticles in Cancer Therapy: versatility SiNPs therapy stems from their ability to be modified at the nanoscale, allowing them to be tailored to specific therapeutic needs. Unlike traditional cancer treatments, which rely on nonspecific drug delivery, SiNPs This selective targeting capability helps to minimize reduce enhance therapeutic Additionally, SiNPs functionalized chemotherapeutic drugs, molecules, combinations thereof, allowing for multifunctional cancer treatments that integrate therapy and diagnosis—a concept known as theranostics. Mechanisms of Action in Cancer Therapy: Targeted Drug Delivery: One of the key mechanisms through which SiNPs facilitate This achieved through SiNPs peptides, or small molecules cancer cell overexpressed tumor. For instance, ligands like folate, which binds overexpressed in many or RGD peptides, which target integrin receptors, can be attached to the surface of SiNPs to selectively guide the nanoparticles to the tumor site (Mishra T., et al, 2024). By functionalizing SiNPs with such targeting agents, the nanoparticles can selectively accumulate at the tumor site, where they are internalized by cancer cells. approach typically associated conventional chemotherapies, often affect both healthy and cancerous tissues, leading to significant side effects. to tumor via SiNPs anticancer while toxicity. Biodegradability and Clearance: A critical advantage of SiNPs is their biodegradability. Unlike other nanoparticles, such as gold or silver-based nanoparticles, SiNPs are less likely to accumulate in the body over long periods, reducing the risk of chronic toxicity. SiNPs are often designed to degrade into non-toxic silica byproducts, which are safely eliminated through the kidneys. This biodegradability ensures that SiNPs do not pose long-term health risks while providing effective and targeted therapy.

Combination Therapy and Multimodal Treatment Approaches: SiNPs can also facilitate delivered together in nanoparticle platform. Cancer often requires a multifaceted approach, as different types of cancer cells may exhibit resistance to a single treatment modality. SiNPs can be loaded with chemotherapeutic drugs, immunotherapeutic agents, such as CRISPR-Cas9, allowing for a synergistic approach to attacking cancer cells from multiple angles. For instance, SiNPs can simultaneously carry and RNA molecules to inhibit oncogene expression or enhance immune responses (Ma X., et al, 2024). This combination therapy the helps overcome resistance mechanisms that may develop when using a single treatment modality. Additionally, SiNPs can be functionalized with immunotherapeutic improving. Photothermal and Photodynamic Therapy: SiNPs have also gained attention for their potential use In SiNPs can be loaded with materials that allowing selective cancer cells specific wavelengths of. Silicon nanowires and silica-coated gold nanoparticles are commonly used in PTT because they efficiently absorb light and generate heat, which induces localized hyperthermia in the tumor tissue, causing tumor cell death. In photodynamic therapy, SiNPs can be loaded with photosensitizers, which are produce cancer cells. The combination of silicon nanostructures with photosensitizers offers the potential for highly localized and precise tumor treatment, as the therapeutic effect is only triggered in the presence of light at the tumor site. Imaging and Diagnostic Capabilities: A significant advantage of SiNPs in cancer therapy is their multifunctionality, as not only but also for and diagnosis (Sugumaran K M D., et al. 2024). SiNPs loaded or radionuclides, enabling monitoring nanoparticles' location, distribution, therapeutic efficacy. Fluorescence Imaging: SiNPs can be conjugated with fluorophores, allowing for non-invasive fluorescence imaging of tumors. This helps to track SiNPs tumor assess effectiveness delivery, monitor tumor response during therapy. Magnetic Resonance Imaging (MRI): SiNPs can be functionalized with magnetic nanoparticles, enabling their use in MRI for high-resolution imaging of tumor tissue. This is especially useful in tracking the biodistribution of SiNPs and determining how well the nanoparticles are reaching the tumor and surrounding tissues (Yu X., et al, 2024). Furthermore, biodegradability SiNPs ensures that they do not accumulate in further Once the SiNPs have delivered their therapeutic payload, they degrade into harmless byproducts, which are eliminated via the kidneys. Enhanced Drug Delivery Efficiency: SiNPs allow for the improving efficiency Drugs encapsulated within SiNPs or conjugated to their surface, ensuring a slow and sustained release of the therapeutic payload sustained enhances efficacy the drug by levels tumor thus increasing chances tumor cell death while minimizing systemic toxicity.

THE DELIVERY OF THERAPEUTIC AGENTS WITH SILICON-BASED NANOPARTICLES

The treatment significant nanotechnology, particularly the use of silicon-based nanoparticles (SiNPs). These nanoparticles have emerged as promising and precisely harm tissues, represents SiNPs offer unique advantages in cancer treatment and potential for multifunctionality. However, despite their promise, SiNPs involves a range strategies and challenges that need to be addressed for effective therapeutic outcomes. This article explores the various strategies using SiNPs and the to maximize utility. The Importance of Targeting in Cancer Therapy: Traditional suffer from lack specificity, which leads to significant toxicity and side effects in healthy tissues (Yan J., et al, 2024). Reduced Toxicity: By directing therapeutic agents specifically to cancer cells, SiNPs can reduce systemic toxicity treatments. Increased Efficacy: Targeting cancer cells directly ensures in higher concentrations, improving the therapeutic efficacy of treatments (Wang J., et al, 2024). The targeting of cancer cells with SiNPs nanoparticles to allow for selective interaction with tumor cells (Duan W., et al, 2024). This allows passive genes, or imaging tumor, reducing need for active targeting. Short peptides, such as the RGD (Arg-Gly-Asp) sequence, internes peptides have the advantage of being smaller and more cost-effective compared to antibodies while still offering high specificity for their targets. Aptamers: can into three-dimensional structures tightly to offer specificity selectivity, similar to antibodies, but are smaller, more stable, and easier to synthesize (Johri S., et al. 2025). A more advanced approach to targeting cancer cells nanoparticles. nanoparticles therapeutic cargo stimuli that are characteristic that can be used for responsive targeting include: pH-Responsive Release: with extracellular pH values ranging from 6.0 to 7.0 compared to SiNPs functionalized with linkers or coatings therapeutic conditions in. This only, reducing systemic toxicity. Enzyme-Responsive Release: Tumors specific or cathepsins, which play key roles in tumor progression and metastasis (Tiwari R., et al, 2024).

THE USE OF SILICON-BASED NANOPARTICLES IN BIOIMAGING

The target with has been most modern Over last few decades, nanotechnology has provided significant advancements in this area, offering new opportunities genes, among various nanomaterials being explored for drug delivery, silicon-based nanoparticles (SiNPs) have their physical, chemical, biological SiNPs biocompatibility, biodegradability, surface modification, multiple simultaneously, making systems. However, while potential SiNPs in drug delivery is immense, there are still several challenges and limitations these nanoparticles fully realized (Zuo Y., et al, 2021). This article will explore the opportunities and limitations associated with the use of SiNPs. Opportunities for the Delivery of Therapeutic Agents with Silicon-Based Nanoparticles: Targeted Delivery and Site-Specific Accumulation SiNPs deliver therapeutic agent's specific sites in the body. often hindered a lack of specificity, which results in widespread affecting both healthy and diseased tissues. This lack of precision reduced therapeutic efficacy. SiNPs, however, tissues or several mechanisms: Passive Targeting via the Enhanced Permeability and Retention (EPR) Effect: nanoparticles to accumulate preferentially in these areas, where the target tissue. SiNPs take advantage of this effect to accumulate more readily in tumors compared to normal tissues, enhancing (Alagarasan J K., et al, 2023). Active Targeting Using Surface Functionalization: SiNPs specific making folate-functionalized SiNPs an effective way to selectively target and treat tumors. Similarly, SiNPs can be conjugated with monoclonal antibodies or peptides to target other cancer-related antigens, therapeutic only cancer while off-target effects. For example, SiNPs release the drug response to tumors or inflamed tissues. Similarly, incorporated into SiNPs trigger drug under hyperthermic conditions, making the system responsive to external stimuli. This type of release mechanism can enhance therapeutic effect drug by ensuring that it is only released in the desired location.

Biodegradability and Reduced Toxicity: Another significant opportunity for SiNPs in drug delivery is their biodegradability. Unlike many other nanoparticles made from materials like gold, silver, or carbon, SiNPs are primarily composed of silicon dioxide (SiO_2), which is naturally biocompatible and breaks down into non-toxic byproducts, such as silicic acid. This biodegradability minimizes the risk of accumulation ensures that they are safely eliminated. ability SiNPs to degrade into non-toxic components also reduces the potential for chronic toxicity, which is a major concern for many nanoparticle-based drug delivery systems (Pudakalakatti S., et al, 2021). After the drug has been released at the target site, SiNPs can undergo gradual degradation and be excreted via the kidneys, ensuring that the body is not burdened with residual nanoparticles. Limitations of Silicon-Based Nanoparticles in Drug Delivery: While SiNPs hold tremendous promise for drug delivery, and limitations addressed fully issues related to biodistribution, immune system interactions, scalability, regulatory concerns, and patient-specific variations in response. Heterogeneity of Tumors and Targeting Efficiency: One of the primary challenges in SiNPs is heterogeneity tumors (Sreenivasulu A., et al. 2022). SiNPs, like other nanoparticles, may trigger an immune response in the body. When nanoparticles are introduced into the bloodstream, they can be recognized by the reticuloendothelial system (RES), which is responsible for clearing foreign particles from the body (Bertão, 2024).

THE IMAGING OF CANCER CELLS WITH SILICON-BASED NANOPARTICLES

Bioimaging modern medicine, non-invasive visualization biological processes in living organisms. This technology has revolutionized diagnostics, therapeutic monitoring, and drug discovery by providing real-time images of tissues, organs, and molecular interactions. In recent years, silicon-based nanoparticles (SiNPs) have garnered significant attention as promising materials for bioimaging applications. These nanoparticles, made primarily from silicon dioxide (SiO_2) or silicon-based compounds, offer distinct advantages over traditional imaging agents such as organic dyes and metals, suitable imaging modalities, delves into principles behind the use of SiNPs in bioimaging, explores across the opportunities associated with their use in medical and preclinical research (Chu B., et al, 2021). Principles of Silicon-Based Nanoparticles in Bioimaging: use SiNPs in bioimaging relies on can be tailored optimize imaging performance. These properties include their size, surface characteristics, optical properties, and magnetic properties, all of which can be fine-tuned to enhance image quality, sensitivity, and specificity. Size and Surface Modifications: SiNPs their use in bioimaging. Nanoparticles typically range size range penetrate barriers, while remaining small enough rapid immune SiNPs in this size range can also facilitate deep tissue penetration and are small enough to accumulate at inflamed. Surface SiNPs can be modified with various functional groups, which allow biomolecules, drugs, or fluorescent dyes. These modifications enable the nanoparticles to be targeted to specific tissues or cells, thereby specificity Surface modifications can also help to enhance the biocompatibility of SiNPs, reducing the risk of immune system activation and promoting longer circulation times in vivo (Karaman D Ş., et al, 2021).

Magnetic Properties: While SiNPs are not inherently magnetic, magnetic nanoparticles can be synthesized by incorporating superparamagnetic iron oxide nanoparticles (SPIONs) or magnetic metals into SiNPs. This modification enhances the magnetic resonance imaging (MRI) contrast of SiNPs, enabling more precise imaging of tissues and organs. Magnetic SiNPs are used in MRI contrast agents to improve image resolution, contrast, and sensitivity, particularly in soft tissue imaging, where distinguishing between different types of tissues can be challenging. The superparamagnetic of these particles means that they can be manipulated using external magnetic fields, which can assist in the localization and targeting of SiNPs ischemic regions (Huang S., et al, 2024). SiNPs great imaging due to be tailored for tumor-specific targeting and their enhanced optical properties. Drug Delivery Monitoring: exciting SiNPs bioimaging their role monitoring as carriers for chemotherapeutic agents or gene therapy, SiNPs can be tracked in real-time to monitor the distribution and release of drugs in vivo (Wei D., et al, 2024). For example, SiNPs loaded with chemotherapeutic drugs can be conjugated with fluorescent probes enabling researchers' clinicians to track the nanoparticles' biodistribution and determine whether the drug reaches the intended target, such as a tumor site. This approach can also be used to monitor assess effectiveness systems, adjust dosing schedules accordingly.

THE USE OF SILICON-BASED NANOPARTICLES IN OPTICAL IMAGING

The imaging of cancer cells plays a critical role diagnosis, staging, cancer progression, evaluating response to interventions. Traditional scans, and have made significant contributions to the management of cancer; however, they often lack the sensitivity, specificity, and resolution necessary for detecting early-stage tumors or monitoring dynamic changes in tumor microenvironments. To address these limitations, the integration of nanotechnology into cancer imaging has become a revolutionary approach (Lian Y., et al, 2023). Among the various nanomaterials being explored for this purpose, silicon-based nanoparticles (SiNPs) stand out due to their biocompatibility, ease of functionalization, and optical properties, making them highly effective for both early detection and real-time tracking of cancer cells. SiNPs are unique in their ability to be engineered wide techniques, positron emission (PET), fluorescence The versatility SiNPs stems from their customizable allowing them to be tailored for different imaging modalities. The of SiNPs can be modified with various functional groups that enable them to biomarkers receptors overexpressed cancer cells. antibodies, aptamers, functionalize SiNPs, allowing them to selectively target specific types of cancer cells (Mohajer F., et al, 2023)..

Positron Emission Tomography (PET) and Radioactive Labeling: addition to their optical and magnetic properties, SiNPs can also be functionalized with radioactive isotopes to enable positron emission tomography (PET). PET imaging offers high sensitivity for detecting small tumors or metastatic lesions in preclinical and clinical settings, making it an essential cancer treatment. Radioactive isotopes, fluorine-18 or copper-64, can be conjugated to SiNPs, allowing for in vivo PET imaging of cancer cells (Long Z., et al, 2025). This technique provides quantitative data on concentration, metabolism SiNPs valuable information about tumor biology and the effectiveness of therapeutic interventions. The use of SiNPs in combination with PET can help track drug delivery to cancer cells, assess tumor size and growth, and monitor the response to therapy. The ability to combine PET imaging with optical or magnetic imaging further enhances the diagnostic capabilities of SiNPs, offering a more complete picture of tumor dynamics (Aragani, V M., et al. 2025). These benefits include increased sensitivity, targeted imaging, non-invasive monitoring, and multimodal imaging capabilities, all of which improve the detection, staging, and treatment of cancer. Enhanced Sensitivity and Specificity: SiNPs is improve sensitivity and specificity cancer cell imaging. Traditional imaging modalities may struggle to identify early-stage cancers or small tumor deposits due to their limited resolution. SiNPs, however, can be engineered to interact specifically with cancer cell receptors, ensuring that they accumulate at the tumor site and provide high contrast in imaging studies. This targeted approach improves the detection of cancer cells even at very low concentrations, making SiNPs particularly useful for early diagnosis of cancer (Thapa A., et al, 2025).

THE CHALLENGES AND LIMITATIONS OF SILICON-BASED NANOPARTICLES IN CANCER THERAPY AND BIOIMAGING

Optical imaging has emerged as one of the most powerful and widely used techniques for non-invasive, high-resolution, real-time visualization of biological systems. It allows for the observation of cellular and molecular processes with high spatial and temporal resolution. Among the materials being increasingly explored for optical imaging, silicon-based nanoparticles (SiNPs) significant remarkable properties, biocompatibility, potential for functionalization. SiNPs, particularly those with photoluminescent properties, hold promise biomedical molecular imaging to delivery tracking and therapeutic monitoring. However, despite their potential, there remain harness advantages SiNPs in optical imaging. This article explores the opportunities and silicon-based nanoparticles optical imaging, providing their advancing medical imaging technologies (Bharadwaj K K., et al, 2021). Optical Properties of Silicon-Based Nanoparticles: optical properties SiNPs, particularly their ability to exhibit photoluminescence, make them highly attractive for optical imaging applications. These properties are largely dictated by shape, characteristics specific chemical physical modifications applied during synthesis. The key features of SiNPs that make them suitable for optical imaging include their size-tunable fluorescence, high quantum yield, low photobleaching, and biocompatibility. Size-Tunable Fluorescence: One of the most important characteristics of SiNPs is their ability to exhibit size-tunable fluorescence. The NIR range is particularly valuable, as light in this region can penetrate deeper into biological tissues with reduced scattering and absorption, offering the potential for deep tissue imaging. NIR fluorescence can also reduce the background signal interference from endogenous tissue autofluorescence, improving the signal-to-noise ratio and enhancing image clarity (Bahadur P S., et al, 2024).

Opportunities in Optical Imaging with Silicon-Based Nanoparticles: SiNPs offer a wide range of opportunities in optical imaging, spanning across diagnostic applications, real-time monitoring, multimodal imaging, and both imaging therapy nanoparticle system. These opportunities are largely driven by the unique properties of SiNPs, which combine fluorescence with biocompatibility and functionalization versatility (Omidian H., et al, 2024). Cancer Diagnosis and Monitoring: promising SiNPs in optical cancer and monitoring. Cancerous cells often overexpress specific biomarkers on their surface, which can be targeted by SiNPs functionalized with corresponding targeting ligands. This selective binding allows tumors or metastatic lesions, even at early stages when the tumor mass is too small to be detected by conventional imaging techniques. For example, SiNPs conjugated with antibodies targeting tumor-associated antigens can be used in in vivo fluorescence imaging to monitor tumor growth, angiogenesis, and tumor response to therapy. Additionally, NIR fluorescence imaging can be employed for real-time tracking of SiNPs in animal models, providing insights into tumor behavior and the effectiveness of therapeutic interventions (Mahadev Madgule, et al. 2023). Drug Delivery and Tracking: SiNPs are also being investigated for use systems carry therapeutic agents deliver them directly to target tissues. By loading SiNPs with chemotherapeutic drugs, gene therapies, or small-molecule inhibitors, researchers can monitor the biodistribution and drug release in real-time using optical imaging. clinicians and track location concentration drug-loaded SiNPs in vivo, thereby optimizing dosing regimens and enhancing treatment efficacy. Furthermore, SiNPs targeting ligands target cancer infected specifically, thereby systemic typically traditional chemotherapy (Sadighian S., 2025).

THE IMPORTANCE OF BIOCOMPATIBILITY AND BIODEGRADABILITY IN SILICON-BASED NANOPARTICLES

Silicon-based nanoparticles (SiNPs) attention fields therapy and bioimaging remarkable optical properties, ease of functionalization, and potential for targeted delivery. These nanoparticles have shown great promise for enhancing cancer diagnosis, improving the specificity of enabling tumor dynamics. Despite their potential, the widespread clinical application of SiNPs faces a number of challenges and limitations. Among the most critical concerns are biocompatibility, safety SiNPs in living organisms, especially when used for prolonged periods or at high concentrations. As SiNPs move from laboratory studies to clinical trials, issues essential developing effective, reliable SiNP-based technologies (Bellanthudawa & Nawalage, 2023). Size-Dependent Toxicity: SiNPs crucial toxicity profile. are generally preferred for medical applications because they can efficiently enter cells and tissues. However, smaller nanoparticles (below 10 nm) can be more increased potential for in various organs. Their small size allows them, including and cell membranes, but it also increases the likelihood of renal toxicity as they are cleared by the kidneys (Shi D., et al, 2024). The size-dependent toxicity is particularly concerning in prolonged treatments where SiNPs may accumulate over time, increasing toxic. Surface Charge and Functionalization Effects: SiNPs, surface the functional groups used for surface modification, significantly impact their toxicity. PEGylation other biocompatible coatings are commonly used to reduce immune recognition, but the long-term impact of these modifications on immune tolerance and response is still an area of ongoing research (Da Cruz Schneid & Albuquerque, 2022). Additionally, inflammatory responses triggered by SiNPs may complicate their use in cancer therapy, as inflammation can promote tumor progression and metastasis. Research on long-term biocompatibility and clearance of SiNPs While shown that SiNPs can be excreted through urine or feces, others have indicated that they may persist in tissues, leading to potential long-term adverse effects (Ostroff E., et al, 2022). Thus, the development of biodegradable SiNPs that degrade into non-toxic products is an important area of research prolonged.

FUTURE DIRECTIONS IN SILICON-BASED NANOPARTICLES FOR CANCER THERAPY AND BIOIMAGING

Silicon (SiNPs) has significant attention in fields particularly bioimaging, cancer therapy. combination ease of functionalization, optical characteristics, makes them highly promising for these applications. However, in the SiNPs is biodegradability. The of nanoparticles humans requires them to be biocompatible, meaning they must not induce harmful effects when introduced into the body, and biodegradable, meaning they must degrade into non-toxic products and be eliminated without causing. As such, biocompatibility and biodegradability of SiNPs are fundamental requirements that must be carefully considered and optimized these nanoparticles medical applications. article explores the importance of these two properties in SiNPs, the requirements for their clinical use, and the challenges that arise in ensuring their safe and effective application (Wegner K D., et al, 2024). Biocompatibility: The Foundation of Safe Nanoparticles: Biocompatibility desired inducing harmful effects on the biological system. For SiNPs, biocompatibility is critical, as these particles will often come into direct contact with cells, tissues, and organs, and any adverse reactions could lead to complications in the treatment or imaging process. interaction SiNPs to minimize such as toxicity, inflammation, immune responses, and organ damage. Cellular Interaction and Toxicity: The biocompatibility of SiNPs is largely determined by their Upon introduction body, SiNPs interact with and as well as circulating proteins. The size, modifications of SiNPs critical they interact with Positively charged SiNPs, for instance, may interact more readily potentially greater cellular and increased risk of cytotoxicity. Similarly, SiNPs with unmodified or hydrophobic surfaces might aggregate or bind to plasma proteins, which could trigger the complement system or phagocytosis, leading to immune activation and inflammatory responses. The surface functionalization of SiNPs is a key strategy for improving their biocompatibility. By attaching biocompatible polymers like poly-ethylene glycol (PEG), or using surface coatings such as silane-based agents (Zhou

Z., et al, 2024), SiNPs reduce non-specific binding, minimize immune recognition, and prevent the nanoparticles from aggregating.

Degradation Mechanisms of Silicon Nanoparticles: SiNPs are generally considered to be biodegradable, especially when compared to silver. The degradation of SiNPs typically involves the release of silica or silicic acid, which are generally regarded as non-toxic to the body at low concentrations. Silicon itself is an abundant and essential element in living organisms, and silicic acid is naturally found in small quantities in the human body. The breakdown of SiNPs usually occurs through processes such as oxidation, hydrolysis, and dissolution in biological environments, leading to the release of non-toxic silica and its excretion via the urinary system (Dave S., et al, 2024). While the degradation of SiNPs into soluble silica is generally considered safe, the rate of degradation is a critical factor. SiNPs must degrade at a controlled rate to ensure that they are cleared from the body before they accumulate in tissues and cause harm. Too slow a degradation rate could buildup liver, kidneys, while too rapid a degradation rate could result in the premature release of toxic byproducts or drug payloads. Therefore, the degradation kinetics of SiNPs must be carefully engineered to match the specific application, whether it involves drug release or imaging, while ensuring minimal risk of toxicity. Challenges in Biodegradability: Despite the advantages of biodegradable SiNPs, there are several challenges in ensuring their safe and effective breakdown in the body. One of the primary challenges is which can the nanoparticle's size, surface area, and functionalization (Ruan Y., et al, 2022).

RECOMMENDATIONS FOR RESEARCHERS AND CLINICIANS

Silicon-based nanoparticles (SiNPs) are at the forefront of transformative advancements in nanomedicine, especially in the realms of cancer therapy and bioimaging. designing, synthesizing, and characterizing SiNPs for a variety of biomedical applications. As research continues to evolve, new trends, technologies, and opportunities are emerging that promise to further enhance the potential of SiNPs in targeted cancer therapy, diagnostics, and real-time bioimaging. This article aims to explore these exciting developments, shedding light on the cutting-edge approaches and the future directions in SiNP-based cancer treatment and bioimaging (Sun D., 2024). Emerging Trends in Silicon-Based Nanoparticles for Cancer Therapy: use SiNPs steadily gained traction due to their ability to be engineered for targeted drug delivery, photothermal therapy, radiotherapy enhancement, and gene delivery. Moving forward, several emerging trends are shaping the development of SiNPs as efficient, precise, and less invasive cancer therapies. Precision and Targeted Therapy through Surface Functionalization: One of the most promising directions in SiNPs for cancer therapy lies in targeted drug delivery. Over the past decade, the ability to conjugate to SiNP surfaces has significantly improved the precision of therapeutic delivery. These modifications enable SiNPs cancer improving the bioavailability of chemotherapeutic agents while minimizing side effects in healthy tissues (Su Y., et al, 2021). In the future, there is a growing emphasis on the development of multifunctional SiNPs that can simultaneously deliver drugs, monitor disease progression, and potentially even stimulate immune responses.

Advancements in Bioimaging with Silicon-Based Nanoparticles: In addition to their potential in therapy, SiNPs bioimaging, allowing for enhanced Over the past decade, significant strides have been made in improving the optical properties of SiNPs, making them more suitable techniques, fluorescence, luminescence, photoacoustic. Multifunctional Imaging and Theranostics: One of the key trends in bioimaging nanoparticles performing both imaging therapeutic functions simultaneously. These theranostic while delivering drugs or performing other therapeutic activities. SiNPs applications versatility in surface functionalization, optical properties, and biocompatibility. The development of multimodal imaging systems that combine different is another exciting direction. For example, SiNPs that are functionalized with paramagnetic agents could enhance MRI contrast, while simultaneously emitting fluorescent signals for optical imaging, allowing for comprehensive tumor imaging with high spatial resolution (Sobhanan J., et al, 2023). This multimodal approach would provide more accurate and detailed information about tumor morphology, location, and response to treatment. Quantum Dots and Photonic SiNPs for Enhanced Imaging: Recent advancements have development SiNPs, which can emit light at specific wavelengths depending on their size. These SiNPs exhibit superior fluorescence properties, including high brightness, photostability, and the ability to be excited by a single light source. Quantum dot-based SiNPs can provide enhanced imaging signals, tumors. In addition to fluorescence, photoacoustic imaging (PAI) is becoming an increasingly important technique for deep tissue imaging. SiNPs have the potential to act as photoacoustic agents, absorbing laser light and converting it into ultrasound waves that can be used for non-invasive imaging of tumors (Li R., et al, 2025).

CONCLUSION

silicon-based (SiNPs) bioimaging holds tremendous potential for advancing modern medicine. SiNPs offer unique advantages, including biocompatibility, optical properties, surface functionalization, from targeted and gene to real-time imaging of tumor sites. However, the clinical translation of these nanomaterials requires meticulous design, rigorous testing, and a deep understanding of both scientific and regulatory complexities. In order to maximize the therapeutic and diagnostic potential of SiNPs, researchers and clinicians must adhere to best practices and established guidelines that ensure both the efficacy and safety of these novel technologies. This article outlines key recommendations for researchers and clinicians' application SiNPs for therapy bioimaging. Ensuring Safe and Effective Nanoparticle Design: The foundation of any SiNP-based approach lies in the design and synthesis of the nanoparticles themselves. Researchers must carefully consider several factors that influence both the therapeutic performance and biocompatibility of SiNPs. These include chemistry, surface SiNPs biological interactions and tissue distribution. Smaller nanoparticles generally have better cellular uptake and enhanced tumor penetration, while larger nanoparticles may be more suited for imaging applications or for carrying larger payloads of therapeutic agents. In Vitro Safety Testing: Before progressing to animal studies, researchers should conduct thorough in vitro assays to assess the cytotoxicity and biocompatibility of SiNPs. This includes evaluating cell viability, apoptosis induction, cellular uptake, and immune activation. Standard assays such as the MTT assay (for cell viability) and flow cytometry (for apoptosis) should be performed to determine the impact of SiNPs on different cell lines, including cancer cells, immune cells, and healthy tissues. The interaction of SiNPs with serum proteins and their potential to induce inflammatory responses or activate the complement system should also be evaluated using appropriate models. In addition, researchers must carefully assess the surface charge of SiNPs, as positively charged particles are more likely to cause cellular membrane disruption, leading to cytotoxicity. Neutral or slightly negatively charged SiNPs are often preferred for biocompatible applications, as they tend to minimize non-specific interactions with healthy cells. In Vivo Toxicity and Biodistribution Studies: After initial testing, are critical understanding biodistribution, clearance kinetics, and long-term toxicity of SiNPs. These studies involve administering SiNPs to animal models and tracking their distribution throughout the body, including. Researchers should assess both acute and chronic toxicity by evaluating biomarkers of organ function, histological tissue changes, and long-term survival. or optical imaging, be employed track SiNPs their clearance from body. Scalability and Manufacturing of SiNPs: Once the safety and efficacy of SiNPs have been demonstrated in preclinical studies, the next step is to scale up production for clinical trials and eventual commercialization. The reproducibility of SiNPs must be ensured at large scales, as uniformity in size, shape, and surface properties is critical for consistent therapeutic outcomes and imaging performance. Researchers should focus on developing cost-effective, reproducible, and scalable synthesis methods that can produce SiNPs in large quantities with high purity and biocompatibility. Standardization of Synthesis Methods: Researchers should prioritize the use of well-defined synthesis protocols to ensure batch-to-batch consistency in the properties of SiNPs. laser ablation, microemulsion synthesis are common techniques for producing SiNPs, each with its own advantages and limitations in terms of size control, scalability, and reproducibility.

REFERENCES

Alagarasan, J. K., Shasikala, S., & Ganesan, S. (2023). Silicon Nanoparticles as a Fluorometric Probe for Sensitive Detection of Cyanide Ion and Its Application in C. elegans Bio-Imaging. *Environmental Research*, *224*, 115402. DOI: 10.1016/j.envres.2023.115402

Aragani, V. M., Anumolu, V. R., & Selvakumar, P. (2025). Democratization in the Age of Algorithms: Navigating Opportunities and Challenges. In Wongmahesak, K., Wekke, I., Seftyono, C., & Nurdin, N. (Eds.), *Democracy and Democratization in the Age of AI* (pp. 39–56). IGI Global Scientific Publishing., DOI: 10.4018/979-8-3693-8749-8.ch003

Bahadur, P. S., Shameer, P. M., Sharma, N., & Manjunath, T. C. (2024). "Nano Technology: The Future of Innovation." Books.Google. *Com*, *1*, 567–589. DOI: 10.1007/google.nanotechnology2024

Bellanthudawa, B. K. A., & Nawalage, N. (2023). A Perspective on Biodegradable and Non-Biodegradable Nanoparticles in Industrial Sectors: Applications, Challenges, and Future Prospects. *Nanotechnology for Environmental Engineering*, *8*(4), 975–1013. DOI: 10.1007/s41204-023-00344-7

Bertão, A. R. C. (2024). Silicon-Based Nanostructures as Drug Delivery Systems for Biological and Medical Applications. *Universidade do Minho Repository*, *7*, 92402. DOI: 10.1822/92402

Bharadwaj, K. K., Rabha, B., Pati, S., & Sarkar, T. (2021). Green Synthesis of Gold Nanoparticles Using Plant Extracts as Beneficial Prospect for Cancer Theranostics. *Molecules (Basel, Switzerland)*, *26*(4), 123–145. DOI: 10.3390/mdpi.goldnanoparticles2021

Chaudhari, K. N., Rajeev, K. K., Kim, S., & Nersisyan, H. (2022). Performance Enhancement of Carbon-Coated Si Nanoparticles for Lithium-Ion Batteries Through the Generation of Lithophilic Sites by a Simple Oxidation Process. *Applied Surface Science*, *601*, 154361. DOI: 10.1016/j.apsusc.2022.154361

Chen, L., Zhang, S., Duan, Y., Song, X., & Chang, M. (2024). Silicon-Containing Nanomedicine and Biomaterials: Materials Chemistry, Multi-Dimensional Design, and Biomedical Application. *Chemical Society Reviews*, *53*(3), 567–589. DOI: 10.1039/d1cs01022k

Cheng, Z., Jiang, H., Zhang, X., & Cheng, F. (2023). Fundamental Understanding and Facing Challenges in Structural Design of Porous Si-Based Anodes for Lithium-Ion Batteries. *Advanced Functional Materials*, *33*(4), 789–812. DOI: 10.1002/adfm.202300456

Chu, B., Wang, H., & He, Y. (2021). Fluorescent Silicon-Based Nanomaterials Imaging Technology in Diseases. *Chemical Research in Chinese Universities*, *37*(2), 123–145. DOI: 10.1007/springer.nanomaterials2021

Da Cruz Schneid, A., & Albuquerque, L. J. C. (2022). Colloidal Stability and Degradability of Silica Nanoparticles in Biological Fluids: A Review. *Journal of Sol-Gel Science and Technology*, *102*(1), 41–62. DOI: 10.1007/s10971-021-05695-8

Daulay, A., & Gea, S. (2022). Synthesis and Application of Silicon Nanoparticles Prepared from Rice Husk for Lithium-Ion Batteries. *Case Studies in Chemical and Environmental Engineering*, *6*, 100256. DOI: 10.1016/j.cscee.2022.100256

Dave, S., Das, J., & Sillanpää, M. (2024). "Nanomaterials and Point of Care Technologies." Books.Google. *Com*, *1*, 567–589. DOI: 10.1007/google.nanotechnology2024

Debnath, P., Deshmukh, S. M., Laroia, M., Selvakumar, P., Manjunath, T. C., & Bhattacharya, S. (2025). Challenges Faced by Countries in Renewable Energy Adoption. In Özsungur, F., Chaychi Semsari, M., & Küçük Bayraktar, H. (Eds.), *Geopolitical Landscapes of Renewable Energy and Urban Growth* (pp. 295–320). IGI Global Scientific Publishing., DOI: 10.4018/979-8-3693-8814-3.ch010

Díaz-Rodríguez, T. G., Antunez, E. E., & Celaya, C. A. (2025). Exploring Contaminants and Analyte Adsorption on Functionalized Porous Silicon: Insights from a Combined Theoretical and Experimental Approach. *Surfaces and Interfaces*, *30*, 100456. DOI: 10.1016/j.surfin.2025.100456

Duan, W., Zhao, J., Gao, Y., Xu, K., Huang, S., & Zeng, L. (2024). Porous Silicon-Based Sensing and Delivery Platforms for Wound Management Applications. *Journal of Controlled Release*, *350*, 123–145. DOI: 10.1016/j.jconrel.2024.012345

Gavas, S., Quazi, S., & Karpiński, T. M. (2021). Nanoparticles for Cancer Therapy: Current Progress and Challenges. *Nanoscale Research Letters*, *16*(1), 173–190. DOI: 10.1186/s11671-021-03628-6

Ge, M., Cao, C., Biesold, G. M., & Sewell, C. D. (2021). Recent Advances in Silicon-Based Electrodes: From Fundamental Research Toward Practical Applications. *Advanced Materials*, *33*(12), 1234–1250. DOI: 10.1002/adma.202100123

Huang, S., & Huang, G. (2024). The Utilization of Quantum Dot Labeling as a Burgeoning Technique in the F ield of Biological Imaging. *RSC Advances*, *14*(3), 789–812. DOI: 10.1039/rsc.quantumdots2024

Jin, H., Wang, L., & Bernards, R. (2023). Rational Combinations of Targeted Cancer Therapies: Background, Advances and Challenges. *Nature Reviews. Drug Discovery*, *22*(3), 215–230. DOI: 10.1038/s41573-022-00615-z

Johri, S., Selvakumar, P., Satyanarayana, P., Das, A., Gupta, S., & Bhattacharya, S. (2025). Urbanization and Energy: The Nexus of Global Challenges. In Özsungur, F., Chaychi Semsari, M., & Küçük Bayraktar, H. (Eds.), *Geopolitical Landscapes of Renewable Energy and Urban Growth* (pp. 453–478). IGI Global Scientific Publishing., DOI: 10.4018/979-8-3693-8814-3.ch016

Karaman, D. Ş., Kaasalainen, M., & Kettiger, H. (2021). Opportunities and Challenges of Silicon-Based Nanoparticles for Drug Delivery and Imaging. *Characterization of Nanoparticles*, *12*(4), 567–589. DOI: 10.1002/wiley.nanoparticles2021

Kim, T., & Lee, J. (2023). Silicon Nanoparticles: Fabrication, Characterization, Application and Perspectives. *Micro and Nano Systems Letters*, *11*(3), 45–67. DOI: 10.1007/springer.nanoparticles2023

Labrie, M., Brugge, J. S., & Mills, G. B. (2022). Therapy Resistance: Opportunities Created by Adaptive Responses to Targeted Therapies in Cancer. *Nature Reviews. Cancer*, *22*(6), 323–339. DOI: 10.1038/s41568-022-00454-5

Lai, Y., Li, H., Yang, Q., Li, H., Liu, Y., & Song, Y. (2022). Revisit the Progress of Binders for a Silicon-Based Anode from the Perspective of Designed Binder Structure and Special Sized Silicon Nanoparticles. *Industrial & Engineering Chemistry Research*, *61*(12), 567–589. DOI: 10.1021/acs.iecr.2c01234

Li, R., Xu, J., Mu, X., & Zeng, F. (2025). A Comprehensive Review on the Synthesis Methods and Applications of Silicon Quantum Dots (SiQDs). *Next Nanotechnology*, *10*(2), 567–589. DOI: 10.1016/j.nextnano.2025.123456

Lian, Y., Sun, J., & Jiang, L. (2023). Probing Electron and Lattice Dynamics by Ultrafast Electron Microscopy: Principles and Applications. *International Journal of Mechanical Sciences*, *14*(2), 567–589. DOI: 10.1002/wiley.ultrafast2023

Liu, J., Yao, Y., Li, X., & Zhang, Z. (2021). Fabrication of Advanced Polydimethylsiloxane-Based Functional Materials: Bulk Modifications and Surface Functionalizations. *Chemical Engineering Journal*, *408*, 127262. DOI: 10.1016/j.cej.2020.127262

Long, Z., Zhou, Y., Ding, Y., Qiu, X., & Poddar, S. (2025). Biomimetic Optoelectronics with Nanomaterials for Artificial Vision. *Nature Reviews. Materials, 10*(1), 123–145. DOI: 10.1038/nature.optoelectronics2025

Ma, X., Tian, Y., Yang, R., Wang, H., & Allahou, L. W. (2024). Nanotechnology in Healthcare, and Its Safety and Environmental Risks. *Journal of Nanotechnology, 15*(3), 345–367. DOI: 10.1007/springer.nano2024

Madgule, M., Vinayaka, N., Sonkhaskar, Y. M., Patel, D. R., Karthikeyan, R., & Selvakumar, P. (2023). Mechanical properties and microstructure of activated TIG welded similar joints of Inconel alloys by desirability approaches. *Materials Today: Proceedings, 77*, 528–533.

Mishra, T., Sengupta, P., & Basu, S. (2024). Biomaterials for Targeting Endoplasmic Reticulum in Cancer. *Chemistry, an Asian Journal, 19*(2), 567–589. DOI: 10.1002/asia.202400123

Mohajer, F., & Mirhosseini-Eshkevari, B. (2023). Advanced Nanosystems for Cancer Therapeutics: A Review. *ACS Applied Nano Materials, 5*(4), 789–812. DOI: 10.1021/acs.nano2023

Mohammadi, J., Hheidari, A., Sardari, S., & Nouri, M. (2024). A Comprehensive Review of Lessons Learned from Quantum Dots in Cancer Therapy. *Biomedical Materials (Bristol, England), 19*(1), 1–25. DOI: 10.1088/1748-605X/19/1/012345

Omidian, H., & Wilson, R. L. (2024). Polydopamine Applications in Biomedicine and Environmental Science. *Materials (Basel), 15*(3), 123–145. DOI: 10.3390/mdpi.polydopamine2024

Ostroff, E., Parekh, K., & Prominski, A. (2022). Biocompatible and Nanoenabled Technologies for Biological Modulation. *Advanced Materials Technologies, 7*(4), 567–589. DOI: 10.1002/admt.202100216

Patel, G. M., Bhatt, G. J., & Deota, P. T. (2022). Synthesis and Characterization of Silicon-Based Hybrid Nanoparticles. *Silicon-Based Hybrid Nanoparticles, 12*(3), 345–360. DOI: 10.1016/elsevier.hybrid2022

Pudakalakatti, S., & Enriquez, J. S. (2021). Hyperpolarized MRI with Silicon Micro and Nanoparticles: Principles and Applications. *Wiley Online Library, 408*, 127–161. DOI: 10.1007/978-3-662-56333-5_3

Rajput, V. D., Minkina, T., Feizi, M., Kumari, A., & Khan, M. (2021). Effects of Silicon and Silicon-Based Nanoparticles on Rhizosphere Microbiome, Plant Stress and Growth. *Biology (Basel), 10*(3), 567–589. DOI: 10.3390/biology10030567

Ramakrishnan, T., Mohan Gift, M. D., Chitradevi, S., Jegan, R., Hency Jose, P. S., Nagaraja, H. N., & Hailegiorgis, S. M. (2022). Study of numerous resins used in polymer matrix composite materials. *Advances in Materials Science and Engineering*, *2022*(1), 1088926.

Ruan, Y., Sohail, M., Zhao, J., Hu, F., & Li, Y. (2022). Applications of Material-Binding Peptides: A Review. *ACS Biomaterials Science & Engineering*, *8*(3), 567–589. DOI: 10.1021/acsbiomaterials.2c01234

Sadighian, S. (2025). Metal and Metal Oxide-Based Nanostructures for Drug Delivery. *Medical Nanobiotechnology*, *20*(1), 567–589. DOI: 10.1016/elsevier.nanobiotechnology2025

Shende, A. D., Tekade, S. A., Deshmukh, A. A., Tembhurkar, S. P., & Selvakumar, P. (2025). Water Quality Monitoring and Control With AI. In Ruiz-Vanoye, J., & Díaz-Parra, O. (Eds.), *Smart Water Technology for Sustainable Management in Modern Cities* (pp. 51–80). IGI Global Scientific Publishing., DOI: 10.4018/979-8-3693-8074-1.ch003

Shi, D., Narayanan, S., & Woeppel, K. (2024). Improving the Biocompatibility and Functionality of Neural Interface Devices with Silica Nanoparticles. *Accounts of Chemical Research*, *57*(3), 567–589. DOI: 10.1021/acs.accounts.4c00160

Sobhanan, J., Rival, J. V., Anas, A., & Shibu, E. S. (2023). Luminescent Quantum Dots: Synthesis, Optical Properties, Bioimaging and Toxicity. *Advanced Drug Delivery Reviews*, *191*, 567–589. DOI: 10.1016/j.addr.2023.123456

Somavarapu, S., Ramesh, B., & Venkatrayulu, C. (2021). Nanotechnology—A New Frontier in Medical Microbiology. *Nanotechnology for Biomedical Applications*, *12*(2), 45–67. DOI: 10.1007/springer.nanomed2021

Sreenivasulu, A., & Manimozhi, T. Satheesh kumar P, Selvakumar P. and Singh P, (2022). "The Model of Smart Sensing Device For Sensitive Nanoclusters Modification in Sensing Properties," 2022 2nd International Conference on Advance Computing and Innovative Technologies in Engineering (ICACITE), Greater Noida, India, 2022, pp. 1043-1047, doi: DOI: 10.1109/ICACITE53722.2022.9823846

Su, Y., Yu, B., & Wang, S. Cong H, Shen Y, (2021). "NIR-II Bioimaging of Small Organic Molecules." Biomaterials, Vol. 276, pp. 123–145. DOI: DOI: 10.1016/j.biomaterials.2021.123456

Sugumaran K M D, Pandiyarajan V Y B S, Selvakumar P. and M. S, (2024). "Analysis of Electromyography Signals for Control of Mechanical Prosthesis using Machine Learning Techniques," 2024 2nd International Conference on Intelligent Data Communication Technologies and Internet of Things (IDCIoT), Bengaluru, India, 2024, pp. 1059-1067, doi: .DOI: 10.1109/IDCIoT59759.2024.10467707

Sun, D. (2024). Development of Fluorescent-Based Nanoparticles for Bioimaging, Gene Regulation Expression, and Photo-Induced Synergistic Cancer Therapy. *ProQuest Dissertations & Theses*, *1*, 567–589. DOI: 10.1000/proquest.nanoparticles2024

Thapa, A., & Gao, H. (2025). High Areal Loading Silicon Nanoparticle-Based Lithium-Ion Batteries. *ACS Applied Materials & Interfaces*, *17*(3), 567–589. DOI: 10.1021/acs.materials2025

Tiwari, R., Kolli, M., Chauhan, S., & Yallapu, M. M. (2024). Tabletized Nanomedicine: From the Current Scenario to Developing Future Medicine. *ACS Nano*, *18*(4), 789–812. DOI: 10.1021/acs.nano.2024.012345

Vijayakumar, G., Rajkumar, M., Chandar, N. R., Selvakumar, P., & Duraisamy, R. (2024). Environmentally friendly TDS removal from waste water by electrochemical ion exchange batch-type recirculation (EIR) technique. *Environmental Science. Water Research & Technology*, *10*(4), 826–835.

Wang, J., Yu, Y., & Li, Y. (2024). Molecular Modeling in Drug Delivery. *Pharmaceutics–AI and Modeling in Pharma*, *18*(3), 567–589. DOI: 10.1002/wiley.pharma2024

Wang, Y., Zhou, Y., Wang, J., Zhang, L., Liu, C., & Guo, D. (2025). Nucleolin-Targeted Silicon-Based Nanoparticles for Enhanced Chemo-Sonodynamic Therapy of Diffuse Large B-Cell Lymphoma. *International Journal of Nanomedicine*, *20*, 123–145. DOI: 10.2147/IJN.S123456

Wegner, K. D., & Hildebrandt, N. (2024). Near Infrared Quantum Dots for Biosensing and Bioimaging. *Trends in Analytical Chemistry*, *157*, 123–145. DOI: 10.1016/j.trac.2024.123456

Wei, D., Sun, H., Zhang, M., Zhao, Y., & Yuan, H. (2024). Mapping the Technological Trajectory of Inorganic Nanomaterials in the Cancer Field. *Journal of Nanoparticle Research*, *26*(1), 123–145. DOI: 10.1007/springer.nanotechnology2024

Yan, J., Siwakoti, P., Shaw, S., Bose, S., & Kokil, G. (2024). Porous Silicon and Silica Carriers for Delivery of Peptide Therapeutics. *Drug Delivery and Translational Research*, *14*(2), 345–360. DOI: 10.1007/springer.drug2024

Yang, D., Cui, Z., Wen, Z., Piao, Z., He, H., & Wei, X. (2023). Recent Updates on Functionalized Silicon Quantum-Dot-Based Nanoagents for Biomedical Applications. *ACS Materials Letters*, *5*(4), 789–812. DOI: 10.1021/acs.matlett.3c01234

Yeom, S. J., Wi, T. U., Ko, S., & Park, C. (2022). Nitrogen Plasma-Assisted Functionalization of Silicon/Graphite Anodes to Enable Fast Kinetics. *ACS Applied Materials & Interfaces*, *14*(12), 5678–5690. DOI: 10.1021/acsami.2c01234

Yu, X., & Zhu, L. (2024). Nanoparticles for the Treatment of Bone Metastasis in Breast Cancer: Recent Advances and Challenges. *International Journal of Nanomedicine*, *19*(1), 567–589. DOI: 10.2147/IJN.S123457

Zhang, F., Zhu, W., Li, T., Yuan, Y., Yin, J., & Jiang, J. (2022). Advances of Synthesis Methods for Porous Silicon-Based Anode Materials. *Frontiers in Chemistry*, *10*(2), 123–145. DOI: 10.3389/fchem.2022.012345

Zhong, L., Li, Y., Xiong, L., Wang, W., & Wu, M. (2021). Small Molecules in Targeted Cancer Therapy: Advances, Challenges, and Future Perspectives. *Signal Transduction and Targeted Therapy*, *6*(1), 201–220. DOI: 10.1038/s41392-021-00572-w

Zhou, Z., Shi, P., Wang, C., Sun, Y., & Gao, C. (2024). Recent Updates in Nanoscale Delivery Systems of Platinum (IV) Antitumor Prodrugs. *Coordination Chemistry Reviews*, *472*, 567–589. DOI: 10.1016/j.ccr.2024.123456

Zuo, Y., Liang, X., Yin, J., Gou, Z., & Lin, W. (2021). Understanding the Significant Role of SiOSi Bonds: Organosilicon Materials as Powerful Platforms for Bioimaging. *Coordination Chemistry Reviews*, *447*, 214166. DOI: 10.1016/j.ccr.2021.214166

Compilation of References

Abbas A, Zehra S, Aslam R, Mobin M, (2023). "Functionalized Carbon Nanotubes for Biomedical Imaging: The Recent Advances". DOI: DOI: 10.1002/9783527835874.ch6

Acharya, R., Patil, T. V., Dutta, S. D., & Lee, J. (2024). Single-Walled Carbon Nanotube-Based Optical Nano/Biosensors for Biomedical Applications: Role in Bioimaging, Disease Diagnosis, and Biomarkers Detection. *Advanced Materials*. Advance online publication. DOI: 10.1002/adma.202308281

Adeniyi, M., & Awosan, W. (2025). Nanoparticle-enhanced near-infrared fluorescence probes: A breakthrough in cancer imaging techniques. *International Journal For Multidisciplinary Research*, *7*(1). Advance online publication. DOI: 10.36948/ijfmr.2025.v07i01.34883

Adhyam, M., & Gupta, A. K. (2012). A review on the clinical utility of PSA in cancer prostate. *Indian Journal of Surgical Oncology*, *3*(2), 120–129. DOI: 10.1007/S13193-012-0142-6

Ahmad I, Parween T, Khandare L, (2023). "Applications of Functionalized Carbon Nanotubes in Cancer Therapy and Diagnosis". DOI: DOI: 10.1002/9783527835874.ch5

Ahmad, F., Ghazal, H., Rasheed, F., Shahid, M., Vasantham, S. K., Rafiq, W., & Atiq, S. (2024). Graphene and its derivatives in medical applications: A comprehensive review. *Synthetic Metals*, •••, 117594.

Akmal, M. H., Kalashgrani, M. Y., Mousavi, S. M., Rahmanian, V., Sharma, N., Gholami, A., & Chiang, W. H. (2024). Recent advances in synergistic use of GQD-based hydrogels for bioimaging and drug delivery in cancer treatment. *Journal of Materials Chemistry. B*, *12*(21), 5039–5060. PMID: 38716622

Alagarasan, J. K., Shasikala, S., & Ganesan, S. (2023). Silicon Nanoparticles as a Fluorometric Probe for Sensitive Detection of Cyanide Ion and Its Application in C. elegans Bio-Imaging. *Environmental Research, 224*, 115402. DOI: 10.1016/j.envres.2023.115402

Alharbi, A. L., Alreshidi, M. S., Alanazi, A., Almutairi, K. F., & Alotibi, F. S. (2021). Advancements and challenges in nanotheranostics for cancer: A comprehensive review. [IJHS]. *International Journal of Health Sciences, 5*(S1), 1436–1443. DOI: 10.53730/ijhs.v5ns1.15352

Ali, A., Zafar, H., & Zia, M.. (2021). Synthesis, characterization, applications, and challenges of iron oxide nanoparticles. *Nanotechnology, Science and Applications, 14*, 49–76. PMID: 27578966

Allamyradov, Y., Yosef, J. B., Annamuradov, B., Ateyeh, M., Street, C., Whipple, H., & Er, A. (2024). Photodynamic therapy review: Past, present, future, opportunities and challenges. *Photochem, 4*(4), 434–461. DOI: 10.3390/photochem4040027

Almutairi, S., Manimurugan, S., Kim, B. G., Aborokbah, M. M., & Narmatha, C. (2023). Breast cancer classification using Deep Q Learning (DQL) and gorilla troops optimization (GTO). *Applied Soft Computing, 142*, 110292.

Al-Thani, A. N., Jan, A. G., Abbas, M., & Geetha, M. (2024). Nanoparticles in cancer theragnostic and drug delivery: A comprehensive review. *Life Sciences*. Advance online publication. DOI: 10.1016/j.lfs.2024.122511

Amaral, M. N., Kumar, P., Faísca, P., Ferreira, H. A., Coelho, J. M. P., Gaspar, M. M., & Reis, C. P. (2025). Gold nanoparticle-mediated photothermal therapy: Expanding the frontiers of cancer treatment and theragnostics. *Biomedicine & `Pharmacotherapy, 190*, 118399. https://doi.org/DOI: 10.1016/j.biopha.2025.118399

American Cancer Society. *Cancer Facts & Figures 2025*. (n.d.). Retrieved on June 29, 2025, from https://www.cancer.org/research/cancer-facts-statistics/

Amethiya, Y., Pipariya, P., Patel, S., & Shah, M. (2022). Comparative analysis of breast cancer detection using machine learning and biosensors. *Intelligent Medicine, 2*(2), 69–81.

Anandasabapathy, S., Asirwa, C., Grover, S., & Mungo, C. (2024). Cancer burden in low-income and middle-income countries. *Nature Reviews. Cancer, 24*, 167–170. DOI: 10.1038/s41568-023-00659-2

Ananthakumar, B., Nagaraja, B. G., Umesh, S. M., & Raghavan, R. (2025). Predictive Medicine: A New Era of Pain Management for Assistive Augmentative Rehabilitation Systems. In *Predictive Algorithms for Rehabilitation and Assistive Systems* (pp. 291–314). IGI Global Scientific Publishing. https://doi.org/DOI: 10.4018/979-8-3373-0194-5.ch011

Ananthakumar, B., Nanjundeshwara, N., Lakshmaiah, A., & Raghavan, R. (2024). Synthetic Gene Circuits as a Promising Approach in Cancer Immunotherapy. *Critical Developments in Cancer Immunotherapy*, 391–418. https://doi.org/DOI: 10.4018/979-8-3693-3976-3.ch011

Ananthakumar, B., Nanjundeshwara, N., Lakshmaiah, A., & Raghavan, R. (2025). Nanocarriers for Delivering Nucleic Acids in Plants: Mechanisms and Applications. In *Nanocarriers in Plant Science and Agriculture* (pp. 207–234). IGI Global Scientific Publishing. https://doi.org/DOI: 10.4018/979-8-3693-9869-2.ch008

Appidi, T., Sivasankaran, R. P., Chinchulkar, S. A., Patra, P., Murugaiyan, K., Veeresh, B., & Rengan, A. K. (2024). A lipo-polymeric hybrid nanosystem with metal enhanced fluorescence for targeted imaging of metastatic breast cancer. *Nanotheranostics*, *8*(2), 239–246. DOI: 10.7150/ntno.92410

Aragani, V. M., Anumolu, V. R., & Selvakumar, P. (2025). Democratization in the Age of Algorithms: Navigating Opportunities and Challenges. In Wongmahesak, K., Wekke, I., Seftyono, C., & Nurdin, N. (Eds.), *Democracy and Democratization in the Age of AI* (pp. 39–56). IGI Global Scientific Publishing., DOI: 10.4018/979-8-3693-8749-8.ch003

Arakawa, T., Niikura, T., Kita, Y., & Akuta, T. (2024). Sodium dodecyl sulfate analogs as a potential molecular biology reagent. *Current Issues in Molecular Biology*, *46*(1), 621–633. PMID: 38248342

Arias, L, J., Clares, B., E Morales, M., Gallardo, V., & A Ruiz, M. (2011). Lipid-based drug delivery systems for cancer treatment. *Current Drug Targets*, *12*(8), 1151–1165. PMID: 21443475

Arnold, A. M., Singh, J., & Sydlik, S. A. (2025). The Role and Future of Functional Graphenic Materials in Biomedical and Human Health Applications. *Biomacromolecules*, *26*(4), 2015–2042. PMID: 40101190

Asghar, K., Qasim, M., Dharmapuri, G., & Das, D. (2017). Investigation on a smart nanocarrier with a mesoporous magnetic core and thermo-responsive shell for co-delivery of doxorubicin and curcumin: A new approach towards combination therapy of cancer. *RSC Advances*, *7*(46), 28802–28818. DOI: 10.1039/c7ra03735j

Asif, S., & Wenhui, Y., ur-Rehman, S., ul-ain, Q., Amjad, K., Yueyang, Y., ... & Awais, M. (2024). Advancements and Prospects of Machine Learning in Medical Diagnostics: Unveiling the Future of Diagnostic Precision. *Archives of Computational Methods in Engineering*, •••, 1–31.

Assali, M., Kittana, N., & Khiar, N. (2021). Dual covalent functionalization of single-walled carbon nanotubes for effective targeted cancer therapy. *Nanotechnology*. Advance online publication. DOI: 10.1088/1361-6528/ac1d3d

Atre, P., & Rizvi, S. A. (2024). A brief overview of quality by design approach for developing pharmaceutical liposomes as nano-sized parenteral drug delivery systems. *RSC Pharmaceutics*.

Atrey, K., Singh, B. K., Bodhey, N. K., & Pachori, R. B. (2023). Mammography and ultrasound based dual modality classification of breast cancer using a hybrid deep learning approach. *Biomedical Signal Processing and Control, 86*, 104919.

Attri, S., Sharma, S., Sharma, U., Srivastava, M., Sonkar, S. C., Singh, M., & Bhukya, P. (2021a). Use of lipids, polymers, and peptides for drug delivery and targeting to cancer cells or specific organs. In *Handbook of Research on Advancements in Cancer Therapeutics* (pp. 276–289). IGI Global Scientific Publishing.

Attri, S., Sharma, S., Sharma, U., Srivastava, M., Sonkar, S. C., Singh, M., & Bhukya, P. (2021b). *Use of Lipids*. Polymers, and Peptides for Drug Delivery and Targeting to Cancer Cells or Specific Organs.

Ayreen, Z., Khatoon, U., Kirti, A., Sinha, A., Gupta, A., Lenka, S. S., & Verma, S. K. (2024). Perilous paradigm of graphene oxide and its derivatives in biomedical applications: Insight to immunocompatibility. *Biomedicine and Pharmacotherapy, 176*, 116842. PMID: 38810404

Aziz, A. A., Gayed, I., Heakel, F., & El Nashar, R. M. (2025). Synthesis, characterization, and versatile applications of metal oxides and mixed metal oxide nanoparticles. *Journal of the Electrochemical Society, 172*, 023503. DOI: 10.1149/1945-7111/adada4

Aziz, S. H., & Abdullah, S. (2022). Toxicity of metal oxide nanoparticles in freshwater fish. In Pal, K., & Zaheer, T. (Eds.), *Nanomaterials in the Battle against Pathogens and Disease Vectors* (pp. 257–282). CRC Press., DOI: 10.1201/9781003126256-10

Bahadur, P. S., Shameer, P. M., Sharma, N., & Manjunath, T. C. (2024). "Nano Technology: The Future of Innovation." Books.Google. *Com, 1*, 567–589. DOI: 10.1007/google.nanotechnology2024

Bai, C., & Tang, M. (2020). Toxicological study of metal and metal oxide nanoparticles in zebrafish. *Journal of Applied Toxicology*, *40*(1), 37–63. DOI: 10.1002/jat.3910

Barick, K. C., Singh, S., Bahadur, D., Lawande, M. A., Patkar, D., & Hassan, P. A. (2014). Carboxyl decorated Fe3O4 nanoparticles for MRI diagnosis and localized hyperthermia. *Journal of Colloid and Interface Science*, *418*, 120–125. DOI: 10.1016/j.jcis.2013.11.076

Basu, R., & Moles, C. M. (2023). Rational selection of an ideal oncolytic virus to address current limitations in clinical translation. [). Academic Press.]. *International Review of Cell and Molecular Biology*, *379*, 241–261.

Basu, S., & Hendler-Neumark, A. (2024). Dynamic Tracking of Biological Processes Using Near-Infrared Fluorescent Single-Walled Carbon Nanotubes. *ACS Applied Materials & Interfaces*. Advance online publication. DOI: 10.1021/acsami.3c18157

Behyar, M. B., Nilghaz, A., Bahari, H., Hasanzadeh, M., & Shadjou, N. (2024). Recent advances on nanomaterials for personalized diagnostics in multiplex point-of-care testing. *Trends in Analytical Chemistry*, •••, 117761.

Beitzinger, B., Draphoen, B., Hald, J., Krämer, M., Rudolf, T., Schmitt, F., Schwarz, P., & Lindén, M. (2025). On the cellular selectivity of targeting peptide-nanoparticle conjugates – a nanoparticle perspective. *Expert Opinion on Drug Delivery, ahead-of-print*(ahead-of-print), 1–14. https://doi.org/DOI: 10.1080/17425247.2025.2522245

Bellanthudawa, B. K. A., & Nawalage, N. (2023). A Perspective on Biodegradable and Non-Biodegradable Nanoparticles in Industrial Sectors: Applications, Challenges, and Future Prospects. *Nanotechnology for Environmental Engineering*, *8*(4), 975–1013. DOI: 10.1007/s41204-023-00344-7

Bento, C., Katz, M., Santos, M. M., & Afonso, C. A. (2024). Striving for uniformity: A review on advances and challenges to achieve uniform polyethylene glycol. *Organic Process Research & Development*, *28*(4), 860–890. PMID: 38660381

Bera, I., & Klauda, J. B. (2017). Molecular simulations of mixed lipid bilayers with sphingomyelin, glycerophospholipids, and cholesterol. *The Journal of Physical Chemistry B*, *121*(20), 5197–5208. PMID: 28447449

Bertão, A. R. C. (2024). Silicon-Based Nanostructures as Drug Delivery Systems for Biological and Medical Applications. *Universidade do Minho Repository*, *7*, 92402. DOI: 10.1822/92402

Bharadwaj, K. K., Rabha, B., Pati, S., & Sarkar, T. (2021). Green Synthesis of Gold Nanoparticles Using Plant Extracts as Beneficial Prospect for Cancer Theranostics. *Molecules (Basel, Switzerland)*, *26*(4), 123–145. DOI: 10.3390/mdpi.goldnanoparticles2021

Bhattacharya, D. (2023). *Folate receptor mediated targeting of oral cancer: Therapeutic and mechanistic investigations against tumour cells and tumour associated Macrophages* BITS PILANI, Hyderabad campus].

Bhattacharya, R., Avdieiev, S. S., Bukkuri, A., Whelan, C. J., Gatenby, R. A., Tsai, K. Y., & Brown, J. S. (2025). The hallmarks of cancer as eco-evolutionary processes. *Cancer Discovery*, *15*(4), 685–701. DOI: 10.1158/2159-8290.cd-24-0861

Bhattacharya, S., Belemkar, S., & Prajapati, B. G. (2024). The application of graphene oxide nanoarchitectures in the treatment of cancer: Phototherapy, immunotherapy, and the development of vaccines. *Current Medicinal Chemistry*, *31*(27), 4320–4339. DOI: 10.2174/0109298673288750240117115141 PMID: 38549530

Bhupathi, P., Elhassan A-Elgadir, T. M., Mohammed Ali, R. H., Sanaan Jabbar, H., Gulnoza, D., Joshi, S. K., ... & Alsaalamy, A. (2025). Fluorescence resonance energy transfer (FRET)-based sensor for detection of foodborne pathogenic bacteria: A review. *Critical Reviews in Analytical Chemistry*, *55*(2), 233–250. PMID: 37917532

Bhuvana, G. N., Umesh, S. M., & Raghavan, R. (2025). Expanding the frontiers of Schistosomiasis Treatment: The promise of Nanoparticle-Based Therapies. *Journal of Vector Borne Diseases, 10*(4103/jvbd.jvbd_180_23). https://doi.org/DOI: 10.4103/jvbd.jvbd_180_23

Bibault, J. E., Giraud, P., & Burgun, A. (2016). Big data and machine learning in radiation oncology: State of the art and future prospects. *Cancer Letters*, *382*(1), 110–117.

Bi, J., Mo, C., Li, S., Huang, M., Yuan, P., Liu, Z., Jia, B., & Xu, S. (2023). Immunotoxicity of metal and metal oxide nanoparticles: From toxic mechanisms to metabolism and outcomes. *Biomaterials Science*, *11*(12), 4151–4183. DOI: 10.1039/d3bm00271c

Biradar, S., Suresh, C., Nanjundeshwara, N., & Raghavan, R. (2025). Spatially Variable Genes. *Spatially Variable Genes in Cancer*, 1–28. https://doi.org/DOI: 10.4018/979-8-3693-7728-4.ch001

Bischof, J. J., Presley, C. J., & Caterino, J. M. (2019). Addressing new diagnostic and treatment challenges associated with a new age of cancer treatment. *Annals of Emergency Medicine*, *73*(1), 88–90. DOI: 10.1016/j.annemergmed.2018.08.421

Biswas, H. S., Bala, S., Kundu, A. K., Saha, I., Poddar, S., Sarkar, S., & Mandal, P. (2022). Tuned synthesis and designed characterization of graphene oxide thin film. *Inorganic Chemistry Communications*, *139*, 109356.

Bjerk, T. R., Severino, P., Jain, S., Marques, C., Silva, A. M., Pashirova, T., & Souto, E. B. (2021). Biosurfactants: Properties and applications in drug delivery, biotechnology and ecotoxicology. *Bioengineering (Basel, Switzerland)*, *8*(8), 115. PMID: 34436118

Bjørklund, G., Đorđević, A. B., Hamdan, H., Wallace, D. R., & Peana, M. (2024). Metal-induced autoimmunity in neurological disorders: A review of current understanding and future directions. *Autoimmunity Reviews*, *23*(3), 103509. DOI: 10.1016/j.autrev.2023.103509

Boire, A., Burke, K., Cox, T., Guise, T., Jamal-Hanjani, M., Janowitz, T., Kaplan, R., Lee, R., Swanton, C., Heiden, M. G. V., & Sahai, E. (2024). Why do patients with cancer die? *Nature Reviews. Cancer*, *24*(8), 578–589. DOI: 10.1038/s41568-024-00708-4

Botequim, D., Oliveira-Silva, R., Vaz Serra, V., Zijlstra, P., Prazeres, D. M. F., & Paulo, P. M. R. (2025). Nanoplasmonics for enhanced fluorescence detection of nucleic acids: From fundamentals to boosting cancer management. *Advanced NanoBiomed Research*, *5*(2), 2400088. DOI: 10.1002/anbr.202400088

Bray, F., Laversanne, M., Sung, H., Ferlay, J., Siegel, R. L., Soerjomataram, I., & Jemal, A. (2024). Global cancer statistics 2022: GLOBOCAN estimates of incidence and mortality worldwide for 36 cancers in 185 countries. *CA: a Cancer Journal for Clinicians*, *74*(3), 229–263. DOI: 10.3322/caac.21834

Brindhadevi, K., Garalleh, H A L., & Alalawi, A. (2023). Carbon nanomaterials: Types, synthesis strategies and their application as drug delivery system for cancer therapy. *Biochemical Engineering Journal*. Advance online publication. DOI: 10.1016/j.bej.2023.108934

Bukowski, K., Kciuk, M., & Kontek, R. (2020). Mechanisms of multidrug resistance in cancer chemotherapy. *International Journal of Molecular Sciences*, *21*(9), 3233. DOI: 10.3390/ijms21093233

Bulte, J. W., & Kraitchman, D. L. (2004). Iron oxide MR contrast agents for molecular and cellular imaging. *NMR in Biomedicine*, *17*(7), 484–499. PMID: 15526347

Bwatanglang, I. B., Mohammad, F., Yusof, N. A., Abdullah, J., Hussein, M. Z., Alitheen, N. B., & Abu, N. (2016). Folic acid targeted Mn:ZnS quantum dots for theranostic applications of cancer cell imaging and therapy. *International Journal of Nanomedicine*, *11*, 413–428. DOI: 10.2147/IJN.S90198 PMID: 26858524

Byalahunasishetter, B. I. N, P. V., Savitha, C., Moger, S. S., & Raghavan, R. (2025). Spatial Omics and Bioprinting: Tissue Engineering and Precision Medicine. In *Innovations in Precision Medicine and Genomics* (pp. 85–110). IGI Global Scientific Publishing. https://doi.org/DOI: 10.4018/979-8-3693-5787-3.ch003

Cabral, H., Li, J., Miyata, K., & Kataoka, K. (2024). Controlling the biodistribution and clearance of nanomedicines. *Nature Reviews Bioengineering*, *2*(3), 214–232.

Cai, C., Zhang, Q., Ye, J., Yao, S., Li, Q., Fan, Z., & Ding, Y. (2025). Tumor microenvironment modulation innovates combinative cancer therapy via versatile graphene oxide nanosystem. *Biomaterials Science*. PMID: 40314424

Campos-Carrillo, A., Weitzel, J. N., Sahoo, P., Rockne, R. C., Mokhnatkin, J. V., Murtaza, M., Gray, S. W., Goetz, L. H., Goel, A., Schork, N. J., Schork, N. J., & Slavin, T. P. (2020). Circulating tumor DNA as an early cancer detection tool. *Pharmacology & Therapeutics*, *207*, 107458. DOI: 10.1016/j.pharmthera.2019.107458

Cavalcanti, I. D. L., & Soares, J. C. S. (2021). *Advances in cancer treatment: From systemic chemotherapy to targeted therapy*. Springer., DOI: 10.1007/978-3-030-68334-4

Chacko, R. T., Ventura, J., Zhuang, J., & Thayumanavan, S. (2012). Polymer nanogels: A versatile nanoscopic drug delivery platform. *Advanced Drug Delivery Reviews*, *64*(9), 836–851. PMID: 22342438

Chandana, K. V., Gupta, N. V., & Kanna, S. (2019). Nanostructured lipid carriers: The frontiers in drug delivery. *Asian Journal of Pharmaceutical and Clinical Research*, *12*(7), 8–12.

Chary, P. S., Bhawale, R., Vasave, R., & Rajana, N. (2023). A review on emerging role of multifunctional carbon nanotubes as an armament in cancer therapy, imaging and biosensing. *Journal of Drug Delivery Science and Technology*. Advance online publication. DOI: 10.1016/j.jddst.2023.104455

Chaudhari, K. N., Rajeev, K. K., Kim, S., & Nersisyan, H. (2022). Performance Enhancement of Carbon-Coated Si Nanoparticles for Lithium-Ion Batteries Through the Generation of Lithophilic Sites by a Simple Oxidation Process. *Applied Surface Science*, *601*, 154361. DOI: 10.1016/j.apsusc.2022.154361

Chaudhary, S., Pathak, A. K., & Ahmad, S. (2020). Nano-based materials and their synthesis. In I. Bhushan, V. Singh & D. Tripathi, D. (Eds.), *Nanomaterials and Environmental Biotechnology* (pp. 127-140). Springer, Cham. https://doi.org/DOI: 10.1007/978-3-030-34544-0_8

Chen, D.-D., Xi, T.-F., & Bai, J. (2007). Biological effects induced by nanosilver particles: In vivo study. *Biomedical Materials (Bristol, England)*, *2*(3), S126. DOI: 10.1088/1748-6041/2/3/S08

Cheng, Z., Jiang, H., Zhang, X., & Cheng, F. (2023). Fundamental Understanding and Facing Challenges in Structural Design of Porous Si-Based Anodes for Lithium-Ion Batteries. *Advanced Functional Materials*, *33*(4), 789–812. DOI: 10.1002/adfm.202300456

Chen, L., Zhang, S., Duan, Y., Song, X., & Chang, M. (2024). Silicon-Containing Nanomedicine and Biomaterials: Materials Chemistry, Multi-Dimensional Design, and Biomedical Application. *Chemical Society Reviews*, *53*(3), 567–589. DOI: 10.1039/d1cs01022k

Chen, M., Tang, S., Guo, Z., Wang, X., Mo, S., Huang, X., Liu, G., & Zheng, N. (2014a). Core–Shell Pd@ Au nanoplates as theranostic agents for in-vivo photoacoustic imaging, CT imaging, and photothermal therapy. *Advanced Materials*, *26*(48), 8210–8216. DOI: 10.1002/adma.201404013

Chen, S. H., Bell, D. R., & Luan, B. (2022). Understanding interactions between biomolecules and two-dimensional nanomaterials using in silico microscopes. *Advanced Drug Delivery Reviews*, *186*, 114336. PMID: 35597306

Chen, Z., Tian, Y., Yang, J., Wu, F., Liu, S., Cao, W., Xu, W., Hu, T., Siegwart, D. J., & Xiong, H. (2023). Modular Design of Biodegradable Ionizable Lipids for Improved mRNA Delivery and Precise Cancer Metastasis Delineation In Vivo. *Journal of the American Chemical Society*, *145*(44), 24302–24314. DOI: 10.1021/jacs.3c09143 PMID: 37853662

Chen, Z., Wang, Y., Ba, T., Li, Y., Pu, J., Chen, T., Song, Y., Gu, Y., Qian, Q., Yang, J., & Jia, G. (2014b). Genotoxic evaluation of titanium dioxide nanoparticles in vivo and in vitro. *Toxicology Letters*, *226*(3), 314–319. DOI: 10.1016/j.toxlet.2014.02.020

Chime, S. A., Akpa, P. A., & Attama, A. A. (2019). The utility of lipids as nanocarriers and suitable vehicle in pharmaceutical drug delivery. *Current Nanomaterials*, *4*(3), 160–175.

Choi, H. S., Liu, W., Misra, P., Tanaka, E., Zimmer, J. P., Ipe, B. I., Bawendi, M. G., & Frangioni, J. V. (2007). Renal clearance of nanoparticles. *Nature Biotechnology*, *25*(10), 1165–1170. DOI: 10.1038/nbt1340

Chu, B., Wang, H., & He, Y. (2021). Fluorescent Silicon-Based Nanomaterials Imaging Technology in Diseases. *Chemical Research in Chinese Universities*, *37*(2), 123–145. DOI: 10.1007/springer.nanomaterials2021

Conde, J., Doria, G., & Baptista, P. (2012). Noble metal nanoparticles applications in cancer. *Journal of Drug Delivery*, *751075*. Advance online publication. DOI: 10.1155/2012/751075

Contreras-Torres, F. F., & Salas-Trevino, D. (2022). Carbon nanotubes in tumor-targeted chemotherapeutic formulations: A review of opportunities and challenges. *ACS Applied Nano Materials*. Advance online publication. DOI: 10.1021/acsanm.1c03971

Correa-Soto, C. E., Gao, Y., Indulkar, A. S., Zhang, G. G. Z., & Taylor, L. S. (2022). Role of surfactants in improving release from higher drug loading amorphous solid dispersions. *International Journal of Pharmaceutics*, *625*, 122120. DOI: 10.1016/j.ijpharm.2022.122120 PMID: 35987321

Cullion, K., Ostertag-Hill, C. A., Pan, M., Timko, B. P., Boscolo, E., & Kohane, D. S. (2023). Ablation of venous malformations by photothermal therapy with intravenous gold nanoshells. *Nano Letters*, *23*(15), 7092–7099. DOI: 10.1021/acs.nanolett.3c01945

Da Cruz Schneid, A., & Albuquerque, L. J. C. (2022). Colloidal Stability and Degradability of Silica Nanoparticles in Biological Fluids: A Review. *Journal of Sol-Gel Science and Technology*, *102*(1), 41–62. DOI: 10.1007/s10971-021-05695-8

Das, S., Mazumdar, H., Khondakar, K. R., & Kaushik, A. (2025). Machine learning integrated graphene oxide-based diagnostics, drug delivery, analytical approaches to empower cancer diagnosis. *BMEmat*, *3*(1), e12117.

Das, S., Roy, S., Dinda, S. C., Bose, A., & Mahapatra, C. (2025). Carbon nanotubes in brain targeted drug delivery: A comprehensive review. *Results in Chemistry*. Advance online publication. DOI: 10.1016/j.rechem.2024.101401

Daulay, A., & Gea, S. (2022). Synthesis and Application of Silicon Nanoparticles Prepared from Rice Husk for Lithium-Ion Batteries. *Case Studies in Chemical and Environmental Engineering*, *6*, 100256. DOI: 10.1016/j.cscee.2022.100256

De Stefani, C., Lodovichi, J., Albonetti, L., Salvatici, M. C., Quintela, J. C., Bilia, A. R., & Bergonzi, M. C. (2022). Solubility and Permeability Enhancement of Oleanolic Acid by Solid Dispersion in Poloxamers and gamma-CD. *Molecules (Basel, Switzerland)*, *27*(9). Advance online publication. DOI: 10.3390/molecules27093042 PMID: 35566392

Debnath, P., Deshmukh, S. M., Laroia, M., Selvakumar, P., Manjunath, T. C., & Bhattacharya, S. (2025). Challenges Faced by Countries in Renewable Energy Adoption. In Özsungur, F., Chaychi Semsari, M., & Küçük Bayraktar, H. (Eds.), *Geopolitical Landscapes of Renewable Energy and Urban Growth* (pp. 295–320). IGI Global Scientific Publishing., DOI: 10.4018/979-8-3693-8814-3.ch010

Deng, B., Kong, Y., Ma, Y., Zhan, Y., Sun, Y., Wang, R., Huang, P., & Liu, L. (2025). A novel carrier-free nanoparticle with stable distinctive three-dimensional structure for tumor-targeted precision chemoimmunotherapy. *Journal of Nanobiotechnology*, *23*(1), 1–15. DOI: 10.1186/s12951-025-03568-8 PMID: 40598510

Dhande, H. S., Taksande, J. B., Awandekar, N. B., Burle, S. S., Umekar, M. J., & Lade, S. N. (2025). Pharmacodynamic and Physicochemical Evaluation of Intranasal Zolmitriptan-Loaded Nanostructured Lipid Carriers: For Enhanced Antimigraine Potential. *Assay and Drug Development Technologies*. Advance online publication. DOI: 10.1089/adt.2025.016 PMID: 40415589

Díaz-Rodríguez, T. G., Antunez, E. E., & Celaya, C. A. (2025). Exploring Contaminants and Analyte Adsorption on Functionalized Porous Silicon: Insights from a Combined Theoretical and Experimental Approach. *Surfaces and Interfaces*, *30*, 100456. DOI: 10.1016/j.surfin.2025.100456

Dillekås, H., Rogers, M. S., Straume, O., & Straume, O. (2019). Are 90% of deaths from cancer caused by metastases? *Cancer Medicine*, *8*(12), 5574–5576. DOI: 10.1002/cam4.2474

Ding, H., Dai, W., Ju, H., Lu, H., Wang, S., Xu, L., Zhou, S. F., Zhang, Y., & Zhang, X. (2015). Multifunctional poly(L-lactide)-polyethylene glycol-grafted graphene quantum dots for intracellular microRNA imaging and combined specific-gene-targeting agents delivery for improved therapeutics. *ACS Applied Materials & Interfaces*, *7*(20), 11015–11023. DOI: 10.1021/acsami.5b02803 PMID: 25942410

Ding, L., Liu, Z., Aggrey, M. O., Li, C., Chen, J., & Tong, L. (2015). Nanotoxicity: The toxicity research progress of metal and metal-containing nanoparticles. *Mini-Reviews in Medicinal Chemistry*, *15*(7), 529–542. DOI: 10.2174/1389557515071 50424104334

Divband, B., Gharehaghaji, N., & Hassani, S. (2022). Fe3O4/graphene-based nanotheranostics for bimodal magnetic resonance/fluorescence imaging and cancer therapy. *Journal of Inorganic and Organometallic Polymers and Materials*, *32*(12), 4443–4460.

Dochez, V., Caillon, H., Vaucel, E., Dimet, J., Winer, N., & Ducarme, G. (2019). Biomarkers and algorithms for diagnosis of ovarian cancer: CA125, HE4, RMI and ROMA, a review. *Journal of Ovarian Research*, *12*(1), 28. DOI: 10.1186/S13048-019-0503-7

Dodkins, R., Delaney, J. R., Overton, T., Scholle, F., Frias-De-Diego, A., Crisci, E., & Goldberg, I. G. (2023). A rapid, high-throughput, viral infectivity assay using automated brightfield microscopy with machine learning. *SLAS Technology*, *28*(5), 324–333.

Dong, H., Dong, C., Ren, T., Li, Y., & Shi, D. (2014). Surface-engineered graphene-based nanomaterials for drug delivery. *Journal of Biomedical Nanotechnology*, *10*(9), 2086–2106. DOI: 10.1166/jbn.2014.1989 PMID: 25992450

Duan, W., Zhao, J., Gao, Y., Xu, K., Huang, S., & Zeng, L. (2024). Porous Silicon-Based Sensing and Delivery Platforms for Wound Management Applications. *Journal of Controlled Release*, *350*, 123–145. DOI: 10.1016/j.jconrel.2024.012345

Dutta, A. K. (2019). Introductory chapter: surfactants in household and personal care formulations-An overview. *Surfactants and detergents*.

Dutz, S., & Hergt, R. (2014). Magnetic nanoparticle heating and heat transfer on a microscale: Basic principles, realities, and physical limitations of hyperthermia for tumor therapy. *International Journal of Hyperthermia*, *29*(8), 790–800. PMID: 23968194

Dutz, S., & Hergt, R. (2014). Magnetic nanoparticle heating and heat transfer on a microscale: Basic principles, realities, and physical limitations of hyperthermia for tumor therapy. *Nanotechnology*, *25*(45), 452001. PMID: 25337919

Dutz, S., & Hergt, R. (2014). Magnetic particle hyperthermia—A promising tumour therapy? *Nanotechnology*, *25*(45), 452001. PMID: 25337919

Dyagala, S., Paul, M., Aswal, V. K., Biswas, S., & Saha, S. K. (2023). Compaction of Calf Thymus DNA by a Potential One-Head-Two-Tail Surfactant: Properties of Nanomaterials and Biological Testing for Gene Delivery. *ACS Applied Bio Materials*, *6*(9), 3848–3862. DOI: 10.1021/acsabm.3c00470 PMID: 37647161

Dy, G. K., & Adjei, A. A. (2013). Understanding, recognizing, and managing toxicities of targeted anticancer therapies. *CA: a Cancer Journal for Clinicians*, *63*(4), 249–279. DOI: 10.3322/caac.21184

Eda, G., Fanchini, G., & Chhowalla, M. (2008). Large-area ultrathin films of reduced graphene oxide as a transparent and flexible electronic material. *Nature Nanotechnology*, *3*(5), 270–274. DOI: 10.1038/nnano.2008.83 PMID: 18654522

El Majzoub, I., Qdaisat, A., Thein, K. Z., Win, M. A., Han, M. M., Jacobson, K., Chaftari, P., Prejean, M., Reyes-Gibby, C. C., & Yeung, S. C. J. (2019). Adverse effects of immune checkpoint therapy in cancer patients visiting the emergency department of a comprehensive cancer center. *Annals of Emergency Medicine*, *73*(1), 79–87. DOI: 10.1016/j.annemergmed.2018.04.019

ElBaz, N., Chauhan, R., James, K. T., Malik, D. A., Zhu, M., ElBaz, A., Keynton, R. S., Ng, C. K., Bates, P. J., Malik, T., & O'Toole, M. G. (2018). Gold nanoparticles conjugate cancer-targeting aptamer and gadolinium chelate for MR cancer imaging. In *2018 IEEE International Symposium on Signal Processing and Information Technology* (pp. 520-525). Louisville, USA. https://doi.org/DOI: 10.1109/isspit.2018.8642635

Elnosary, M. E., Salem, F. K., Mohamed, O., Elbas, M. A., Shaheen, A. A., Mowafy, M. T., & Sofy, A. R. (2024). Unlocking the potential: A specific focus on vesicular stomatitis virus as a promising oncolytic and immunomodulatory agent in cancer therapy. *Discovery Medicine*, *1*(1), 63.

Enclopedia, S. c. (2023). *Zwitterionics Surfactants* https://encyclopedia.pub/entry/41390

Erickson, K., Erni, R., Lee, Z., Alem, N., Gannett, W., & Zettl, A. (2010). Determination of the local chemical structure of graphene oxide and reduced graphene oxide. *Advanced Materials*, *22*(40), 4467–4472. DOI: 10.1002/adma.201000732 PMID: 20717985

Estelrich, J., Escribano, E., Queralt, J., & Busquets, M. A. (2015). Iron oxide nanoparticles for magnetically guided and magnetically responsive drug delivery. *International Journal of Nanomedicine*, *10*, 1727–1741. PMID: 25834422

Estelrich, J., Escribano, E., Queralt, J., & Busquets, M. A. (2015). Iron oxide nanoparticles for magnetically-guided and magnetically-responsive drug delivery. *International Journal of Nanomedicine*, *10*, 1727–1741. PMID: 25834422

Fatfat, Z., Fatfat, M., & Gali-Muhtasib, H. (2022). Micelles as potential drug delivery systems for colorectal cancer treatment. *World Journal of Gastroenterology, 28*(25), 2867. PMID: 35978871

Faulkner, S., Gunnlaugsson, T., & Máille, G. Ó. (2022). Supramolecular. *Chemical & Biomedical Imaging.*

Fernández-Borbolla, A., García-Hevia, L., & Fanarraga, M. L. (2024). Cell Membrane-Coated Nanoparticles for Precision Medicine: A Comprehensive Review of Coating Techniques for Tissue-Specific Therapeutics. *International Journal of Molecular Sciences, 25*(4), 2071. DOI: 10.3390/ijms25042071 PMID: 38396747

Fortin, J. P., Wilhelm, C., Servais, J., Ménager, C., Bacri, J. C., & Gazeau, F. (2007). Size-sorted anionic iron oxide nanomagnets as colloidal mediators for magnetic hyperthermia. *Journal of the American Chemical Society, 129*(9), 2628–2635. PMID: 17266310

Fu, B., Dang, M., Tao, J., Li, Y., & Tang, Y. (2020). Mesoporous platinum nanoparticle-based nanoplatforms for combined chemo-photothermal breast cancer therapy. *Journal of Colloid and Interface Science, 570*, 197–204. DOI: 10.1016/J.JCIS.2020.02.051

Fujita, K., & Urano, Y. (2024). Activity-Based Fluorescence Diagnostics for Cancer. *Chemical Reviews, 124*(7), 4021–4078. DOI: 10.1021/acs.chemrev.3c00612

Furrer, D., Jacob, S., Caron, C., Sanschagrin, F., Provencher, L., & Diorio, C. (2017). Concordance of HER2 immunohistochemistry and fluorescence in situ hybridization using tissue microarray in breast cancer. *Anticancer Research, 37*(6), 3323–3329. DOI: 10.21873/anticanres.11701

Gao, C., Guo, M., Liu, Y., Zhang, D., Gao, F., Sun, L., & Li, J. (2023). Surface modification methods and mechanisms in carbon nanotubes dispersion. *Carbon.* Advance online publication. DOI: 10.1016/j.carbon.2023.01.045

Gao, F., Sun, Z., Zhao, L., Chen, F., & Stenzel, M. (2021). Bioactive engineered photothermal nanomaterials: From theoretical understanding to cutting-edge application strategies in anti-cancer therapy. *Materials Chemistry Frontiers.* Advance online publication. DOI: 10.1039/D0QM00953A

Gao, S., Xu, B., Sun, J., & Zhang, Z. (2024). Nanotechnological advances in cancer: Therapy a comprehensive review of carbon nanotube applications. *Frontiers in Bioengineering and Biotechnology.* Advance online publication. DOI: 10.3389/fbioe.2024.1324568

García-Torra, V., Cano, A., Espina, M., Ettcheto, M., Camins, A., Barroso, E., Vazquez-Carrera, M., García, M. L., Sánchez-López, E., & Souto, E. B. (2021). State of the art on toxicological mechanisms of metal and metal oxide nanoparticles and strategies to reduce toxicological risks. *Toxics*, *9*(8), 195. DOI: 10.3390/toxics9080195

Gato-Diaz, U., Concheiro, A., Alvarez-Lorenzo, C., & Blanco-Fernandez, B. (2025). 4D printed hydrogels for precision delivery of bioactive molecules in cancer. *Expert Opinion on Drug Delivery*, *0*(0), 1–17. DOI: 10.1080/17425247.2025.2539959 PMID: 40739734

Gavas, S., Quazi, S., & Karpiński, T. M. (2021). Nanoparticles for Cancer Therapy: Current Progress and Challenges. *Nanoscale Research Letters*, *16*(1), 173–190. DOI: 10.1186/s11671-021-03628-6

Gayathri, K., & Vidya, R. (2024). Carbon nanomaterials as carriers for the anti-cancer drug doxorubicin: A review on theoretical and experimental studies. *Nanoscale Advances*, *6*(16), 3992–4014. PMID: 39114152

Ge, M., Cao, C., Biesold, G. M., & Sewell, C. D. (2021). Recent Advances in Silicon-Based Electrodes: From Fundamental Research Toward Practical Applications. *Advanced Materials*, *33*(12), 1234–1250. DOI: 10.1002/adma.202100123

Ge, X., Chen, B., Liu, T., Wei, L., Tong, L., Ma, Q., Gao, S., & Song, J. (2019). Active targeting drug-gold nanorod hybrid nanoparticles for amplifying photoacoustic signal and enhancing anticancer efficacy. *RSC Advances*, *9*(24), 13494–13502. DOI: 10.1039/c9ra01547g

Ghorbian, M., & Ghorbian, S. (2024). Comprehensive review of reinforcement learning in lung cancer diagnosis and treatment: Taxonomy, challenges and recommendations. *Computers in Biology and Medicine*, *183*, 109326. DOI: 10.1016/j.compbiomed.2024.109326

Ghulam, A. N., Dos Santos, O. A., Hazeem, L., Pizzorno Backx, B., Bououdina, M., & Bellucci, S. (2022). Graphene oxide (GO) materials—Applications and toxicity on living organisms and environment. *Journal of Functional Biomaterials*, *13*(2), 77. PMID: 35735932

Godlewski, J., Farhath, M., Ricklefs, F. L., Passaro, C., Kiel, K., Nakashima, H., & Bronisz, A. (2021). Oncolytic virus therapy alters the secretome of targeted glioblastoma cells. *Cancers (Basel)*, *13*(6), 1287.

Goleij, P., Babamohamadi, M., Rezaee, A., Majma Sanaye, P., Tabari, M., & Sadreddini, S., ArefNezhad, R., & Motedayyen, H. (2023). Types of RNA therapeutics. *Progress in Molecular Biology and Translational Science*, *203*, 41–63. DOI: 10.1016/bs.pmbts.2023.12.022

Gong, L., Zhang, Y., Liu, C., Zhang, M., & Han, S. (2021). Application of radiosensitizers in cancer radiotherapy. *International Journal of Nanomedicine*, *16*, 1083–1102. DOI: 10.2147/ijn.s290438

Goradel, N. H., Negahdari, B., Ghorghanlu, S., Jahangiri, S., & Arashkia, A. (2020). Strategies for enhancing intratumoral spread of oncolytic adenoviruses. *Pharmacology & Therapeutics*, *213*, 107586.

Gospodinova, Z., Hristova-Panusheva, K., Kamenska, T., Antov, G., & Krasteva, N. (2025). Insights into cellular and molecular mechanisms of graphene oxide nanoparticles in photothermal therapy for hepatocellular carcinoma. *Scientific Reports*, *15*(1), 15541. PMID: 40319094

Gostaviceanu, A., Gavrilaş, S., Copolovici, L., & Copolovici, D. M. (2024). Graphene-oxide peptide-containing materials for biomedical applications. *International Journal of Molecular Sciences*, *25*(18), 10174. PMID: 39337659

Govender, S., & Swart, P. (2008). Surfactant formulations for multifunctional surface modification. *Colloids and Surfaces. A, Physicochemical and Engineering Aspects*, *331*(1-2), 97–102.

Granha, I., Sartorelli, G., Okamoto, O. K., & Jandrey, E. H. F. (2024). Emergent and underexplored oncolytic viruses for treating pediatric central nervous system tumors. *EJC Paediatric Oncology*, *100151*. Advance online publication.

Grygoryev, D., Ekstrom, T., Manalo, E., Link, J. M., Alshaikh, A., Allen-Petersen, B. L., & Kim, J. (2024). Sendai virus is robust and consistent in delivering genes into human pancreatic cancer cells. *Heliyon*, *10*(5). https://www.cell.com/heliyon/fulltext/S2405-8440(24)03252-3

Guan, X. L., Souza, C. M., Pichler, H., Dewhurst, G., Schaad, O., Kajiwara, K., Wakabayashi, H., Ivanova, T., Castillon, G. A., & Piccolis, M. (2009). Functional interactions between sphingolipids and sterols in biological membranes regulating cell physiology. *Molecular Biology of the Cell*, *20*(7), 2083–2095. PMID: 19225153

Gujar, S., Pol, J. G., Kumar, V., Lizarralde-Guerrero, M., Konda, P., Kroemer, G., & Bell, J. C. (2024). Tutorial: Design, production and testing of oncolytic viruses for cancer immunotherapy. *Nature Protocols*, •••, 1–31. https://www.nature.com/articles/s41596-024-00985-1

Guo, S., Garaj, S., Bianco, A., & Ménard-Moyon, C. (2022). Controlling covalent chemistry on graphene oxide. *Nature Reviews. Physics*, *4*(4), 247–262.

Gupta, A. K., & Gupta, M. (2005). Synthesis and surface engineering of iron oxide nanoparticles for biomedical applications. *Biomaterials*, *26*(18), 3995–4021. PMID: 15626447

GVS, A. C., Ananthakumar, B., & Raghavan, R. (2025). Enhancing Assistive Technologies With Neuromorphic Computing. *Revolutionizing AI with Brain-Inspired Technology: Neuromorphic Computing*, 207–232. https://doi.org/DOI: 10.4018/979-8-3693-6303-4.ch009

GVS, A. C., Biradar, S., & Raghavan, R. (2025). Neuroscience-Driven Solutions for Speech Impairment: Assistive Technologies for Speech Rehabilitation. *Advancing Medical Research Through Neuroscience*, 119–146. https://doi.org/DOI: 10.4018/979-8-3693-5464-3.ch006

Hajipour Keyvani, A., Mohammadnejad, P., Pazoki-Toroudi, H., Perez Gilabert, I., Chu, T., Manshian, B. B., & Sohrabi, B. (2025). Advancements in Cancer Treatment: Harnessing the Synergistic Potential of Graphene-Based Nanomaterials in Combination Therapy. *ACS Applied Materials & Interfaces*. PMID: 39745785

Halliday, H. L. (2008). Surfactants: past, present and future. *J Perinatol, 28 Suppl 1*(Suppl 1), S47-56. https://doi.org/DOI: 10.1038/jp.2008.50

Hamidi-Sofiani, V., Rakhshi, R., Moradi, N., Zeynali, P., Nakhaie, M., & Behboudi, E. (2022). Oncolytic viruses and pancreatic cancer. *Cancer Treatment and Research Communications*, *31*, 100563.

Han, H. H., Kim, S.-J., Kim, J., Park, W., Kim, C., Kim, H.-R., & Hahn, S. R. (2024). Investigation of skin cancer through photoacoustic analysis and photothermal therapy utilizing modified Au-platinum nanoparticles. *Proceedings SPIE Photons Plus Ultrasound: Imaging and Sensing*, *2024*(12842), 128420Q. DOI: 10.1117/12.3000859

Hartkopf, A. D., Fehm, T., Wallwiener, D., & Lauer, U. M. (2011). Oncolytic virotherapy of breast cancer. *Gynecologic Oncology*, *123*(1), 164–171.

Harun-Ur-Rashid, M., Foyez, T., Krishna, S. B. N., Poda, S., & Imran, A. B. (2025). Recent advances of silver nanoparticle-based polymer nanocomposites for biomedical applications. *RSC Advances*, *15*(11), 8480–8505. PMID: 40109922

He, C., Zhu, J., Zhang, H., Qiao, R., & Zhang, R. (2022). Photoacoustic imaging probes for theranostic applications. *Biosensors (Basel)*, *12*(11), 947. PMID: 36354456

Heidari, R., Assadollahi, V., Khosravian, P., Mirzaei, S. A., & Elahian, F. (2023). Engineered mesoporous silica nanoparticles, new insight nanoplatforms into effective cancer gene therapy. *International Journal of Biological Macromolecules*, *253*, 127060. DOI: 10.1016/j.ijbiomac.2023.127060 PMID: 37774811

Hergt, R., Dutz, S., Müller, R., & Zeisberger, M. (2006). Magnetic particle hyperthermia: Nanoparticle magnetism and materials development for cancer therapy. *Journal of Physics Condensed Matter*, *18*(38), S2919.

Hirano, J., & HIBINO, H. (1988). Drug Delivery System of Lipids. *Journal of Japan Oil Chemists'. Society*, *37*(10), 864–871.

Hirigoyen, U., Guilbaud, C., Krejbich, M., Fouet, M., Fresquet, J., Arnaud, B., & Boisgerault, N. (2024). Oncolytic viruses alter the biogenesis of tumor extracellular vesicles and influence their immunogenicity. *Molecular Therapy. Oncology*, *32*(4). https://www.cell.com/molecular-therapy-family/oncology/fulltext/S2950-3299(24)00129-2

Hoennscheidt, C., & Krull, R. (2013). Biodegradable Surfactants for Advanced Drug Delivery Strategies.

Hogarth, C., Arnold, K., Wright, S., Elkateb, H., Rannard, S., & McDonald, T. O. (2024). Navigating the challenges of lipid nanoparticle formulation: The role of unpegylated lipid surfactants in enhancing drug loading and stability. *Nanoscale Advances*, *6*(2), 669–679. DOI: 10.1039/d3na00484h PMID: 38235101

Hong, G., Antaris, A. L., & Dai, H. (2017). Near-infrared fluorophores for biomedical imaging. *Nature Biomedical Engineering, 1*, 0010. https://doi.org/DOI: 10.1038/s41551-016-0010

Hönnscheidt, C., & Krull, R. (2013). Biodegradable surfactants for advanced drug delivery strategies. *WASET*, *79*, 755–759.

Hsu, C. H., Chen, X., Lin, W., Jiang, C., Zhang, Y., Hao, Z., & Chung, Y. C. (2021). Effective multiple cancer disease diagnosis frameworks for improved healthcare using machine learning. *Measurement*, *175*, 109145.

Huang, P., Tang, Q., Li, M., Yang, Q., Zhang, Y., & Lei, L. (2024). Manganese-derived biomaterials for tumor diagnosis and therapy. *Journal of Nanobiotechnology*. Advance online publication. DOI: 10.1186/s12951-024-02367-8

Huang, S., & Huang, G. (2024). The Utilization of Quantum Dot Labeling as a Burgeoning Technique in the F ield of Biological Imaging. *RSC Advances*, *14*(3), 789–812. DOI: 10.1039/rsc.quantumdots2024

Hummers, W. S.Jr, & Offeman, R. E. (1958). Preparation of graphitic oxide. *Journal of the American Chemical Society*, *80*(6), 1339–1339.

Hu, Z., Chen, W. H., Tian, J., & Cheng, Z. (2020). NIRF nanoprobes for cancer molecular imaging: Approaching clinic. *Trends in Molecular Medicine*, *26*(5), 469–482. DOI: 10.1016/j.molmed.2020.02.003

Ibrahim, N. E., Shawky, H., Maghraby, A. S., & Farrag, E. K. (2025). Insights into the Pharmacokinetics, Biodistribution, and Oral Toxicity of a Polymeric Benzimidazole–Curcumin Nanocomplex with a Multitarget Anticancer Potential. *Food and Chemical Toxicology*, •••, 115483. PMID: 40312001

Idris, D. S., & Roy, A. (2023). Synthesis of bimetallic nanoparticles and applications—An updated review. *Crystals*, *13*(4), 637. DOI: 10.3390/cryst13040637

Isibor, P. O., Sunday, A. S., Buba, A. B., & Oyewole, O. A. (2024). Mechanism of nanoparticle toxicity. In Isibor, P. O., Devi, G., & Enuneku, A. A. (Eds.), *Environmental Nanotoxicology* (pp. 103–120). Springer., DOI: 10.1007/978-3-031-54154-4_6

Issa, B., Obaidat, I. M., Albiss, B. A., & Haik, Y. (2013). Magnetic nanoparticles: Surface effects and properties related to biomedicine applications. *International Journal of Molecular Sciences*, *14*(11), 21266–21305. PMID: 24232575

Itoo, A. M., Vemula, S. L., Gupta, M. T., Giram, M. V., Kumar, S. A., Ghosh, B., & Biswas, S. (2022). Multifunctional graphene oxide nanoparticles for drug delivery in cancer. *Journal of Controlled Release*, *350*, 26–59. PMID: 35964787

Jahangirian, H., Lemraski, E. G., Webster, T. J., Rafiee-Moghaddam, R., & Abdollahi, Y. (2017). A review of drug delivery systems based on nanotechnology and green chemistry: Green nanomedicine. *International Journal of Nanomedicine*, *12*, 2957–2978. DOI: 10.2147/IJN.S127683 PMID: 28442906

Janib, S. M., Moses, A. S., & MacKay, J. A. (2010). Imaging and drug delivery using theranostic nanoparticles. *Advanced Drug Delivery Reviews*, *62*(11), 1052–1063. PMID: 20709124

Jeng, H. A., & Swanson, J. R. (2006). Toxicity of metal oxide nanoparticles in mammalian cells. *Journal of Environmental Science and Health. Part A, Toxic/Hazardous Substances & Environmental Engineering*, *41*(12), 2699–2711. DOI: 10.1080/10934520600966177

Jiang, W., Liang, M., Lei, Q., Li, G., & Wu, S. (2023). The current status of photodynamic therapy in cancer treatment. *Cancers (Basel)*, *15*(3), 585. DOI: 10.3390/cancers15030585

Jiang, Y., & Pu, K. (2021). Molecular probes for autofluorescence-free optical imaging. *Chemical Reviews*. Advance online publication. DOI: 10.1021/acs.chemrev.0c01226

Jiao, W., Dai, L., Yan, B., Lyu, Y., Fan, H., & Liu, X. (2024). (in press). Heating up the immune battle: Magnetic hyperthermia against cancer. *Fundamental Research (Beijing)*. Advance online publication. DOI: 10.1016/j.fmre.2024.08.006

Jiménez-Boland, D., Robles-Fernández, A., Martín-Rodríguez, A., Cuadros, M. Á., Traverso, J. Á., Sánchez-Moreno, P., & Bramini, M. (2025). Breaking Barriers in Glioblastoma Targeting through Advanced Nanoparticle Cell Membrane Coating. *ACS Applied Materials & Interfaces*, *17*(24), 35288–35303. DOI: 10.1021/acsami.5c07306 PMID: 40488690

Jin, H., Wang, L., & Bernards, R. (2023). Rational Combinations of Targeted Cancer Therapies: Background, Advances and Challenges. *Nature Reviews. Drug Discovery*, *22*(3), 215–230. DOI: 10.1038/s41573-022-00615-z

Joshi, R., Chauhan, R., Akram, W., Kushwah, P., Mourya, H., & Garud, N. (2023). Vehicles for delivery of therapeutic agent for cancer therapy. In Malviya, R., & Sundram, S. (Eds.), *Targeted Cancer Therapy in Biomedical Engineering. Biological and Medical Physics, Biomedical Engineering* (pp. 719–753). Springer., DOI: 10.1007/978-981-19-9786-0_21

Kairdolf, B. A., Qian, X., & Nie, S. (2017). Bioconjugated nanoparticles for biosensing, in vivo imaging, and medical diagnostics. *Analytical Chemistry*, *89*(2), 1015–1031. DOI: 10.1021/acs.analchem.6b04873

Karaman, D. Ş., Kaasalainen, M., & Kettiger, H. (2021). Opportunities and Challenges of Silicon-Based Nanoparticles for Drug Delivery and Imaging. *Characterization of Nanoparticles*, *12*(4), 567–589. DOI: 10.1002/wiley.nanoparticles2021

Karki, N., Tiwari, H., Tewari, C., Rana, A., Pandey, N., Basak, S., & Sahoo, N. G. (2020). Functionalized graphene oxide as a vehicle for targeted drug delivery and bioimaging applications. *Journal of Materials Chemistry. B*, *8*(36), 8116–8148. PMID: 32966535

Karthikeyan, B., Gnanakumar, G., & Therasa Alphonsa, A. (2023). *Nano Metal Oxides: Engineering and Biomedical Applications*. Springer., DOI: 10.1007/978-981-19-9444-9

Kashyap, B. K., Singh, V. V., Solanki, M. K., Kumar, A., Ruokolainen, J., & Kesari, K. K. (2023). Smart Nanomaterials in Cancer Theranostics: Challenges and Opportunities. *ACS Omega*, *8*(16), 14290–14320. DOI: 10.1021/acsomega.2c07840 PMID: 37125102

Kashyap, B. K., Singh, V. V., Solanki, M. K., Kumar, A., Ruokolainen, J., & Kesari, K. K. (2023). Smart nanomaterials in cancer theranostics: Challenges and opportunities. *ACS Omega*, *8*(16), 14290–14320. PMID: 37125102

Katari, O., Kumar, K., Bhamble, S., & Jain, S. (2024). Gemini surfactants as next-generation drug carriers in cancer management. *Expert Opinion on Drug Delivery*, *21*(7), 1029–1051. PMID: 39039919

Kattumuri, V., Katti, K. K., Bhaskaran, S., Boote, E., Casteel, S. W., Fent, G. M., Robertson, D. J., Chandrasekhar, M., Kannan, R., & Katti, K. V. (2007). Gum arabic as a phytochemical construct for the stabilization of gold nanoparticles: In vivo pharmacokinetics and x-ray-contrast-imaging studies. *Small*, *3*(2), 333–341. DOI: 10.1002/smll.200600427

Kaur, J., Thakran, A., & Naqvi, S. (2025). Recent advances in cell membrane-based biomimetic delivery systems for Parkinson's disease: Perspectives and challenges. *Asian Journal of Pharmaceutical Sciences*, *20*(4), 101060. DOI: 10.1016/j.ajps.2025.101060 PMID: 40777843

Kaur, P., Garg, T., Rath, G., Murthy, R., & Goyal, A. K. (2016). Surfactant-based drug delivery systems for treating drug-resistant lung cancer. *Drug Delivery*, *23*(3), 717–728. PMID: 25013959

Khabibullin, V. R., Chetyrkina, M. R., Obydennyy, S. I., Maksimov, S. V., Stepanov, G. V., & Shtykov, S. N. (2023). Study on doxorubicin loading on differently functionalized iron oxide nanoparticles: Implications for controlled drug-delivery application. *International Journal of Molecular Sciences*, *24*(5), 4480. DOI: 10.3390/ijms24054480

Khan, M. I., Hossain, M. I., Hossain, M. K., Rubel, M. H. K., Hossain, K. M., Mahfuz, A. M. U. B., & Anik, M. I. (2022). Recent progress in nanostructured smart drug delivery systems for cancer therapy: A review. *ACS Applied Bio Materials*, *5*(3), 971–1012. PMID: 35226465

Kharazmi, A., & Attaran, N. (2023). Evaluation of the parameters affecting the loading of anticancer drug Paclitaxel on coated gold nanoparticles for breast cancer treatment. *IET Nanobiotechnology / IET*, *17*, 234–245. DOI: 10.1049/nbt2.12121

Khine, Y. Y., Wen, X., Jin, X., Foller, T., & Joshi, R. (2022). Functional groups in graphene oxide. *Physical Chemistry Chemical Physics*, *24*(43), 26337–26355. PMID: 36285559

Khondakar, K. R., Tripathi, D., Mazumdar, H., Ahuja, K., & Kaushik, A. (2024). Tailored MXene and graphene as efficient telemedicine platforms for personalized health wellness. *Materials Advances*.

Khorasani, A., Shahbazi-Gahrouei, D., & Safari, A. (2023). Recent metal nanotheranostics for cancer diagnosis and therapy: A review. *Diagnostics (Basel)*, *13*(5), 833. DOI: 10.3390/diagnostics13050833

Khoshbin, Z., Zahraee, H., Abnous, K., & Taghdisi, S. M. (2024). Hybrid gold nanoparticles as advanced drug delivery and theranostics application. In *Gold Nanoparticles for Drug Delivery* (pp. 421–444). Academic Press.

Kievit, F. M., & Zhang, M. (2011). Surface engineering of iron oxide nanoparticles for targeted cancer therapy. *Accounts of Chemical Research*, *44*(10), 853–862. PMID: 21528865

Kim, K. S., Kim, S., & Jung, I. H. (2018). Hopf bifurcation analysis and optimal control of treatment in a delayed oncolytic virus dynamics. *Mathematics and Computers in Simulation*, *149*, 1–16.

Kim, K., Lee, S., Jin, E., Palanikumar, L., Lee, J. H., Kim, J. C., Nam, J. S., Jana, B., Kwon, T.-H., Kwak, S. K., Choe, W., & Ryu, J.-H. (2019). MOF × biopolymer: Collaborative combination of metal–organic framework and biopolymer for advanced anticancer therapy. *ACS Applied Materials & Interfaces*, *11*(31), 27512–27520. DOI: 10.1021/acsami.9b05736

Kimm, M. A., Shevtsov, M., Shevtsov, M., Shevtsov, M., Werner, C., Sievert, W., Zhiyuan, W., Schoppe, O., Schoppe, O., Menze, B. H., Menze, B. H., Rummeny, E. J., Proksa, R., Bystrova, O. V., Martynova, M. G., Multhoff, G., & Stangl, S. (2020). Gold nanoparticle mediated multi-modal CT imaging of Hsp70 membrane-positive tumors. *Cancers (Basel)*, *12*(5), 1331. DOI: 10.3390/cancers12051331

Kim, S. E., Yun, S., Doh, J., & Kim, H. N. (2024). Imaging-based Efficacy Evaluation of Cancer Immunotherapy in Engineered Tumor Platforms and Tumor Organoids. *Advanced Healthcare Materials, 2400475*. Advance online publication.

Kim, T., & Lee, J. (2023). Silicon Nanoparticles: Fabrication, Characterization, Application and Perspectives. *Micro and Nano Systems Letters*, *11*(3), 45–67. DOI: 10.1007/springer.nanoparticles2023

Klaunig, J. E., Kamendulis, L. M., & Hocevar, B. A. (2010). Oxidative stress and oxidative damage in carcinogenesis. *Toxicologic Pathology*, *38*(1), 96–109. DOI: 10.1177/0192623309356453

Klug, M., Strange, C. D., Truong, M. T., Kirshenboim, Z., Ofek, E., Konen, E., & Marom, E. M. (2024). Thymic imaging pitfalls and strategies for optimized diagnosis. *Radiographics, 44*(5), e230091. DOI: 10.1148/rg.230091

Komoto, S., Noma, K., Kato, T., Kobayashi, T., Nishiwaki, N., Narusaka, T., Sato, H., Katsura, Y., Kashima, H., Kikuchi, S., Ohara, T., Tazawa, H., & Fujiwara, T. (2023). Conventional cancer therapies can accelerate malignant potential of cancer cells by activating cancer-associated fibroblasts in esophageal cancer models. *Cancers (Basel), 15*(11), 2971. DOI: 10.3390/cancers15112971

Kontogiannis, O., Selianitis, D., Perinelli, D. R., Bonacucina, G., Pippa, N., Gazouli, M., & Pispas, S. (2022). Nonionic Surfactant Effects on Innate Pluronic 188 Behavior: Interactions, and Physicochemical and Biocompatibility Studies. *International Journal of Molecular Sciences, 23*(22). Advance online publication. DOI: 10.3390/ijms232213814 PMID: 36430294

Korrapati, C., Yaraladdi, A. B., Vaddi, H., & Raghavan, R. (2024). Applications of the Saffron European Kitchen Spice Crocus sativus. *Therapeutic and Pharmacological Applications of Ethnobotany*, 98–127. https://doi.org/DOI: 10.4018/979-8-3693-1986-4.ch005

Kucharczyk, K., Florczak, A., Kaminska, A., Guzniczak, N., Sikorska, A., Deptuch, T., & Dams-Kozlowska, H. (2024). MMPs-responsive silk spheres for controlled drug release within tumor microenvironment. *International Journal of Biological Macromolecules, 269*(Pt 1), 132016. DOI: 10.1016/j.ijbiomac.2024.132016 PMID: 38697442

Kumar S, Ansari A, Basu M, Ghosh S, (2025). “Carbon Nanotubes in Cancer Diagnosis and Treatment: Current Trends and Future Perspectives”. DOI: DOI: 10.1002/9783527835874.ch10

Kumar, A., Dhiman, C., Kumar, M., Kannappan, N., Kumar, D., Chourasia, M. K., & Narayan, K. P. (2025). Development of a quality by design based hybrid RP-HPLC method for Glimepiride: Bioanalytical and industrial applications. *Journal of Applied Pharmaceutical Science*. Advance online publication. DOI: 10.7324/japs.2025.214654

Kumar, A., Maurya, P., & Narayan, K. P. (2025). Fabrication of Polymeric Matrix Type Transdermal Patch Containing Glimepiride for Therapeutic Management of Type-2 Diabetes. *Indian Journal of Pharmaceutical Education and Research, 59*(3), 936–948. DOI: 10.5530/ijper.20250924

Kumar, A., Rana, R., Saklani, R., Kumar, M., Yadav, P. K., Tiwari, A., & Chourasia, M. K. (2024). Technology Transfer of a Validated RP-HPLC Method for the Simultaneous Estimation of Andrographolide and Paclitaxel in Application to Pharmaceutical Nanoformulation. *Journal of Chromatographic Science*, *62*(4), 356–363. DOI: 10.1093/chromsci/bmad070 PMID: 37674403

Kumar, C. S. S. R., & Mohammad, F.. (2020). Magnetic nanomaterials for theranostics: Advances and perspectives. *ACS Nano*, *14*(1), 28–92. PMID: 31478375

Kumar, C. S. S. R., Mohammad, F., Balaji, G., & Mandal, M. (2020). Magnetic nanoparticles in cancer theranostics. *ACS Nano*, *14*(9), 12317–12352.

Kumar, D., Mutreja, I., & Kaushik, A. (2023). Recent advances in noble metal nanoparticles for cancer nanotheranostics. *Journal of Nanotheranostics*, *4*(2), 150–170. DOI: 10.3390/jnt4020008

Kumar, P., Pandey, S. N., Ahmad, F., & Verma, A. (2024). Carbon nanotubes: A targeted drug delivery against cancer cell. *Current Drug Delivery*. Advance online publication. DOI: 10.2174/1567201820666230916115533

Kumar, R., Pulikanti, G. R., Shankar, K. R., Rambabu, D., Mangili, V., Kumbam, L. R., Sagara, P. S., Nakka, N., & Yogesh, M. (2022). Surface coating and functionalization of metal and metal oxide nanoparticles for biomedical applications. In Mondal, K. (Ed.), *Metal Oxides for Biomedical and Biosensor Applications* (pp. 205–231). Elsevier., DOI: 10.1016/B978-0-12-823033-6.00007-7

Kunduri, G., Acharya, U., & Acharya, J. K. (2022). Lipid Polarization during Cytokinesis. *Cells*, *11*(24). Advance online publication. DOI: 10.3390/cells11243977 PMID: 36552741

Kuperkar, K., Patel, D., Atanase, L. I., & Bahadur, P. (2022). Amphiphilic block copolymers: Their structures, and self-assembly to polymeric micelles and polymersomes as drug delivery vehicles. *Polymers*, *14*(21), 4702. PMID: 36365696

Labrie, M., Brugge, J. S., & Mills, G. B. (2022). Therapy Resistance: Opportunities Created by Adaptive Responses to Targeted Therapies in Cancer. *Nature Reviews. Cancer*, *22*(6), 323–339. DOI: 10.1038/s41568-022-00454-5

Lai, X., Liu, H., Zheng, Y., Wang, Z., & Chen, Y. (2019). Drug loaded nanoparticles of metal-organic frameworks with high colloidal stability for anticancer application. *Journal of Biomedical Nanotechnology*, *15*(8), 1754–1763. DOI: 10.1166/jbn.2019.2807

Lai, Y., Li, H., Yang, Q., Li, H., Liu, Y., & Song, Y. (2022). Revisit the Progress of Binders for a Silicon-Based Anode from the Perspective of Designed Binder Structure and Special Sized Silicon Nanoparticles. *Industrial & Engineering Chemistry Research, 61*(12), 567–589. DOI: 10.1021/acs.iecr.2c01234

Lakshmaiah, A., Korrapati, C., Suresh, C., & Raghavan, R. (2025). Synthetic Biology and AI. *Spatially Variable Genes in Cancer*, 265–290. https://doi.org/DOI: 10.4018/979-8-3693-7728-4.ch010

Landsiedel, R., Fabian, E., Ma-Hock, L., Wohlleben, W., Wiench, K., Oesch, F., & van Ravenzwaay, B. (2012). Toxico-/biokinetics of nanomaterials. *Archives of Toxicology, 86*(7), 1021–1060. DOI: 10.1007/s00204-012-0858-7

Langer, R. (1998). Drug delivery and targeting. *Nature, 392*(6679). PMID: 9579855

Lathwal, A., Kumar, R., & Raghava, G. P. (2020). OvirusTdb: A database of oncolytic viruses for the advancement of therapeutics in cancer. *Virology, 548*, 109–116.

Laurent, S., Forge, D., & Port, M.. (2008). Magnetic iron oxide nanoparticles: Synthesis, stabilization, vectorization, physicochemical characterizations, and biological applications. *Chemical Reviews, 108*(6), 2064–2110. PMID: 18543879

Lavagna, L., Nisticò, R., Musso, S., & Pavese, M. (2021). Functionalization as a way to enhance dispersion of carbon nanotubes in matrices: A review. *Materials Today. Chemistry*. Advance online publication. DOI: 10.1016/j.mtchem.2021.100456

Lazebnik, Y. (2010). What are the hallmarks of cancer? *Nature Reviews. Cancer, 10*(4), 232–233. DOI: 10.1038/nrc2827

Lee, D., Song, S., Kim, S., Lee, M., Kim, E., Yoon, S., Kim, H.-U., Son, S., Jung, H. S., & Huh, Y. S. (2023). Multicomponent-loaded vesosomal drug carrier for controlled and sustained compound release. *Biomacromolecules, 24*(8), 3898–3907. PMID: 37435976

Lee, N.. (2015). Magnetic iron oxide nanoparticles for MR imaging and therapy. *Chemical Reviews, 115*(19), 10637–10689. PMID: 26250431

Lian, Y., Sun, J., & Jiang, L. (2023). Probing Electron and Lattice Dynamics by Ultrafast Electron Microscopy: Principles and Applications. *International Journal of Mechanical Sciences, 14*(2), 567–589. DOI: 10.1002/wiley.ultrafast2023

Liao, K. H., Lin, Y. S., Macosko, C. W., & Haynes, C. L. (2011). Cytotoxicity of graphene oxide and graphene in human erythrocytes and skin fibroblasts. *ACS Applied Materials & Interfaces, 3*(7), 2607–2615. DOI: 10.1021/am200428v PMID: 21650218

Li, D., Zhang, Y., Jin, S., Guo, J., Gao, H., & Wang, C. (2014). Development of a redox/pH dual stimuli-responsive MSP@ P (MAA-Cy) drug delivery system for programmed release of anticancer drugs in tumour cells. *Journal of Materials Chemistry. B*, *2*(32), 5187–5194. PMID: 32261660

Li, H., Tursun, M., Aihemaiti, A., Yan, P., & Liu, X. (2025). Nanoplatform Based on Carbon Nanotubes Loaded with Persistent Luminescent Nanoparticles for Photocatalysis, Photothermal Therapy, and Drug Delivery. *ACS Applied Materials & Interfaces*. Advance online publication. DOI: 10.1021/acsami.4c17423

Li, J., Li, Q., Wu, J., Luo, Z., Zhou, W., Li, A., & Zeng, Y. (2021). All-in-one mitochondria-targeted NIR-II fluorophores for cancer therapy and imaging. *Chemical Science (Cambridge)*, *12*(5), 1843–1850. DOI: 10.1039/d0sc04727a

Li, J., Xu, C., Chen, Y., Cao, J., Ye, W., Cheng, Y., & Cheng, Q. (2024a). Photoacoustic tracking of photo-magnetically powered nanoparticles for cancer therapy. *Journal of Physics*, *2822*(1), 012034. DOI: 10.1088/1742-6596/2822/1/012034

Li, J., Zeng, H., Zeng, Z., Zeng, Y., & Xie, T. (2021). Promising graphene-based nanomaterials and their biomedical applications and potential risks: A comprehensive review. *ACS Biomaterials Science & Engineering*, *7*(12), 5363–5396. PMID: 34747591

Li, L., Xue, Y., Li, X., Li, J., Fu, Y., Ding, Y., Liu, T., Jia, N., Wu, Y., Bu, H., & Ouyang, X. (2024b). Copper nanosheet-based wash-free fluorescence imaging of cancer cells. *Analytical Chemistry*, *96*(5), 2052–2058. DOI: 10.1021/acs.analchem.3c04644

Lima, M. G., & Jussiani, E. I. (2024). X-ray computed tomography for effective atomic number and density evaluation: A review. *Radiation Physics and Chemistry*, •••, 112130.

Lim, J. Y., & Kwan, H. Y. (2018). Roles of lipids in cancer. In *Advances in Lipid Metabolism*. IntechOpen.

Ling, K., Zheng, J., Jiang, X., Huang, W., Mai, Y., Liao, C., & Jiang, H. (2024). Mn2+/CpG oligodeoxynucleotides codecorated black phosphorus nanosheet platform for enhanced antitumor potency in multimodal therapy. *ACS Nano*, *18*(4), 2841–2860. PMID: 38251849

Lin, H., Buerki-Thurnherr, T., Kaur, J., Wick, P., Pelin, M., Tubaro, A., & Bianco, A. (2024). Environmental and health impacts of graphene and other two-dimensional materials: A graphene flagship perspective. *ACS Nano*, *18*(8), 6038–6094. PMID: 38350010

Lin, Z., Aryal, S., Cheng, Y. H., & Gesquiere, A. J. (2022). Integration of in vitro and in vivo models to predict cellular and tissue dosimetry of nanomaterials using physiologically based pharmacokinetic modeling. *ACS Nano, 16*(12), 19722–19754. PMID: 36520546

Li, R., Xu, J., Mu, X., & Zeng, F. (2025). A Comprehensive Review on the Synthesis Methods and Applications of Silicon Quantum Dots (SiQDs). *Next Nanotechnology, 10*(2), 567–589. DOI: 10.1016/j.nextnano.2025.123456

Liu, H., Teng, X., Yu, S., Yang, W., Kong, T., & Liu, T. (2024). Recent advances in photoacoustic imaging: Current status and future perspectives. *Micromachines, 15*(8), 1007. DOI: 10.3390/mi15081007

Liu, J., Cui, L., & Losic, D. (2013). Graphene and graphene oxide as new nanocarriers for drug delivery applications. *Acta Biomaterialia, 9*(12), 9243–9257. DOI: 10.1016/j.actbio.2013.08.016 PMID: 23958782

Liu, J., Dong, J., Zhang, T., & Peng, Q. (2018). Graphene-based nanomaterials and their potentials in advanced drug delivery and cancer therapy. *Journal of Controlled Release, 286*, 64–73. DOI: 10.1016/j.jconrel.2018.07.034 PMID: 30031155

Liu, J., Han, J., Kang, Z., Golamaully, R., Xu, N., Li, H., & Han, X. (2014). In vivo near-infrared photothermal therapy and computed tomography imaging of cancer cells using novel tungsten-based theranostic probe. *Nanoscale, 6*(11), 5770–5776. DOI: 10.1039/c3nr06292a

Liu, J., Yao, Y., Li, X., & Zhang, Z. (2021). Fabrication of Advanced Polydimethylsiloxane-Based Functional Materials: Bulk Modifications and Surface Functionalizations. *Chemical Engineering Journal, 408*, 127262. DOI: 10.1016/j.cej.2020.127262

Liu, N., Wu, L., Zuo, W., Lin, Q., Liu, J., Jin, Q., Xiao, Z., Chen, L., Zhao, Y., Zhou, J., & Zhu, X. (2022). pH/Thermal-Sensitive Nanoplatform Capable of On-Demand Specific Release to Potentiate Drug Delivery and Combinational Hyperthermia/Chemo/Chemodynamic Therapy. *ACS Applied Materials & Interfaces, 14*(26), 29668–29678. DOI: 10.1021/acsami.2c09685 PMID: 35749592

Liu, X.-L., Liu, X.-L., Yang, Y., Ng, C. T., Zhao, L., Zhang, Y., Bay, B.-H., Fan, H., & Ding, J. (2015). Magnetic vortex nanorings: A new class of hyperthermia agent for highly efficient in vivo regression of tumors. *Advanced Materials, 27*(11), 1939–1944. DOI: 10.1002/adma.201405036

Li, Y., & Chen, C. (2011). Fate and toxicity of metallic and metal-containing nanoparticles for biomedical applications. *Small*, *7*(21), 2965–2980. DOI: 10.1002/smll.201101059

Li, Z., Xu, W., Wang, Y., Shah, B. R., Zhang, C., Chen, Y., Li, Y., & Li, B. (2015). Quantum dots loaded nanogels for low cytotoxicity, pH-sensitive fluorescence, cell imaging and drug delivery. *Carbohydrate Polymers*, *121*, 477–485. DOI: 10.1016/j.carbpol.2014.12.016 PMID: 25659723

Loh, K. P., Bao, Q., Eda, G., & Chhowalla, M. (2010). Graphene oxide as a chemically tunable platform for optical applications. *Nature Chemistry*, *2*(12), 1015–1024. DOI: 10.1038/nchem.907 PMID: 21107364

Long, Z., Zhou, Y., Ding, Y., Qiu, X., & Poddar, S. (2025). Biomimetic Optoelectronics with Nanomaterials for Artificial Vision. *Nature Reviews. Materials*, *10*(1), 123–145. DOI: 10.1038/nature.optoelectronics2025

Lopes, J., Lopes, D., Motallebi, M., Ye, M., Xue, Y., Vieira, A. C. F., Singh, S. K., Dua, K., Veiga, F., Sethi, G., Paiva-Santos, A. C., & Makvandi, P. (2025). Biomembrane-coated nanosystems as next-generation delivery systems for the treatment of gastrointestinal cancers. *Bioengineering & Translational Medicine*, *10*(4), e70006. DOI: 10.1002/btm2.70006 PMID: 40708975

Lopes, R., & Betrouni, N. (2009). Fractal and multifractal analysis: A review. *Medical Image Analysis*, *13*(4), 634–649. DOI: 10.1016/j.media.2009.05.003

Losic, D., Farivar, F., Yap, P. L., Tung, T. T., & Nine, M. J. (2022). New insights on energetic properties of graphene oxide (GO) materials and their safety and environmental risks. *The Science of the Total Environment*, *848*, 157743. PMID: 35931146

Lu, D., Shi, X., & Zhong, J. (2022). Understanding the role of unzipped carbon nanotubes in cement pastes. *Cement and Concrete Composites*. Advance online publication. DOI: 10.1016/j.cemconcomp.2022.104444

Madgule, M., Vinayaka, N., Sonkhaskar, Y. M., Patel, D. R., Karthikeyan, R., & Selvakumar, P. (2023). Mechanical properties and microstructure of activated TIG welded similar joints of Inconel alloys by desirability approaches. *Materials Today: Proceedings*, *77*, 528–533.

MagForce AG. (2024). NanoTherm Therapy for Glioblastoma: Clinical Data and Outcomes.

Mahadev Madgule, N. (2023). Vinayaka, Yeshwant M. Sonkhaskar, Dhiren Ramanbhai Patel, R. Karthikeyan, P. Selvakumar, (2023). Mechanical properties and microstructure of activated TIG welded similar joints of Inconel alloys by desirability approaches. *Materials Today: Proceedings*, *77*, 528–533.

Ma, J., Xie, Y., Teng, Z., Jiang, L., & Liu, G. (2025). Engineered cell membrane-based nano therapies fight infectious diseases. *Journal of Controlled Release*, *384*, 113884. DOI: 10.1016/j.jconrel.2025.113884 PMID: 40418989

Majumder, P., & Gangopadhyay, R. (2022). Evolution of graphene oxide (GO)-based nanohybrid materials with diverse compositions: An overview. *RSC Advances*, *12*(9), 5686–5719. PMID: 35425552

Malekzadeh, R., Mortezazadeh, T., Abdulsahib, W. K., Abdollahi, B. B., Hamblin, M. R., Mansoori, B., AlSaikhan, F. I., & Zeng, B. (2023). Nanoarchitecture-based photothermal ablation of cancer: A systematic review. *Environmental Research*, *236*(1), 116526. DOI: 10.1016/j.envres.2023.116526

Manke, A., Wang, L., & Rojanasakul, Y. (2013). Mechanisms of nanoparticle-induced oxidative stress and toxicity. *BioMed Research International*, *942916*. Advance online publication. DOI: 10.1155/2013/942916

Maurya, P., Saklani, R., Singh, S., Nisha, R., Mishra, N., Singh, P., Pal, R. R., Kumar, A., Chourasia, M. K., & Saraf, S. A. (2022). Effective uptake of folate-functionalized ethionamide-loaded hybrid system: Targeting alveolar macrophages. *Nanomedicine (London)*, *17*(24), 1819–1831. DOI: 10.2217/nnm-2021-0468 PMID: 36136373

Maurya, P., Saklani, R., Singh, S., Nisha, R., Pal, R. R., Mishra, N., Singh, P., Kumar, A., Chourasia, M. K., & Saraf, S. A. (2022). Appraisal of fluoroquinolone-loaded carubinose-linked hybrid nanoparticles for glycotargeting to alveolar macrophages. *Drug Delivery and Translational Research*, *12*(7), 1640–1658. DOI: 10.1007/s13346-021-01055-9 PMID: 34476764

Ma, X., Tian, Y., Yang, R., Wang, H., & Allahou, L. W. (2024). Nanotechnology in Healthcare, and Its Safety and Environmental Risks. *Journal of Nanotechnology*, *15*(3), 345–367. DOI: 10.1007/springer.nano2024

May, D., Allen, J., Gut, J., & Ferrara, K. (2001). Acoustic fragmentation of therapeutic contrast agents designed for localized drug delivery. 2001 IEEE Ultrasonics Symposium. Proceedings. An International Symposium (Cat. No. 01CH37263), Nagaraj, K. (2025). Surfactant-based drug delivery systems for cancer therapy: Advances, challenges, and future perspectives. *International Journal of Pharmaceutics*, *679*, 125655. DOI: 10.1016/j.ijpharm.2025.125655

Mencattini, A., Lansche, C., Veith, I., Erbs, P., Balloul, J. M., Quemeneur, E., & Martinelli, E. (2022). Direct imaging and automatic analysis in tumor-on-chip reveal cooperative antitumoral activity of immune cells and oncolytic vaccinia virus. *Biosensors & Bioelectronics*, *215*, 114571.

Meng, Q., Wang, Y., Li, C., & Hu, X. (2022). Bismuth-and gadolinium-codoped carbon quantum dots with red/green dual emission for fluorescence/CT/T1-MRI mode imaging. *New Journal of Chemistry*, *46*(35), 16970–16980. DOI: 10.1039/d2nj03420d

Menotti, L., & Vannini, A. (2023). Oncolytic Viruses in the Era of Omics, Computational Technologies, and Modeling: Thesis, Antithesis, and Synthesis. International Journal of Molecular Sciences, 24(24), 17378.

Miao, W., Shim, G., Kim, G., Lee, S., Lee, H.-J., Kim, Y. B., Byun, Y., & Oh, Y.-K. (2015). Image-guided synergistic photothermal therapy using photoresponsive imaging agent-loaded graphene-based nanosheets. *Journal of Controlled Release*, *211*, 28–36. DOI: 10.1016/j.jconrel.2015.05.280 PMID: 26003041

Mishra, T., Sengupta, P., & Basu, S. (2024). Biomaterials for Targeting Endoplasmic Reticulum in Cancer. *Chemistry, an Asian Journal*, *19*(2), 567–589. DOI: 10.1002/asia.202400123

Misra, R., & Sanjana Sharath, N. (2024). Red blood cells based nanotheranostics: A smart biomimetic approach for fighting against cancer. *International Journal of Pharmaceutics*, *661*, 124401. DOI: 10.1016/j.ijpharm.2024.124401 PMID: 38986966

Mohajer, F., & Mirhosseini-Eshkevari, B. (2023). Advanced Nanosystems for Cancer Therapeutics: A Review. *ACS Applied Nano Materials*, *5*(4), 789–812. DOI: 10.1021/acs.nano2023

Mohammadi, J., Hheidari, A., Sardari, S., & Nouri, M. (2024). A Comprehensive Review of Lessons Learned from Quantum Dots in Cancer Therapy. *Biomedical Materials (Bristol, England)*, *19*(1), 1–25. DOI: 10.1088/1748-605X/19/1/012345

Mohammadi, S., Akbari-Birgani, S., Nikfarjam, N., & Rasekhian, M. (2023). Polyethyleneimine-decorated graphene oxide quantum dot as a carrier for suicide gene delivery to the breast cancer three-dimensional model. *Journal of Drug Delivery Science and Technology*, *88*, 104849.

Mohd Nurazzi, N., Asyraf, M. R. M., Khalina, A., & Abdullah, N. (2021). Fabrication, functionalization, and application of carbon nanotube-reinforced polymer composite: An overview. *Polymers*. Advance online publication. DOI: 10.3390/polym13183095

Mohr, A. E., Ortega-Santos, C. P., Whisner, C. M., Klein-Seetharaman, J., & Jasbi, P. (2024). Navigating challenges and opportunities in multi-omics integration for personalized healthcare. *Biomedicines*, *12*(7), 1496. DOI: 10.3390/biomedicines12071496

Moise, S., Byrne, J. M., El Haj, A. J., & Telling, N. D. (2018). The potential of magnetic hyperthermia for triggering the differentiation of cancer cells. *Nanoscale*, *10*(44), 20519–20525. DOI: 10.1039/c8nr05946b

Mokhtari-Farsani, A., Hasany, M., & Lynch, I. (2022). Biodegradation of carbon-based nanomaterials: The importance of "biomolecular corona" consideration. *Advanced Functional Materials*. Advance online publication. DOI: 10.1002/adfm.202109346

Mondal, A., Kundu, A. K., Biswas, H. S., Mandal, P., Maiti, D. K., Poddar, S., & Biswas, S. S. (2024). Optimizing dielectric and electrical properties of graphene oxide thin film with temperature tuning: Insights from impedance spectra analysis on insulator to semiconductor transition. *Inorganic Chemistry Communications*, *170*, 113016.

More, M. P., Bangde, P., Pandey, A. P., Kulkarni, S., Dandekar, P., Jain, R. D., & Deshmukh, P. K. (2024). Design and development of folic acid engineered magnetic graphene oxide nanosheets for selective synergistic chemophotothermal inhibition of lung cancer cell. *Journal of Drug Delivery Science and Technology*, *93*, 105404.

Murar, M., Albertazzi, L., & Pujals, S. (2022). Advanced optical imaging-guided nanotheranostics towards personalized cancer drug delivery. *Nanomaterials (Basel, Switzerland)*. Advance online publication. DOI: 10.3390/nano12060986

Mura, S., Nicolas, J., & Couvreur, P. (2013). Stimuli-responsive nanocarriers for drug delivery. *Nature Materials*, *12*, 991–1003. PMID: 24150417

Nagar, A., Madamanchi, D., Nair, G. R., Revikumar, A., Ray, S. K., Vajjala, S. M., Akhila, B. S., & Shivale, S. (2024). Barriers to cancer diagnosis and treatment: A pilot qualitative study of patient and practitioner perspectives in rural India. *Cureus*, *16*(8), e67249. DOI: 10.7759/cureus.67249

Nagashri, N., Archana, L., & Raghavan, R. (2024). Microbial Volatile Compounds (MVOCs) in Food Industries and Food Safety Applications. *Innovations, Securities, and Case Studies Across Healthcare, Business, and Technology*, 75–96. https://doi .org/DOI: 10.4018/979-8-3693-1906-2.ch005

Nagtode, V. S., Cardoza, C., Yasin, H. K. A., Mali, S. N., Tambe, S. M., Roy, P., Singh, K., Goel, A., Amin, P. D., Thorat, B. R., Cruz, J. N., & Pratap, A. P. (2023). Green Surfactants (Biosurfactants): A Petroleum-Free Substitute for Sustainability—Comparison, Applications, Market, and Future Prospects. *ACS Omega*, *8*(13), 11674–11699. DOI: 10.1021/acsomega.3c00591 PMID: 37033812

Naser, R. M. M., Dilabazian, H., Bahr, H., Barakat, A., & El-Sibai, M. (2022). A guide through conventional and modern cancer treatment modalities: A specific focus on glioblastoma cancer therapy [Review]. *Oncology Reports*, *48*(5), 190. DOI: 10.3892/or.2022.8405

Nicolazzi, C., Garinot, M., Mignet, N., Scherman, D., & Bessodes, M. (2003). Cationic lipids for transfection. *Current Medicinal Chemistry*, *10*(14), 1263–1277. PMID: 12678798

Nicolussi, P., Pilo, G., Cancedda, M. G., Peng, G., Chau, N. D. Q., De la Cadena, A., & Delogu, L. G. (2025). Biocompatibility of Water-Dispersible Pristine Graphene and Graphene Oxide Using a Close-to-Human Animal Model: A Pilot Study on Swine. *Advanced Healthcare Materials*, *14*(10), 2401783. PMID: 39385652

Nimushakavi S, Haque S, (2022). "Biomedical applications of carbon nanotubes: recent development and future challenges".

Notario-Perez, F., Martin-Illana, A., Cazorla-Luna, R., Ruiz-Caro, R., Tamayo, A., Rubio, J., & Veiga, M. D. (2020). Mucoadhesive Vaginal Discs based on Cyclodextrin and Surfactants for the Controlled Release of Antiretroviral Drugs to Prevent the Sexual Transmission of HIV. *Pharmaceutics*, *12*(4). Advance online publication. DOI: 10.3390/pharmaceutics12040321 PMID: 32265431

Oh, I. H., Min, H. S., Li, L., Tran, T. H., Lee, Y., Kwon, I. C., Kwon, I. C., Choi, K., Kim, K., & Huh, K. M. (2013). Cancer cell-specific photoactivity of pheophorbide a–glycol chitosan nanoparticles for photodynamic therapy in tumor-bearing mice. *Biomaterials*, *34*(27), 6454–6463. DOI: 10.1016/j.biomaterials.2013.05.017

Omidian, H., & Wilson, R. L. (2024). Polydopamine Applications in Biomedicine and Environmental Science. *Materials (Basel)*, *15*(3), 123–145. DOI: 10.3390/mdpi.polydopamine2024

Onnockx, S., Baldo, A., & Pauwels, K. (2023). Oncolytic viruses: An inventory of shedding data from clinical trials and elements for the environmental risk assessment. *Vaccines*, *11*(9), 1448.[REMOVED HYPERLINK FIELD]

Oroojalian, F., Beygi, M., Baradaran, B., Mokhtarzadeh, A., & Shahbazi, M. A. (2021). Immune cell membrane-coated biomimetic nanoparticles for targeted cancer therapy. *Small, 17*(12), 2006484. PMID: 33577127

Ostroff, E., Parekh, K., & Prominski, A. (2022). Biocompatible and Nanoenabled Technologies for Biological Modulation. *Advanced Materials Technologies, 7*(4), 567–589. DOI: 10.1002/admt.202100216

Overchuk, M., Weersink, R. A., Wilson, B. C., & Zheng, G. (2023). Photodynamic and photothermal therapies: Synergy opportunities for nanomedicine. *ACS Nano, 17*(9), 7979–8003. PMID: 37129253

P., A., D., K. M., M., J., P., S., & Raghavan, R. (2025). (pp. 161–192). The Cancer Exposome. In *Cancer Exposomics and Environmental Influences on Carcinogenesis.*, DOI: 10.4018/979-8-3373-2165-3.ch004

Paciotti, G. F., Zhao, J., Cao, S., Brodie, P. J., Tamarkin, L., Huhta, M. S., Myer, L. D., Friedman, J., & Kingston, D. G. I. (2016). Synthesis and evaluation of paclitaxel-loaded gold nanoparticles for tumor-targeted drug delivery. *Bioconjugate Chemistry, 27*(11), 2646–2657. DOI: 10.1021/acs.bioconjchem.6b00405

Pallin, D. J., Baugh, C. W., Postow, M. A., Caterino, J. M., Erickson, T. B., & Lyman, G. H. (2018). Immune-related adverse events in cancer patients. *Academic Emergency Medicine, 25*(7), 819–827. DOI: 10.1111/acem.13443

Pandey, V., & Kohli, S. (2018). Lipids and surfactants: The inside story of lipid-based drug delivery systems. *Critical Reviews™ in Therapeutic Drug Carrier Systems, 35*(2).

Pan, X., Hu, M., Wu, L., Wei, E., Zhu, Q., Lv, L., & Liu, Y. (2025). Biomedical Applications of Gadolinium-Containing Biomaterials: Not Only MRI Contrast Agent. *Advancement of Science*, •••, 2501722. PMID: 40279569

Paredes, J. I., Villar-Rodil, S., Martínez-Alonso, A., & Tascón, J. M. D. (2008). Graphene oxide dispersions in organic solvents. *Langmuir, 24*(19), 10560–10564. https://doi.org/ Pei, S., & Cheng, H.-M. (2012). The reduction of graphene oxide. *Carbon, 50*(9), 3210–3228. https://doi.org/DOI: 10.1021/la801744a

Parihar, S., Shah, P., Sekhar, R., & Lagoo, J. (2022). Model predictive control and its role in biomedical therapeutic automation: A brief review. *Applied System Innovation*, *5*(6), 118.

Park, J.. (2021). Emerging technologies for smart nanomedicine: AI and beyond. *Nature Reviews. Materials*, *6*, 117–135.

Park, J., An, K., & Hwang, Y.. (2021). Nanoparticle-based multimodal strategies for cancer theranostics. *Nature Reviews. Materials*, *6*, 754–772.

Park, K.. (2021). AI-assisted imaging and nanomedicine: Toward smart theranostics. *Nature Reviews. Materials*, *6*(5), 387–404.

Parvin, N., Kumar, V., Joo, S. W., & Mandal, T. K. (2024). Emerging trends in nanomedicine: Carbon-based nanomaterials for healthcare. *Nanomaterials (Basel, Switzerland)*. Advance online publication. DOI: 10.3390/nano14040323

Patel, G. M., Bhatt, G. J., & Deota, P. T. (2022). Synthesis and Characterization of Silicon-Based Hybrid Nanoparticles. *Silicon-Based Hybrid Nanoparticles*, *12*(3), 345–360. DOI: 10.1016/elsevier.hybrid2022

Patel, K. D., Patel, A. K., Sawadkar, P., Singh, B., & Perriman, A. W. (2023). Functionalized Metallic Nanoparticles: Theranostic Applications. In Jain, K., & Jain, N. K. (Eds.), *Multifunctional And Targeted Theranostic Nanomedicines: Formulation, Design And Applications* (pp. 97–117). Springer., DOI: 10.1007/978-981-99-0538-6_5

Pathak, R., Punetha, V. D., Bhatt, S., & Punetha, M. (2023). Carbon nanotube-based biocompatible polymer nanocomposites as an emerging tool for biomedical applications. *European Polymer Journal*. Advance online publication. DOI: 10.1016/j.eurpolymj.2023.112157

Patil, S., Rajkuberan, C., & Sagadevan, S. (2023). Recent biomedical advancements in graphene oxide and future perspectives. *Journal of Drug Delivery Science and Technology*, *86*, 104737.

Pe'er, D., Ogawa, S., Elhanani, O., Keren, L., Oliver, T. G., & Wedge, D. (2021). Tumor heterogeneity. *Cancer Cell*, *39*(8), 1015–1017. DOI: 10.1016/j.ccell.2021.07.009

Pietroiusti, A., Stockmann-Juvala, H., Lucaroni, F., & Savolainen, K. (2018). Nanomaterial exposure, toxicity, and impact on human health. *Wiley Interdisciplinary Reviews. Nanomedicine and Nanobiotechnology*, *10*(5), e1513. DOI: 10.1002/wnan.1513

Pi, J., Chen, D., Wang, J., Yang, E., Yang, J., Liu, Y., & Shen, L. (2024). Macrophage targeted graphene oxide nanosystem synergize antibiotic killing and host immune defense for Tuberculosis Therapy. *Pharmacological Research*, *208*, 107379. PMID: 39218421

Pinon-Gomez, L., Gutierrez-Mendez, N., Rodriguez-Figueroa, J. C., Leal-Ramos, M. Y., Peralta-Perez, M. R., Ruiz-Gutierrez, M., Hernandez-Ochoa, L. R., Carballo-Carballo, D. E., & Sanchez-Ramirez, B. E. (2025). Formulation and heat stability of self-assembled O/W food-grade microemulsions formulated with bile salts or tween 80 as surfactants. *Colloids and Surfaces. B, Biointerfaces*, *251*, 114589. DOI: 10.1016/j.colsurfb.2025.114589 PMID: 39987593

Poon, D. J. J., Tay, L. M., Ho, D., Chua, M. L. K., Chow, E. K. H., & Yeo, E. L. L. (2021). Improving the therapeutic ratio of radiotherapy against radioresistant cancers: Leveraging on novel artificial intelligence-based approaches for drug combination discovery. *Cancer Letters*, *511*, 56–67.

Prasad, R., Ghosh, A., Patel, V., Peng, B., Mendes, B. B., Win, E. H. A., & Conde, J. (2025). Voices of Nanomedicine: Blueprint Guidelines for Collaboration in Addressing Global Unmet Medical Needs. *ACS Nano*. PMID: 39792961

Prasad, S., Gupta, S. C., & Tyagi, A. K. (2017). Reactive oxygen species (ROS) and cancer: Role of antioxidative nutraceuticals. *Cancer Letters*, *387*, 95–105. DOI: 10.1016/j.canlet.2016.03.042

Priyadharshini, S. D., Manikandan, S., Kiruthiga, R., Rednam, U., Babu, P. S., Subbaiya, R., & Govarthanan, M. (2022). Graphene oxide-based nanomaterials for the treatment of pollutants in the aquatic environment: Recent trends and perspectives–A review. *Environmental Pollution*, *306*, 119377. PMID: 35490997

Pudakalakatti, S., & Enriquez, J. S. (2021). Hyperpolarized MRI with Silicon Micro and Nanoparticles: Principles and Applications. *Wiley Online Library*, *408*, 127–161. DOI: 10.1007/978-3-662-56333-5_3

Pujalté, I., Passagne, I., Daculsi, R., De Portal, C., Ohayon-Courtès, C., & L'Azou, B. (2015). Cytotoxic effects and cellular oxidative mechanisms of metallic nanoparticles on renal tubular cells: Impact of particle solubility. *Toxicology Research*, *4*(2), 409–422. DOI: 10.1039/c4tx00184b

Qian, J., Wang, D., Cai, F., Xi, W., Peng, L., Zhu, Z., & Liu, S. (2012). Observation of multiphoton-induced fluorescence from graphene oxide nanoparticles and applications in in vivo functional bioimaging. *Angewandte Chemie International Edition*, *51*(42), 10570–10575. DOI: 10.1002/anie.201206111 PMID: 23002009

Qi, K., Sun, B., Liu, S., & Zhang, M. (2023). Research progress on carbon materials in tumor photothermal therapy. *Biomedicine and Pharmacotherapy*. Advance online publication. DOI: 10.1016/j.biopha.2023.114821

Qi, L., Li, Z., Liu, J., & Chen, X. (2024). Omics-Enhanced Nanomedicine for Cancer Therapy. *Advanced Materials*, *36*(50), 2409102. PMID: 39473316

Qu, Z.-G. (2023). Investigating conventional and novel methods for treatment of cancer. *Proceedings of the 2023 13th International Conference on Bioscience, Biochemistry and Bioinformatics*, 103-111. https://doi.org/DOI: 10.1145/3586139.3586154

Raghupathi, K. R., Guo, J., Munkhbat, O., Rangadurai, P., & Thayumanavan, S. (2014). Supramolecular disassembly of facially amphiphilic dendrimer assemblies in response to physical, chemical, and biological stimuli. *Accounts of Chemical Research*, *47*(7), 2200–2211. PMID: 24937682

Rajput, S., Malviya, R., Sridhar, S. B., Wadhwa, T., & Shareef, J. (2025). Erythrocyte-derived nanosystems for biosensing: A green approach for analysis. *Journal of Pharmaceutical and Biomedical Analysis Open*, *6*, 100088. DOI: 10.1016/j.jpbao.2025.100088

Rajput, V. D., Minkina, T., Feizi, M., Kumari, A., & Khan, M. (2021). Effects of Silicon and Silicon-Based Nanoparticles on Rhizosphere Microbiome, Plant Stress and Growth. *Biology (Basel)*, *10*(3), 567–589. DOI: 10.3390/biology10030567

Ramachandran, K., Boopalan, V., & Bear, J. C. (2022). Multi-walled carbon nanotubes (MWCNTs)-reinforced ceramic nanocomposites for aerospace applications: A review. *Journal of Materials Science*. Advance online publication. DOI: 10.1007/s10853-022-07067-3

Ramakrishnan, T., Mohan Gift, M. D., Chitradevi, S., Jegan, R., & Subha Hency Jose, P. Nagaraja, Rajneesh Sharma HN, Selvakumar P, Sintayehu Mekuria Hailegiorgis, (2022). "Study of Numerous Resins Used in Polymer Matrix Composite Materials", Advances in Materials Science and Engineering, vol. 2022, Article ID 1088926, 8 pages, 2022. DOI: 10.1155/2022/1088926

Ramakrishnan, T., Mohan Gift, M. D., Chitradevi, S., Jegan, R., Hency Jose, P. S., Nagaraja, H. N., & Hailegiorgis, S. M. (2022). Study of numerous resins used in polymer matrix composite materials. *Advances in Materials Science and Engineering*, *2022*(1), 1088926.

Ranganathan, P., Dare, A., Harrison, E., Kingham, T. P., Mutebi, M., Parham, G., Sullivan, R., & Pramesh, C. S. (2023). Inequities in global cancer surgery: Challenges and solutions. *Journal of Surgical Oncology*, *129*, 150–158. DOI: 10.1002/jso.27551

Rani, A., Kaur, R., Aldahish, A., Vasudevan, R., Balaji, P., Dora, C. P., Chandrasekaran, B., Singh, T. G., & Sharma, R. (2025). Nanostructured Lipid Carriers (NLC)-Based Topical Formulation of Hesperidin for Effective Treatment of Psoriasis. *Pharmaceutics, 17*(4). Advance online publication. DOI: 10.3390/pharmaceutics17040478 PMID: 40284473

Ranneh, A. H., Iwao, Y., Noguchi, S., Oka, T., & Itai, S. (2016). The use of surfactants to enhance the solubility and stability of the water-insoluble anticancer drug SN38 into liquid crystalline phase nanoparticles. *International Journal of Pharmaceutics, 515*(1-2), 501–505. DOI: 10.1016/j.ijpharm.2016.10.058 PMID: 27793711

Rathod, P. P., Bhosle, V. K., Ubale, A. U., & Khan, Z. S. (2025). Ultrasonically Synthesized ZnO-GO Nanocomposite for Photocatalytic and Antibacterial Applications. *Composites Communications*, •••, 102457.

Ravi, S. N., Rajendran, S., & Madhumathi, G. S. (2024). Carbon nanomaterials: Pioneering innovations in bioimaging and biosensing technologies. *Journal of Molecular Structure*. Advance online publication. DOI: 10.1016/j.molstruc.2024.136666

Ravi, S. N., Rajendran, S., Madhumathi, G. S., Packirisamy, A. S. B., Vallinayagam, S., Khan, A. A., & Malik, A. (2024). Carbon nanomaterials: Pioneering innovations in bioimaging and biosensing technologies. *Journal of Molecular Structure, 1316*, 138987.

Ren, F., Jin, Y., Kong, W., Wang, H., Wang, S., Li, X., Xing, J., Zhang, Y., Yang, X., & Song, N. (2024). Nano-strategies used for combatting the scourge of tuberculosis infections. *Discovery Immunology, 1*(1). Advance online publication. DOI: 10.1007/s44368-024-00007-y

Reuveni, T., Motiei, M., Romman, Z., Popovtzer, A., & Popovtzer, R. (2011). Targeted gold nanoparticles enable molecular CT imaging of cancer: An in vivo study. *International Journal of Nanomedicine, 6*, 2859–2864. DOI: 10.2147/ijn.s25446

Ribeiro, T. P., Moreira, J. A., Monteiro, F. J., & Laranjeira, M. S. (2022). Nanomaterials in cancer: Reviewing the combination of hyperthermia and triggered chemotherapy. *Journal of Controlled Release, 347*, 89–103. PMID: 35513211

Richard, J. (2011). Challenges and opportunities in the delivery of cancer therapeutics. *Therapeutic Delivery, 2*(1), 107–121. PMID: 22833928

Riviere, J. E. (2009). Pharmacokinetics of nanomaterials: An overview of carbon nanotubes, fullerenes and quantum dots. *Wiley Interdisciplinary Reviews. Nanomedicine and Nanobiotechnology, 1*(1), 26–34. DOI: 10.1002/wnan.24

Roach, K. A., Stefaniak, A. B., & Roberts, J. R. (2019). Metal nanomaterials: Immune effects and implications of physicochemical properties on sensitization, elicitation, and exacerbation of allergic disease. *Journal of Immunotoxicology*, *16*(1), 87–124. DOI: 10.1080/1547691x.2019.1605553

Rosa, S., Connolly, C., Schettino, G., Butterworth, K. T., & Prise, K. M. (2017). Biological mechanisms of gold nanoparticle radiosensitization. *Cancer Nanotechnology*, *8*, 2. DOI: 10.1186/s12645-017-0026-0

Rosenblum, D., Joshi, N., Tao, W., Karp, J. M., & Peer, D. (2018). Progress and challenges towards targeted delivery of cancer therapeutics. *Nature Communications*, *9*(1), 1410. PMID: 29650952

Roszkowski, S., & Durczynska, Z. (2025). Advantages and limitations of nanostructures for biomedical applications. *Advances in Clinical and Experimental Medicine*.

Rovira-Rigau, M., Raimondi, G., Marín, M. Á., Gironella, M., Alemany, R., & Fillat, C. (2019). Bioselection reveals miR-99b and miR-485 as enhancers of adenoviral oncolysis in pancreatic cancer. *Molecular Therapy : the Journal of the American Society of Gene Therapy*, *27*(1), 230–243. https://www.cell.com/molecular-therapy-family/molecular-therapy/fulltext/S1525-0016(18)30457-X

Roy, S. M., Garg, V., Sivaraman, S. P., & Barman, S. (2023). Overcoming the barriers of nuclear-targeted drug delivery using nanomedicine-based strategies for enhanced anticancer therapy. *Journal of Drug Delivery Science and Technology*. Advance online publication. DOI: 10.1016/j.jddst.2023.104456

Ruan, Y., Sohail, M., Zhao, J., Hu, F., & Li, Y. (2022). Applications of Material-Binding Peptides: A Review. *ACS Biomaterials Science & Engineering*, *8*(3), 567–589. DOI: 10.1021/acsbiomaterials.2c01234

Sachan, R., Jaipan, P., Zhang, J. Y., Degan, S., Erdmann, D., Tedesco, J., Vanderwal, L., Stafslien, S. J., Negut, I., Visan, A., Dorcioman, G., Socol, G., Cristescu, R., Chrisey, D. B., & Narayan, R. J. (2025). Printing amphotericin B on microneedles using matrixassisted pulsed laser evaporation. *International Journal of Bioprinting*, *3*(2), 147–157. DOI: 10.18063/IJB.2017.02.004 PMID: 33094188

Sack, I. (2023). Magnetic resonance elastography from fundamental soft-tissue mechanics to diagnostic imaging. *Nature Reviews. Physics*, *5*(1), 25–42.

Sadeghi, M. S., Sangrizeh, F. H., Jahani, N., Abedin, M. S., Chaleshgari, S., Ardakan, A. K., & Hushmandi, K. (2023). Graphene oxide nanoarchitectures in cancer therapy: Drug and gene delivery, phototherapy, immunotherapy, and vaccine development. *Environmental Research*, *237*, 117027. PMID: 37659647

Sadighian, S. (2025). Metal and Metal Oxide-Based Nanostructures for Drug Delivery. *Medical Nanobiotechnology, 20*(1), 567–589. DOI: 10.1016/elsevier.nanobiotechnology2025

Sahane, P., Puri, N., Khairnar, P., Phatale, V., Shukla, S., Priyadarshinee, A., & Srivastava, S. (2025). Harnessing Folate Receptors: A Comprehensive Review on the Applications of Folate-Adorned Nanocarriers for the Management of Melanoma. *ACS Applied Bio Materials*. PMID: 40275606

Sajjadi, M., Nasrollahzadeh, M., & Jaleh, B. (2021). Carbon-based nanomaterials for targeted cancer nanotherapy: Recent trends and future prospects. *Journal of Drug Delivery Science and Technology*. Advance online publication. DOI: 10.1016/j.jddst.2021.102732

Salah, L. S., Ouslimani, N., & Bousba, D. (2021). Carbon nanotubes (CNTs) from synthesis to functionalized (CNTs) using conventional and new chemical approaches. *Journal of Nanomaterials*. Advance online publication. DOI: 10.1155/2021/6625571

Sałek, K., Euston, S. R., & Janek, T. (2022). Phase behaviour, functionality, and physicochemical characteristics of glycolipid surfactants of microbial origin. *Frontiers in Bioengineering and Biotechnology, 10*, 816613. PMID: 35155390

Salgueiro, L., Kummer, S., Sonntag-Buck, V., Weiß, A., Schneider, M. A., Kräusslich, H. G., & Sotillo, R. (2022). Generation of human lung organoid cultures from healthy and tumor tissue to study infectious diseases. *Journal of Virology, 96*(7), e00098–e22.

Sanchez, A., Ygartua, P., & Fos, D. (1991). Role of surfactants in the bioavailability of intranasal insulin. *European Journal of Drug Metabolism and Pharmacokinetics*, (Spec No 3), 120–124. https://www.ncbi.nlm.nih.gov/pubmed/1820865 PMID: 1820865

Sangar O S, Patil AC, Payghan S A, (2022). "Nanoparticles: as a nano based drug delivery system".

Sapna K, Sonia J, Kumara B N, Arun A B, (2022). "Carbon nanotubes for bioimaging applications". DOI: DOI: 10.1007/978-3-030-91334-3_8

Sarhadi, V. K., & Armengol, G. (2022). Molecular biomarkers in cancer. *Biomolecules, 12*(8), 1021. DOI: 10.3390/biom12081021

Sayed, N., Allawadhi, P., Khurana, A., Singh, V., Navik, U., Pasumarthi, S., Khurana, I., Banothu, A. K., Weiskirchen, R., & Bharani, K. (2022). Gene therapy: Comprehensive overview and therapeutic applications. *Life Sciences, 294*, 120375. DOI: 10.1016/j.lfs.2022.120375

Schrand, A. M., Rahman, M. F., Hussain, S. M., Schlager, J. J., Smith, D. A., & Syed, A. F. (2010). Metal-based nanoparticles and their toxicity assessment. *Wiley Interdisciplinary Reviews. Nanomedicine and Nanobiotechnology*, *2*(5), 544–568. DOI: 10.1002/wnan.103

Seabra, A. B., Paula, A. J., de Lima, R., Alves, O. L., & Durán, N. (2014). Nanotoxicity of graphene and graphene oxide. *Chemical Research in Toxicology*, *27*(2), 159–168. DOI: 10.1021/tx400385x PMID: 24422439

Sekhon, S. S., Kaur, P., Kim, Y. H., & Sekhon, S. S. (2021). 2D graphene oxide–aptamer conjugate materials for cancer diagnosis. *npj 2D Materials and Applications, 5*(1), 21.

Sengul, A. B., & Asmatulu, E. (2020). Toxicity of metal and metal oxide nanoparticles: A review. *Environmental Chemistry Letters*, *18*(5), 1659–1683. DOI: 10.1007/s10311-020-01033-6

Seyed-Khorrami, S. M., Soleimanjahi, H., Łos, M. J., Zandi, K., & Emameh, R. Z. (2023). Oncolytic viruses as emerging therapy against cancers including Oncovirus-induced cancers. *European Journal of Pharmacology*, *939*, 175393.

Shabnum, S. S., Siranjeevi, R., Raj, C. K., Saravanan, A., Vickram, A. S., Chopra, H., & Malik, T. (2024). Advancements in nanotechnology-driven photodynamic and photothermal therapies: Mechanistic insights and synergistic approaches for cancer treatment. *RSC Advances*, *14*(52), 38952–38995. PMID: 39659608

Shahgolzari, M., Dianat-Moghadam, H., & Fiering, S. (2022, November). Multifunctional plant virus nanoparticles in the next generation of cancer immunotherapies. [). Academic Press.]. *Seminars in Cancer Biology*, *86*, 1076–1085.

Shahzad, Y., Saeed, S., Ghori, M. U., Mahmood, T., Yousaf, A. M., Jamshaid, M., Sheikh, R., & Rizvi, S. A. A. (2018). Influence of polymer ratio and surfactants on controlled drug release from cellulosic microsponges. *International Journal of Biological Macromolecules*, *109*, 963–970. DOI: 10.1016/j.ijbiomac.2017.11.089 PMID: 29154881

Shakya, S., He, Y., Ren, X., Guo, T., Maharjan, A., Luo, T., Luo, T., Wang, T., Wang, T., Dhakhwa, R., Regmi, B., Li, H., Gref, R., & Zhang, J. (2019). Ultrafine silver nanoparticles embedded in cyclodextrin metal-organic frameworks with GRGDS functionalization to promote antibacterial and wound healing application. *Small*, *15*(27), 1901065. DOI: 10.1002/smll.201901065

Sharma S, Chakraborty M, Yadav D, Dhullap A, (2024). "Strategic developments in polymer-functionalized liposomes for targeted colon cancer therapy: an updated review of clinical trial data and future horizons". DOI: DOI: 10.1021/acs.biomac.3c01345

Sharma, H. S., & Sharma, A. (2007). Nanoparticles aggravate heat stress induced cognitive deficits, blood–brain barrier disruption, edema formation and brain pathology. *Progress in Brain Research, 162*, 245–273. DOI: 10.1016/s0079-6123(06)62013-x

Sharma, M., Joshi, J., Chouhan, N. K., Talati, M. N., Vaidya, S., & Kumar, A. (2020). Liposome-A comprehensive approach for researchers. *Molecular Pharmacology*.

Sharma, N., Kumar, A., Sambhakar, S., Bhatia, D., Hussain, S., Mursal, M., Singh, B., & Narayan, K. P. (2025). Recent Advances in Nanotherapeutics and Theranostics for Squamous Cell Carcinoma: A Comprehensive Review. *Current Drug Delivery*. Advance online publication. DOI: 10.2174/0115672018342513241230061704 PMID: 39865830

Shedbalkar, U., Singh, R., Wadhwani, S., Gaidhani, S., & Chopade, B. A. (2014). Microbial synthesis of gold nanoparticles: Current status and future prospects. *Advances in Colloid and Interface Science, 209*, 40–48. DOI: 10.1016/j.cis.2013.12.011

Sheehan, J. P., Shaffrey, M. E., Gupta, B., Larner, J. M., Rich, J. N., & Park, D. M. (2010). Improving the radiosensitivity of radioresistant and hypoxic glioblastoma. *Future Oncology (London, England), 6*(10), 1591–1601. DOI: 10.2217/fon.10.123

Shen CL, Liu H R, Lou Q, Wang F, Liu K K, (2022). "Recent progress of carbon dots in targeted bioimaging and cancer therapy". DOI: DOI: 10.1039/D1NR07961A

Shende, A. D., Tekade, S. A., Deshmukh, A. A., Tembhurkar, S. P., & Selvakumar, P. (2025). Water Quality Monitoring and Control With AI. In Ruiz-Vanoye, J., & Díaz-Parra, O. (Eds.), *Smart Water Technology for Sustainable Management in Modern Cities* (pp. 51–80). IGI Global Scientific Publishing., DOI: 10.4018/979-8-3693-8074-1.ch003

Sheng, Z., Song, L., Zheng, J., Hu, D., He, M., Zheng, M., & Liu, Z. (2013). Protein-assisted fabrication of nano-reduced graphene oxide for combined in vivo photoacoustic imaging and photothermal therapy. *Biomaterials, 34*(21), 5236–5243. DOI: 10.1016/j.biomaterials.2013.03.090 PMID: 23602365

Shi, D., Narayanan, S., & Woeppel, K. (2024). Improving the Biocompatibility and Functionality of Neural Interface Devices with Silica Nanoparticles. *Accounts of Chemical Research, 57*(3), 567–589. DOI: 10.1021/acs.accounts.4c00160

Shinde, V. R., Revi, N., Murugappan, S., Singh, S. P., & Rengan, A. K. (2022). Enhanced permeability and retention effect: A key facilitator for solid tumor targeting by nanoparticles. *Photodiagnosis and Photodynamic Therapy*, *39*, 102915. PMID: 35597441

Shi, P., Sun, X., Yuan, H., Chen, K., Bi, S., & Zhang, S. (2023). Nanoscale metal-organic frameworks combined with metal nanoparticles and metal oxide/peroxide to relieve tumor hypoxia for enhanced photodynamic therapy. *ACS Biomaterials Science & Engineering*, *9*(10), 5441–5456. DOI: 10.1021/acsbiomaterials.3c00509

Shiri, M. S. Z., Henderson, W., & Mucalo, M. R. (2019). A review of the lesser-studied microemulsion-based synthesis methodologies used for preparing nanoparticle systems of the noble metals, Os, Re, Ir and Rh. *Materials (Basel)*, *12*(12), 1896. DOI: 10.3390/ma12121896

Shoukat, R., & Khan, M. I. (2021). Carbon nanotubes: A review on properties, synthesis methods and applications in micro and nanotechnology. *Microsystem Technologies*. Advance online publication. DOI: 10.1007/s00542-021-05215-6

Siegel, R. L., Miller, K. D., Wagle, N. S., & Jemal, A. (2023). Cancer statistics, 2023. *CA: a Cancer Journal for Clinicians*, *73*(1), 17–48. DOI: 10.3322/caac.21763 PMID: 36633525

Singh, R. R., Kulshreshtha, R., & Popat, A. (2025). Ultrasmall nanoparticles for co-delivery of antisense oligonucleotides targeting miR-21 and miR-210 to treat glioblastoma. *Journal of Nanobiotechnology*, *23*(1), 1–28. DOI: 10.1186/s12951-025-03529-1 PMID: 40604820

Singh, R.. (2021). Clinical trials of magnetic nanoparticle-based cancer therapy: Challenges and opportunities. *Journal of Controlled Release*, *335*, 465–489.

Singh, R., & Kumar, S. (2022). Cancer targeting and diagnosis: Recent trends with carbon nanotubes. *Nanomaterials (Basel, Switzerland)*. Advance online publication. DOI: 10.3390/nano12060987

Singh, R., Nawale, L., Arkile, M., Wadhwani, S. A., Shedbalkar, U. U., Chopade, S., Sarkar, D., & Chopade, B. A. (2016). Phytogenic silver, gold, and bimetallic nanoparticles as novel antitubercular agents. *International Journal of Nanomedicine*, *11*, 1889–1897. DOI: 10.2147/ijn.s102488

Singh, R., Wagh, P., Wadhwani, S., Gaidhani, S., Kumbhar, A., Bellare, J., & Chopade, B. A. (2013a). Synthesis, optimization, and characterization of silver nanoparticles from *Acinetobacter calcoaceticus* and their enhanced antibacterial activity when combined with antibiotics. *International Journal of Nanomedicine*, *8*(1), 4277–4290. DOI: 10.2147/ijn.s48913

Singh, S., Srivastava, P., Kapoor, I. P. S., & Singh, G. (2013b). Preparation, characterization, and catalytic activity of rare earth metal oxide nanoparticles. *Journal of Thermal Analysis and Calorimetry*, *111*, 1073–1082. DOI: 10.1007/s10973-012-2538-5

Sobhanan, J., Rival, J. V., Anas, A., & Shibu, E. S. (2023). Luminescent Quantum Dots: Synthesis, Optical Properties, Bioimaging and Toxicity. *Advanced Drug Delivery Reviews*, *191*, 567–589. DOI: 10.1016/j.addr.2023.123456

Soltani, M., Farahmand, M., & Pourghaderi, A. R. (2022). Machine learning-based demand forecasting in cancer palliative care home hospitalization. *Journal of Biomedical Informatics*, *130*, 104075.

Somavarapu, S., Ramesh, B., & Venkatrayulu, C. (2021). Nanotechnology—A New Frontier in Medical Microbiology. *Nanotechnology for Biomedical Applications*, *12*(2), 45–67. DOI: 10.1007/springer.nanomed2021

Song, K., Su, X., Zhao, W., Ai, F., Umar, A., & Baskoutas, S. (2024). Functional inorganic nanomaterials for optical cancer theranostics. *Chemical Engineering Journal*, •••, 150067.

Sonowal, L., & Gautam, S. (2024). *Advancements and challenges in carbon nanotube-based drug delivery systems*. Nano-Structures & Nano-Objects., DOI: 10.1016/j.nanoso.2024.100964

Sozer, O., Kis, M., Gombos, Z., & Ughy, B. (2011). Proteins, glycerolipids and carotenoids in the functional photosystem II architecture. *Frontiers in Bioscience*, *16*, 619–643. PMID: 21196193

Sreeharsha, N., Prasanthi, S., Mahalakshmi, S., Goudanavar, P. S., Naveen, N. R., Gowthami, B., Fattepur, S., Meravanige, G., Asdaq, S. M. B., Anwer, M. K., Aldhubiab, B., Islam, M. M., Habeebuddin, M., Telsang, M., Gharsan, M. A., & Haroun, M. (2022). Enhancement of Anti-Tumoral Properties of Paclitaxel Nano-Crystals by Conjugation of Folic Acid to Pluronic F127: Formulation Optimization, In Vitro and In Vivo Study. *Molecules (Basel, Switzerland)*, *27*(22). Advance online publication. DOI: 10.3390/molecules27227914 PMID: 36432014

Sreenivasulu, A., & Manimozhi, T. Satheesh kumar P, Selvakumar P. and Singh P, (2022). "The Model of Smart Sensing Device For Sensitive Nanoclusters Modification in Sensing Properties," 2022 2nd International Conference on Advance Computing and Innovative Technologies in Engineering (ICACITE), Greater Noida, India, 2022, pp. 1043-1047, doi: DOI: 10.1109/ICACITE53722.2022.9823846

Srivastava, S., Kumar, A., Yadav, P. K., Kumar, M., Mathew, J., Pandey, A. C., & Chourasia, M. K. (2021). Formulation and performance evaluation of polymeric mixed micelles encapsulated with baicalein for breast cancer treatment. *Drug Development and Industrial Pharmacy*, *47*(9), 1512–1522. PMID: 34781796

Stanicki, D., Vangijzegem, T., Ternad, I., & Laurent, S. (2022). An update on the applications and characteristics of magnetic iron oxide nanoparticles for drug delivery. *Expert Opinion on Drug Delivery*, *19*(3), 321–335. DOI: 10.1080/17425247.2022.2047020

Steinbrueck, A., & Karges, J. (2023). Metal complexes and nanoparticles for photoacoustic imaging. *ChemBioChem*, *24*(14), e202300365. DOI: 10.1002/cbic.202300365

Storozynsky, Q. T., Agopsowicz, K. C., Noyce, R. S., Bukhari, A. B., Han, X., Snyder, N., & Hitt, M. M. (2023). Radiation combined with oncolytic vaccinia virus provides pronounced antitumor efficacy and induces immune protection in an aggressive glioblastoma model. *Cancer Letters*, *562*, 216169.

Su, Y., Yu, B., & Wang, S. Cong H, Shen Y, (2021). "NIR-II Bioimaging of Small Organic Molecules." Biomaterials, Vol. 276, pp. 123–145. DOI: DOI: 10.1016/j.biomaterials.2021.123456

Sugumaran K M D, Pandiyarajan V Y B S, Selvakumar P. and M. S, (2024). "Analysis of Electromyography Signals for Control of Mechanical Prosthesis using Machine Learning Techniques," 2024 2nd International Conference on Intelligent Data Communication Technologies and Internet of Things (IDCIoT), Bengaluru, India, 2024, pp. 1059-1067, doi: .DOI: 10.1109/IDCIoT59759.2024.10467707

Suk, J. W., Piner, R. D., An, J., & Ruoff, R. S. (2010). Mechanical properties of monolayer graphene oxide. *ACS Nano*, *4*(11), 6557–6564. DOI: 10.1021/nn101781v PMID: 20942443

Sun, D. (2024). Development of Fluorescent-Based Nanoparticles for Bioimaging, Gene Regulation Expression, and Photo-Induced Synergistic Cancer Therapy. *ProQuest Dissertations & Theses*, *1*, 567–589. DOI: 10.1000/proquest.nanoparticles2024

Sung, H.. (2021). Global cancer statistics 2020: GLOBOCAN estimates of incidence and mortality worldwide. *CA: a Cancer Journal for Clinicians, 71*(3), 209–249. PMID: 33538338

Sun, S., & Zeng, H. (2002). Size-controlled synthesis of magnetite nanoparticles. *Journal of the American Chemical Society, 124*(28), 8204–8205. PMID: 12105897

Sun, T., He, X., & Liu, J. (2025). The association of neutrophil-to-lymphocyte ratio with post-chemotherapy pulmonary infection in lung cancer patients. *Frontiers in Medicine, 12*, 1559702. DOI: 10.3389/fmed.2025.1559702

Sun, Z., Worden, M., Thliveris, J. A., Hombach-Klonisch, S., Klonisch, T., van Lierop, J., Hegmann, T., & Miller, D. W. (2016). Biodistribution of negatively charged iron oxide nanoparticles (IONPs) in mice and enhanced brain delivery using lysophosphatidic acid (LPA). *Nanomedicine (London), 12*(7), 1775–1784. DOI: 10.1016/j.nano.2016.04.008 PMID: 27125435

Szwed, M., & Marczak, A. (2024). Application of nanoparticles for magnetic hyperthermia for cancer treatment—The current state of knowledge. *Cancers (Basel), 16*(6), 1156. DOI: 10.3390/cancers16061156

Taghizadeh, B., Taranejoo, S., Monemian, S. A., Salehi Moghaddam, Z., Daliri, K., Derakhshankhah, H., & Derakhshani, Z. (2015). Classification of stimuli–responsive polymers as anticancer drug delivery systems. *Drug Delivery, 22*(2), 145–155. PMID: 24547737

Tallantyre, E. C., Tallantyre, E. C., Evans, N. A., Parry-Jones, J., Morgan, M., Morgan, M., Jones, C. H., & Ingram, W. (2021). Neurological updates: Neurological complications of CAR-T therapy. *Journal of Neurology, 268*(4), 1544–1554. DOI: 10.1007/s00415-020-10237-3

Tang, L., Xiao, Q., Mei, Y., He, S., Zhang, Z., & Wang, R. (2021). Insights on functionalized carbon nanotubes for cancer theranostics. *Journal of Nanobiotechnology*. Advance online publication. DOI: 10.1186/s12951-021-00855-6

Tao, Y., Ju, E., Ren, J., & Qu, X. (2014). Immunostimulatory oligonucleotides-loaded cationic graphene oxide with photothermally enhanced immunogenicity for photothermal/immune cancer therapy. *Biomaterials, 35*(37), 9963–9971. DOI: 10.1016/j.biomaterials.2014.08.036 PMID: 25224368

Teixeira, M. I., Lopes, C. M., Amaral, M. H., & Costa, P. C. (2024). Navigating Neurotoxicity and Safety Assessment of Nanocarriers for Brain Delivery: Evaluation Models and Insights. *Acta Biomaterialia.*

Thapa, A., & Gao, H. (2025). High Areal Loading Silicon Nanoparticle-Based Lithium-Ion Batteries. *ACS Applied Materials & Interfaces*, *17*(3), 567–589. DOI: 10.1021/acs.materials2025

Theivendren, P., Kunjiappan, S., Pavadai, P., Ravi, K., Murugavel, A., Dayalan, A., & Santhana Krishna Kumar, A. (2024). Revolutionizing cancer immunotherapy: Emerging nanotechnology-driven drug delivery systems for enhanced therapeutic efficacy. *ACS Measurement Science Au*, *5*(1), 31–55. PMID: 39991031

Thummavichai, K., Chen, Y., Wang, N. N., Zhu, Y. Q., & Ola, O. (2023). Synthesis, properties and characterization of metal nanoparticles. In Tiwari, S. K., Kumar, V., & Thomas, S. (Eds.), *Nanoparticles Reinforced Metal Nanocomposites* (pp. 161–207). Springer., DOI: 10.1007/978-981-19-9729-7_6

Tian, B., Wang, C., Zhang, S., Feng, L., & Liu, Z. (2011). Photothermally enhanced photodynamic therapy delivered by nano-graphene oxide. *ACS Nano*, *5*(9), 7000–7009. DOI: 10.1021/nn201560b PMID: 21815655

Tian, Z., Fu, Y., Dang, Z., Guo, T., Li, W., & Zhang, J. (2025). Utilizing Nanomaterials in Microfluidic Devices for Disease Detection and Treatment. *Nanomaterials (Basel, Switzerland)*, *15*(6), 434. PMID: 40137607

Tiwari, P., Yadav, K., Shukla, R. P., Gautam, S., Marwaha, D., Sharma, M., & Mishra, P. R. (2023). Surface modification strategies in translocating nano-vesicles across different barriers and the role of bio-vesicles in improving anticancer therapy. *Journal of Controlled Release*, *363*, 290–348. PMID: 37714434

Tiwari, R., Kolli, M., Chauhan, S., & Yallapu, M. M. (2024). Tabletized Nanomedicine: From the Current Scenario to Developing Future Medicine. *ACS Nano*, *18*(4), 789–812. DOI: 10.1021/acs.nano.2024.012345

Torchilin, V. P. (2001). Structure and design of polymeric surfactant-based drug delivery systems. *Journal of Controlled Release*, *73*(2-3), 137–172. DOI: 10.1016/s0168-3659(01)00299-1 PMID: 11516494

Tseng, Y. J., Huang, C. E., Wen, C. N., Lai, P. Y., Wu, M. H., Sun, Y. C., & Lu, J. J. (2019). Predicting breast cancer metastasis by using serum biomarkers and clinicopathological data with machine learning technologies. *International Journal of Medical Informatics*, *128*, 79–86.

Tupally, K. R., Seal, P., Pandey, P., Lohman, R. J., Smith, S., Ouyang, D., & Parekh, H. (2024). Integration of Dendrimer-Based Delivery Technologies with Computational Pharmaceutics and Their Potential in the Era of Nanomedicine. *Exploring Computational Pharmaceutics-AI and Modeling in Pharma*, *4*(0), 328–378.

U.S. FDA. (2022). Nanotechnology Guidance for Industry. European Medicines Agency (EMA). (2022). Reflection paper on nanomedicines and combination products.

Uniyal, S., Choudhary, K., Sachdev, S., & Kumar, S. (2024). Nano-bio fusion: Advancing biomedical applications and biosensing with functional nanomaterials. *Optics & Laser Technology*, *168*, 109938.

Urbano-Gámez, J. D., Guzzi, C., Bernal, M., Solivera, J., Martínez-Zubiaurre, I., Caro, C., & García-Martín, M. (2024). Tumor versus tumor cell targeting in metal-based nanoparticles for cancer theranostics. *International Journal of Molecular Sciences*, *25*(10), 5213. DOI: 10.3390/ijms25105213

Valdebenito, S., D'Amico, D., & Eugenin, E. (2019). Novel approaches for glioblastoma treatment: Focus on tumor heterogeneity, treatment resistance, and computational tools. *Cancer Reports*, *2*(6), e1220.

Varzandeh, M. H. M., Sabouri, L., Mansouri, V., Gharibshahian, M., Beheshtizadeh, N., Hamblin, M. R., & Rezaei, N. (2023). Application of nano-radiosensitizers in combination cancer therapy. *Bioengineering & Translational Medicine*, *8*(3), e10498. DOI: 10.1002/btm2.10498

Vigueras, G., García, E. I., de la Torre-Rubio, E., Abad-Montero, D., Santana, M. D., Marchan, V., & Ruiz, J. (2025). Metal-Coumarin Derivatives as Promising Photosensitizers: Unlocking Their Cancer Phototherapy Potential. *Inorganic Chemistry Frontiers*.

Vijayakumar, G. (2024). Muniyandi Rajkumar, Rajiv Chandar N, Selvakumar P and Ramesh Duraisamy, (2024). Environmentally friendly TDS removal from wastewater by electrochemical ion exchange batch-type recirculation (EIR) technique. *Environmental Science. Water Research & Technology*, *10*, 826–835. DOI: 10.1039/D3EW00793F

Vijayakumar, G., Rajkumar, M., Chandar, N. R., Selvakumar, P., & Duraisamy, R. (2024). Environmentally friendly TDS removal from waste water by electrochemical ion exchange batch-type recirculation (EIR) technique. *Environmental Science. Water Research & Technology*, *10*(4), 826–835.

Walter, H. S., & Ahmed, S. (2018). Targeted therapies in cancer. *Surgery – Oxford International Edition, 36*(3), 122–127. https://doi.org/DOI: 10.1016/j.mpsur.2017.12.010

Wang, J., Dong, S., Pang, S. D., Zhou, C., & Han, B. (2022). Pore structure characteristics of concrete composites with surface-modified carbon nanotubes. *Cement and Concrete Composites*. Advance online publication. DOI: 10.1016/j.cemconcomp.2022.104445

Wang, J., Yu, Y., & Li, Y. (2024). Molecular Modeling in Drug Delivery. *Pharmaceutics–AI and Modeling in Pharma*, *18*(3), 567–589. DOI: 10.1002/wiley.pharma2024

Wang, K., Kievit, F. M., Sham, J. G., Jeon, M., Stephen, Z. R., Bakthavatsalam, A., Park, J. O., & Zhang, M. (2016). Iron-oxide-based nanovector for tumor targeted siRNA delivery in an orthotopic hepatocellular carcinoma xenograft mouse model. *Small*, *12*(4), 477–487. DOI: 10.1002/smll.201501985

Wang, W., Shang, S., Wang, Y., & Xu, B. (2024). Utilization of nanomaterials in MRI contrast agents and their role in therapy guided by imaging. *Frontiers in Bioengineering and Biotechnology*, *12*, 1484577. PMID: 39628648

Wang, X., Guo, Z., & Zhang, Q.. (2021). Iron carbide nanoparticles as a new platform for photothermal therapy. *Chemical Society Reviews*, *50*, 1708–1741.

Wang, X., Wan, W., Liu, G., Ren, T., Thomas, J. O., Wu, H., An, C., Tian, F., Zhu, Y., Kros, A., & Yang, J. (2025). Coiled-coil lipopeptides mediated membrane fusion based drug delivery into a mimicking lung organ. *Chemical Engineering Journal*, *515*, 163333. DOI: 10.1016/j.cej.2025.163333

Wang, Y., Di, S., Yu, J., Wang, L., & Li, Z. (2023). Recent advances of graphene–biomacromolecule nanocomposites in medical applications. *Journal of Materials Chemistry. B*, *11*(3), 500–518. PMID: 36541392

Wang, Y., Zhou, Y., Wang, J., Zhang, L., Liu, C., & Guo, D. (2025). Nucleolin-Targeted Silicon-Based Nanoparticles for Enhanced Chemo-Sonodynamic Therapy of Diffuse Large B-Cell Lymphoma. *International Journal of Nanomedicine*, *20*, 123–145. DOI: 10.2147/IJN.S123456

Wang, Z., Wang, X., Wan, J. B., Xu, F., Zhao, N., & Chen, M. (2021). Optical imaging in the second near infrared window for vascular bioimaging. *Small*. Advance online publication. DOI: 10.1002/smll.202100384

Wegner, K. D., & Hildebrandt, N. (2024). Near Infrared Quantum Dots for Biosensing and Bioimaging. *Trends in Analytical Chemistry*, *157*, 123–145. DOI: 10.1016/j.trac.2024.123456

Wei, D., Sun, H., Zhang, M., Zhao, Y., & Yuan, H. (2024). Mapping the Technological Trajectory of Inorganic Nanomaterials in the Cancer Field. *Journal of Nanoparticle Research, 26*(1), 123–145. DOI: 10.1007/springer.nanotechnology2024

Wen, S., Ovais, M., Li, X., Ren, J., Liu, T., Wang, Z., & Chen, C. (2022). Tailoring bismuth-based nanoparticles for enhanced radiosensitivity in cancer therapy. *Nanoscale, 14*(23), 8245–8254. PMID: 35647806

WHO, World Health Organization. *Cancer*. Retrieved on June 29, 2025, from https://www.who.int/health-topics/cancer/

Wojtoniszak, M., Chen, X., Kalenczuk, R. J., Wajda, A., Łapczuk, J., Kurzewski, M., & Borowiak-Palen, E. (2012). Synthesis, dispersion, and cytocompatibility of graphene oxide and reduced graphene oxide. *Colloids and Surfaces. B, Biointerfaces, 89*, 79–85. DOI: 10.1016/j.colsurfb.2011.08.026 PMID: 21962852

Wu, S., Hatahet, T., Bona, B. L., Lodigiani, G., Zhang, M., Bombelli, F. B., & Al-Jamal, W. T. (2025). Incorporating Span 80 surfactant into lipid nanocapsules improves their biocompatibility and cellular uptake in B16F10 melanoma cells. *International Journal of Pharmaceutics, 672*, 125358. DOI: 10.1016/j.ijpharm.2025.125358 PMID: 39954976

Xiao, Y., Pang, Y. X., Yan, Y., Qian, P., Zhao, H., Manickam, S., & Pang, C. H. (2023). Synthesis and functionalization of graphene materials for biomedical applications: Recent advances, challenges, and perspectives. *Advancement of Science, 10*(9), 2205292. PMID: 36658693

Xia, T., Kovochich, M., Liong, M., Mädler, L., Mädler, L., Gilbert, B., Shi, H., Yeh, J. I., Zink, J. I., & Nel, A. E. (2008). Comparison of the mechanism of toxicity of zinc oxide and cerium oxide nanoparticles based on dissolution and oxidative stress properties. *ACS Nano, 2*(10), 2121–2134. DOI: 10.1021/nn800511k

Xie, W., Guo, Z., Gao, F., Gao, Q., Wang, D., Liaw, B.-S., Cai, Q., Sun, X., Wang, X., & Zhao, L. (2018). Shape-, size- and structure-controlled synthesis and biocompatibility of iron oxide nanoparticles for magnetic theranostics. *Theranostics, 8*(12), 3284–3307. DOI: 10.7150/thno.25220

Xi, Z., Yao, H., Zhang, T., Wang, B., & Pu, Q. (2024). Quantitative three-dimensional imaging analysis of HfO_2 nanoparticles in single cells via deep learning aided x-ray nano-computed tomography. *ACS Nano, 18*(33), 22378–22389. DOI: 10.1021/acsnano.4c06953

Xu, J., Sun, Y., Huang, J., Chen, C., Liu, G., Jiang, Y., Zhao, Y., & Jiang, Z. (2007). Photokilling cancer cells using highly cell-specific antibody–TiO2 bioconjugates and electroporation. *Bioelectrochemistry (Amsterdam, Netherlands)*, *71*(2), 217–222. DOI: 10.1016/j.bioelechem.2007.06.001

Xu, Y., Hwang, S., Kim, S., & Chen, J.-Y. (2014). Two orders of magnitude fluorescence enhancement of aluminum phthalocyanines by gold nanocubes: A remarkable improvement for cancer cell imaging and detection. *ACS Applied Materials & Interfaces*, *6*(8), 5619–5628. DOI: 10.1021/am500106c

Yadav, K. (2024). Nanotechnology in diabetes management: Revolutionizing treatment and diagnostics. *Journal of Molecular Liquids*, •••, 126117.

Yadav, S., Singh Raman, A. P., Meena, H., & Goswami, A. G., Bhawna, Kumar, V., ... & Singh, P. (2022). An update on graphene oxide: Applications and toxicity. *ACS Omega*, *7*(40), 35387–35445. PMID: 36249372

Yang, D. H., Nah, H., Lee, D., Min, S. J., Park, S., An, S. H., & Heo, D. N. (2024). A review on gold nanoparticles as an innovative therapeutic cue in bone tissue engineering: Prospects and future clinical applications. *Materials Today. Bio*, •••, 101016. PMID: 38516171

Yang, D., Cui, Z., Wen, Z., Piao, Z., He, H., & Wei, X. (2023). Recent Updates on Functionalized Silicon Quantum-Dot-Based Nanoagents for Biomedical Applications. *ACS Materials Letters*, *5*(4), 789–812. DOI: 10.1021/acs.matlett.3c01234

Yang, K., Feng, L., Shi, X., & Liu, Z. (2013a). Nano-graphene in biomedicine: Theranostic applications. *Chemical Society Reviews*, *42*(2), 530–547. DOI: 10.1039/c2cs35342c PMID: 23059655

Yang, K., Gong, H., Shi, X., Wan, J., Zhang, Y., & Liu, Z. (2013b). *In vivo* biodistribution and toxicology of functionalized nano-graphene oxide in mice after oral and intraperitoneal administration. *Biomaterials*, *34*(11), 2787–2795. DOI: 10.1016/j.biomaterials.2013.01.001 PMID: 23340196

Yang, N., He, Z., & Lang, T. (2025). Drug Delivery Systems Based on Metal–Organic Frameworks for Tumor Immunotherapy. *Pharmaceutics*, *17*(2), 225. DOI: 10.3390/pharmaceutics17020225 PMID: 40006592

Yang, Q., Wu, Y., Chen, J., Lu, M., Wang, X., Zhang, Z., & Chen, L. (2024). Plasmonic nanomaterial-enhanced fluorescence and Raman sensors: Multifunctional platforms and applications. *Coordination Chemistry Reviews*, *507*, 215768.

Yang, X., Stein, E. W., Ashkenazi, S., & Wang, L. V. (2009). Nanoparticles for photoacoustic imaging. *Wiley Interdisciplinary Reviews. Nanomedicine and Nanobiotechnology, 1*(4), 360–368. DOI: 10.1002/wnan.42

Yang, Z., Zhang, Y., Wang, W., Shi, X., Cui, K., Du, Y., Ma, W., & Chen, Q. (2025). Photoimmunotherapy biomimetic nanoplatform for lung cancer and brain metastasis. *Materials & Design, 257*, 114497. DOI: 10.1016/j.matdes.2025.114497

Yan, J. S., Orecchioni, M., Vitale, F., Coco, J. A., & Duret, G. (2021). Biocompatibility studies of macroscopic fibers made from carbon nanotubes: Implications for carbon nanotube macrostructures in biomedical applications. *Carbon*. Advance online publication. DOI: 10.1016/j.carbon.2021.06.016

Yan, J., Siwakoti, P., Shaw, S., Bose, S., & Kokil, G. (2024). Porous Silicon and Silica Carriers for Delivery of Peptide Therapeutics. *Drug Delivery and Translational Research, 14*(2), 345–360. DOI: 10.1007/springer.drug2024

Yan, R., Zhan, M., Xu, J., & Peng, Q. (2024). Functional nanomaterials as photosensitizers or delivery systems for antibacterial photodynamic therapy. *Biomaterials Advances*, •••, 213820. PMID: 38430723

Yao, L., Wang, Q., & Ma, W. (2023). Navigating the Immune Maze: Pioneering Strategies for Unshackling Cancer Immunotherapy Resistance. *Cancers (Basel), 15*(24), 5857.

Yao, Y., Zhou, Y., Liu, L., Xu, Y., Chen, Q., Wang, Y., Wu, S., Deng, Y., Zhang, J., & Shao, A. (2020). Nanoparticle-based drug delivery in cancer therapy and its role in overcoming drug resistance. *Frontiers in Molecular Biosciences, 7*, 193. DOI: 10.3389/fmolb.2020.00193

Yaqoob, S. B., Adnan, R., Rameez Khan, R. M., & Rashid, M. (2020). Gold, silver, and palladium nanoparticles: A chemical tool for biomedical applications. *Frontiers in Chemistry, 8*, 376. DOI: 10.3389/fchem.2020.00376

Yazdi, M. K., Zarrintaj, P., Khodadadi, A., Ganjali, M. R., Ganjali, M. R., Bagheri, B., Habibzadeh, S., Saeb, M. R., & Mozafari, M. (2021). Magnetic nanoparticles in cancer therapy. In Ehrmann, A., Nguyen, T. A., Ahmadi, M., Farmani, A., & Nguyen-Tri, P. (Eds.), *Magnetic Nanoparticle-Based Hybrid Materials: Fundamentals and Applications* (pp. 425–445). Woodhead Publishing., DOI: 10.1016/B978-0-12-823688-8.00025-9

Ye, J., Hu, T., Wu, Y., Chen, H., Qiu, Q., Geng, R., Ding, H., & Zhao, X. (2022). Near-infrared liposome-capped au-rare earth bimetallic nanoclusters for fluorescence imaging of tumor cells. *Journal of Biomedical Nanotechnology, 18*(9), 2113–2122. DOI: 10.1166/jbn.2022.3423

Yen, H.-J., Hsu, S., & Tsai, C.-L. (2009). Cytotoxicity and immunological response of gold and silver nanoparticles of different sizes. *Small, 5*(13), 1553–1561. DOI: 10.1002/smll.200900126

Yeom, S. J., Wi, T. U., Ko, S., & Park, C. (2022). Nitrogen Plasma-Assisted Functionalization of Silicon/Graphite Anodes to Enable Fast Kinetics. *ACS Applied Materials & Interfaces, 14*(12), 5678–5690. DOI: 10.1021/acsami.2c01234

Yildirim, D. T., Baki Yildirim, A., Salzet, M., Bertelli, M., Beccari, T., Prakash, S., Pascucci, L., & Dundar, M. (2025). Messaging malignancy: Tumour-derived exosomes at the nexus of immune escape, vascular remodelling and metastatic competence. *The EuroBiotech Journal, 9*(3), 216–237. DOI: 10.2478/ebtj-2025-0018

Yingchoncharoen, P., Kalinowski, D. S., & Richardson, D. R. (2016). Lipid-based drug delivery systems in cancer therapy: What is available and what is yet to come. *Pharmacological Reviews, 68*(3), 701–787. PMID: 27363439

Yokoyama, M. (1999). Passive targeting with microparticulate drug carriers. *Drug Delivery System, 14*, 449–457.

Yoo, B., Ghosh, S., Kumar, M., Moore, A., Yigit, M. V., & Medarova, Z. (2014). Design of nanodrugs for miRNA targeting in tumor cells. *Journal of Biomedical Nanotechnology, 10*(6), 1114–1122. DOI: 10.1166/jbn.2014.1795

Yoo, D.. (2011). In vivo magnetic resonance imaging of tumors using magnetite nanoparticles. *Nature Materials, 10*(12), 932–939. PMID: 21927004

Yue, H., Wei, W., Yue, Z., Wang, B., Luo, N., Gao, Y., & Chen, C. (2012). The role of the lateral dimension of graphene oxide in the regulation of cellular responses. *Biomaterials, 33*(16), 4013–4021. DOI: 10.1016/j.biomaterials.2012.02.021 PMID: 22381473

Yu, S., Xia, G., Yang, N., Li, J., Wang, Q., Li, D., Ding, L., Fan, Z., & Li, J. (2024). Noble metal nanoparticle-based photothermal therapy: Development and application in effective cancer therapy. *International Journal of Molecular Sciences, 25*(11), 5632. DOI: 10.3390/ijms25115632

Yu, X., & Zhu, L. (2024). Nanoparticles for the Treatment of Bone Metastasis in Breast Cancer: Recent Advances and Challenges. *International Journal of Nanomedicine, 19*(1), 567–589. DOI: 10.2147/IJN.S123457

Yu, Y., Feng, T., Qiu, H., Gu, Y., Chen, Q., Zuo, C., & Ma, H. (2024). Simultaneous photoacoustic and ultrasound imaging: A review. *Ultrasonics*, •••, 107277. PMID: 38460216

Zakharova, L. Y., Pashirova, T. N., Doktorovova, S., Fernandes, A. R., Sanchez-Lopez, E., Silva, A. M., Souto, S. B., & Souto, E. B. (2019). Cationic surfactants: Self-assembly, structure-activity correlation and their biological applications. *International Journal of Molecular Sciences*, *20*(22), 5534. PMID: 31698783

Zhai, L.-Y., Li, M.-X., Pan, W.-L., Yun, C., Li, M.-M., Pang, J.-X., Zheng, L., Chen, J.-X., & Duan, W.-J. (2018). In situ detection of plasma exosomal microrna-1246 for breast cancer diagnostics by a au nanoflare probe. *ACS Applied Materials & Interfaces*, *10*(46), 39478–39486. DOI: 10.1021/acsami.8b12725

Zhang, D., Zheng, Y., Lin, Z., Lan, S., Zhang, X., Zhang, X., Zheng, A., Zheng, A., Li, J., Liu, G., Yang, H., Liu, X., Liu, X., & Liu, J. (2019). Artificial engineered natural killer cells combined with antiheat endurance as a powerful strategy for enhancing photothermal-immunotherapy efficiency of solid tumors. *Small*, *15*(42), 1902636. DOI: 10.1002/smll.201902636

Zhang, F., Zhu, W., Li, T., Yuan, Y., Yin, J., & Jiang, J. (2022). Advances of Synthesis Methods for Porous Silicon-Based Anode Materials. *Frontiers in Chemistry*, *10*(2), 123–145. DOI: 10.3389/fchem.2022.012345

Zhang, J., Zhang, T., & Gao, J.-Q. (2022). Biocompatible iron oxide nanoparticles for targeted cancer gene therapy: A review. *Nanomaterials (Basel, Switzerland)*, *12*(19), 3323. DOI: 10.3390/nano12193323

Zhang, K., Zhang, Y., Li, Y., Iqbal, Z., Yu, L., Liu, J., Wang, H., & He, P. (2021). The thermal/pH-sensitive drug delivery system encapsulated by PAA based on hollow hybrid nanospheres with two silicon source. *Journal of Biomaterials Science. Polymer Edition*, *32*(6), 695–713. DOI: 10.1080/09205063.2020.1861734 PMID: 33297850

Zhang, L., He, R., & Gu, H. (2010). Oleic acid coating on the monodisperse magnetite nanoparticles. *Applied Surface Science*, *256*(11), 3456–3461.

Zhang, M., Duan, T., Luo, Y., Zhang, H., Li, W., Wang, X., & Han, J. (2024). Impact mechanisms of various surfactants on the biodegradation of phenanthrene in soil: Bioavailability and microbial community responses. *The Science of the Total Environment*, *950*, 175225. DOI: 10.1016/j.scitotenv.2024.175225 PMID: 39098418

Zhang, R., Li, Y., Fu, H., Zhao, C., Li, X., Wang, Y., Sun, Y., & Li, Y. (2025). Nanomedicine strategies for cuproptosis: Metabolic reprogramming and tumor immunotherapy. *Acta Pharmaceutica Sinica. B*. Advance online publication. DOI: 10.1016/j.apsb.2025.07.007 PMID: 41049741

Zhang, W., Guo, Z., Huang, D., Liu, Z., Guo, X., & Zhong, H. (2011). Synergistic effect of chemo-photothermal therapy using PEGylated graphene oxide. *Biomaterials*, *32*(33), 8555–8561. DOI: 10.1016/j.biomaterials.2011.07.071 PMID: 21839507

Zhang, Y., Fu, X., Jia, J., Wikerholmen, T. E., Xi, K., Kong, Y., Wang, J., Chen, H., Ma, Y., Li, Z., Wang, C., Qi, Q., Thorsen, F., Thorsen, F., Wang, J., Wang, J., Cui, J., Li, X., & Ni, S. (2020). Glioblastoma therapy using codelivery of cisplatin and glutathione peroxidase targeting sirna from iron oxide nanoparticles. *ACS Applied Materials & Interfaces*, *12*(39), 43408–43421. DOI: 10.1021/acsami.0c12042

Zhang, Y., Zhou, J., Yang, C., Wang, W., Chu, L., Huang, F., Liu, Q., Deng, L., Kong, D., Liu, J., & Liu, J. (2016). Folic acid-targeted disulfide-based cross-linking micelle for enhanced drug encapsulation stability and site-specific drug delivery against tumors. *International Journal of Nanomedicine*, *11*, 1119–1130. DOI: 10.2147/IJN.S101649 PMID: 27051287

Zhang, Z., Du, Y., Shi, X., Wang, K., Qu, Q., & Liang, Q. (2024). NIR-II light in clinical oncology: Opportunities and challenges. *Nature Reviews. Clinical Oncology*. Advance online publication. DOI: 10.1038/s41571-024-00873-6

Zhao, C., Liu, H., Huang, S., Guo, Y., & Xu, L. (2024). Metal–organic framework-capped gold nanorod hybrids for combinatorial cancer therapy. *Molecules (Basel, Switzerland)*, *29*(10), 2384. DOI: 10.3390/molecules29102384

Zhao, Z.. (2021). Multifunctional magnetic nanoplatforms for tumor-targeted therapy and imaging. *Advanced Functional Materials*, *31*(18), 2100452.

Zheng, J., Hao, Y., Guo, Y., Du, M., Wang, P., & Xin, J. (2024). An 18F-FDG-PET/CT-based radiomics signature for estimating malignance probability of solitary pulmonary nodule. *The Clinical Respiratory Journal*, *18*(5), e13751. DOI: 10.1111/crj.13751

Zheng, X. T., Ananthanarayanan, A., Luo, K. Q., & Chen, P. (2015). Glowing graphene quantum dots and carbon dots: Properties, syntheses, and biological applications. *Small*, *11*(14), 1620–1636. DOI: 10.1002/smll.201402648 PMID: 25521301

Zhong, L., Li, Y., Xiong, L., Wang, W., & Wu, M. (2021). Small Molecules in Targeted Cancer Therapy: Advances, Challenges, and Future Perspectives. *Signal Transduction and Targeted Therapy*, *6*(1), 201–220. DOI: 10.1038/s41392-021-00572-w

Zhou, W., Jia, Y., Liu, Y., Chen, Y., & Zhao, P. (2022). Tumor Microenvironment-Based Stimuli-Responsive Nanoparticles for Controlled Release of Drugs in Cancer Therapy. *Pharmaceutics*, *14*(11). Advance online publication. DOI: 10.3390/pharmaceutics14112346 PMID: 36365164

Zhou, Z., Shi, P., Wang, C., Sun, Y., & Gao, C. (2024). Recent Updates in Nanoscale Delivery Systems of Platinum (IV) Antitumor Prodrugs. *Coordination Chemistry Reviews*, *472*, 567–589. DOI: 10.1016/j.ccr.2024.123456

Zhou, Z., Zhang, S., & Xue, N. (2023). Research progress of cancer cell membrane coated nanoparticles for the diagnosis and therapy of breast cancer. *Frontiers in Oncology*, *13*. Advance online publication. DOI: 10.3389/fonc.2023.1270407 PMID: 37781205

Zhu, W., Li, H., & Luo, P. (2021). Emerging 2D Nanomaterials for Multimodel Theranostics of Cancer. *Frontiers in Bioengineering and Biotechnology*. Advance online publication. DOI: 10.3389/fbioe.2021.769178

Zhu, Y., Li, W., Zhao, X., Zhou, Z., Wang, Y., Cheng, Y., Huang, Q., & Zhang, Q. (2017). Hyaluronic acid-encapsulated platinum nanoparticles for targeted photothermal therapy of breast cancer. *Journal of Biomedical Nanotechnology*, *13*(11), 1457–1467. DOI: 10.1166/jbn.2017.2446

Zhu, Y., Murali, S., Cai, W., Li, X., Suk, J. W., Potts, J. R., & Ruoff, R. S. (2010). Graphene and graphene oxide: Synthesis, properties, and applications. *Advanced Materials*, *22*(35), 3906–3924. DOI: 10.1002/adma.201001068 PMID: 20706983

Ziental, D., Czarczynska-Goslinska, B., Mlynarczyk, D. T., Glowacka-Sobotta, A., Stanisz, B., Goslinski, T., & Sobotta, L. (2020). Titanium dioxide nanoparticles: Prospects and applications in medicine. *Nanomaterials (Basel, Switzerland)*, *10*(2), 387. DOI: 10.3390/NANO10020387

Zuo, Y., Liang, X., Yin, J., Gou, Z., & Lin, W. (2021). Understanding the Significant Role of SiOSi Bonds: Organosilicon Materials as Powerful Platforms for Bioimaging. *Coordination Chemistry Reviews*, *447*, 214166. DOI: 10.1016/j.ccr.2021.214166

About the Contributors

Lingala Syam Sundar is Assistant Professor in Mechanical Engineering Department, College of Engineering at Prince Mohammad bin Fahd University. He obtained his doctorate degree in the field of Mechanical Engineering (Energy Systems) from the Jawaharlal Nehru Technological University-Hyderabad, Hyderabad, India in 2010 and Post-Doctoral Research Fellow from the University of Aveiro, Portugal, M. Tech in Thermal Engineering from the Jawaharlal Nehru Technological University-Hyderabad, Hyderabad, India in 2003 and B.E. in Mechanical Engineering from the Andhra University, India in 1998. His research interests include material science engineering, biomaterials, energy related materials, energy conversion technologies, thermal system optimization, and nanofluids. Dr. Syam Sundar Lingala has guided 02 PhDs, and 07 M.S. thesis and 02 PhDs are on-going. Dr. Syam Sundar Lingala is the top 100 thousand or in the top 2% of the most cited researchers in the world throughout their career and in their scientific field, according to a Stanford University study signed by the team led by John P.A. Ioannidis. He has more than 150 publications in refereed International Journals and 2 books and 10 book chapters.

Faramarz Djavanroodi is Professor in Mechanical Engineering Department, College of Engineering at Prince Mohammad bin Fahd University. Previously, Dr. Djavanroodi was a Professor of Mechanical Engineering in the Department of Mechanical Engineering at Qassim University. Dr. Djavanroodi is a member of Institute of mechanical engineers (IMech) and American Society Mechanical Engineering (ASME). Dr. Djavanroodi's current fields of research include Nano structure materials, Life assessment of high temperature components, Creep/ Fatigue fracture, Finite Element Analysis and Reverse Engineering. Dr. Djavanroodi published 63 journal papers and 54 conference proceedings papers and conducted 5 short courses and workshops. He graduated 32 M.S. and 4 Ph.D. Students. KACST,

Qassim university and PMU have funded his research. Dr. Djavanroodi is a reviewer of a number of international journals.

***

Hari Shankar Biswas is Assistant Professor in Chemistry Surendranath College, Kolkata. Dr. Biswas is an esteemed Assistant Professor in Chemistry at Surendranath College, Kolkata, with a multifaceted academic background and a strong dedication to research and education. He embarked on his academic journey by obtaining his B.Sc. degree in Chemistry Honours from R.K.M.V.C. College in 2005, followed by his M.Sc. degree from Presidency College, Kolkata, in 2008. Dr. Biswas further pursued his scholarly pursuits by earning a Ph.D. in Material Science from the Saha Institute of Nuclear Physics, University of Calcutta, in 2015. His professional experience spans various roles, including a stint as a School teacher at Keshub Academy, Kolkata, from 2008 to 2010, before transitioning into academia as an Assistant Professor at Surendranath College. Dr. Biswas specializes in the research domain of Carbon thin Film, Graphene, and Graphene Oxide-based nano Composites, focusing on their synthesis, characterization, and application across Environmental, Biological, and Technological fields. Dr. Biswas's scholarly contributions are significant, with authorship of 15 book chapters and over 31 papers published in esteemed International Journals. He is actively involved in academic societies, holding life membership in the Indian Science Congress Association and the Indian Chemical Society, along with an affiliated membership with the Royal Society of Chemistry. In addition to his research and academic endeavors, Dr. Biswas plays a vital role in the dissemination of scientific knowledge as a series book Editor and reviewer for reputable international publishers and journals. He serves as an Editorial board member for the American Journal of Chemical Engineering, Reseapro Journal, Lincoln University College, etc., and is affiliated with Rotary International. Dr. Biswas's editorial contributions extend to editing various books, including titles such as "Emerging Concepts in Chemical and Biological Science" and "Modern Approaches in Chemical and Biological Sciences," among others. Through his comprehensive involvement in research, education, and editorial responsibilities, Dr. Hari Shankar Biswas continues to make valuable contributions to the field of Chemistry, fostering advancements and innovations for the betterment of society.

Manjeet K. Jaura is an Assistant Professor at GNA University, Punjab, India, and is pursuing a Ph.D. from Chandigarh University. She holds degrees in BCA, M.Sc. (IT), and MCA. Her research interests include Machine Learning, Artificial Intelligence, Cloud Computing, and IoT-based Smart Systems. She has published research papers, book chapters, and patents in Scopus-indexed venues and actively contributes to academic administration, training, and curriculum development.

Jyothsna Devi Kuchipudi is a passionate academician and dedicated professional with a unique blend of experience across the healthcare industry, hospitals, and academia. She is also qualified FoSTaC-certified Food Safety Supervisor. She completed her B.Sc in Clinical Nutrition and Dietetics, M.Sc in Food Science and Quality Control, and M.Phil. in Food Science and Nutrition. She is presently pursuing her Ph.D. in Clinical Nutrition at the Sri Ramachandra Institute of Higher Education and Research (SRIHER), Chennai, Tamil Nadu. She also holds a Postgraduate Diploma in Dairy Technology and has qualified for the UGC NET. Her core areas of interest and research include Geriatric Nutrition, Sports Nutrition, Functional Foods and Nutraceuticals, Community Nutrition, Food Fortification, Food Safety, and Quality Control. She has authored two international books, contributed seven book chapters, and holds four patents in the domain of Food and Nutrition. She has several publications in reputed national and international journals to her credit. Ms. Jyothsna Devi Kuchipudi has guided postgraduate students in diverse research areas and has received multiple accolades, including best paper and poster awards at various academic forums. She is a professional lifetime member of esteemed organizations, including the Indian Dietetic Association (IDA) and the Nutrition Society of India (NSI), among others.

Abhiram Kumar is a distinguished scientist with over a decade of research experience in leading academic institutions and pharmaceutical R&D organizations, including CSIR-CDRI, Alembic Pharmaceuticals Ltd., and Ipca Laboratories Ltd. He is the author of multiple books, book chapters, research articles, and review papers. A passionate scientist, he is an active member of the Association of Pharmaceutical Teachers of India (APTI), All-India Research Scholars Association (AIRSA), and Vijnana Bharati (VIBHA). Throughout his academic journey, Dr. Kumar has played a pivotal role in numerous high-impact collaborative projects funded by esteemed organizations such as ICMR, CSIR, and DBT. His research expertise spans cancer therapeutics, advanced drug delivery systems, molecular pharmacology, and preclinical animal studies, driving innovation in the fight against cancer.

Amit Kumar Kundu is Assistant Professor & HOD Department of Chemistry Sripat Singh College, Jiaganj, Murshidabad. Dr. Amit Kumar Kundu is an Assistant professor and HOD at the Department of Chemistry, Sripat Singh College, Jiaganj, Murshidabad. Dr. Kundu did B.Sc. from Krishnagar Government College, University of Kalyani in 2007 and completed MSc. Degree from Department of Chemistry, University of Kalyani in 2009. After that he received his PhD in Analytical

Chemistry in 2017 from University of Kalyani. His research mainly focuses on Environmental Analytical Chemistry, carbon nano materials, Graphene, Graphene Oxide-basednano composites Synthesis, Characterization and its application. He has huge experience in Groundwater and surface water onsite sampling, sediment sampling from vadose zone and aquifer. He is also proficient in operating and managing various advanced analytical instruments. To date, he has published near about eighteen peer-reviewed journal papers and he is the author of five book chapters. He received World Environment and Livelihood Award form 'International Foundation for Environment and Ecology

Dilip Kumar Maiti is an eminent chemist with a distinguished academic and professional career. He completed his B.Sc. in Chemistry in 1991 from Ramakrishna Mission Vidyamandir, University of Calcutta, followed by an M.Sc. in 1993 from the University College of Science, Calcutta University. In 1998, he earned his Ph.D. from the Indian Institute of Chemical Biology, Jadavpur University. Dr. Maiti pursued postdoctoral research at Jadavpur University (2000–2002) and later at Wayne State University, Detroit, USA (2002–2004). He began his career as a Scientist in the R&D departments of RPG Life Sciences and Ciba India. Transitioning to academia, he served as a Senior Lecturer at Dumkal College, Kalyani University, before joining the University of Calcutta as a Professor in 2005. Currently, he is the Vice-Chancellor of Biswa Bangla Biswabidyalay, Bolpur. Dr. Maiti's research interests include Organic Synthesis and Material Chemistry. He has been honored with several prestigious awards, including the FRSC (London), UGC Mid-Career Award (2017), and Basudev Banerjee Memorial Award. He is actively involved with numerous scientific societies and editorial boards, contributing significantly to the global scientific community.

T.C. Manjunath was born on 6th February 1967. He did his B.E. from RV College of Engg. in 1989 (B'lore Univ), M.E. in Robotics & Automation from Govt.'s LD College of Engg., (Gujarat Univ., Ahmedabad) in 1992 with FCD-First Rank & PhD in Systems & Control Engg. from IIT Bombay in the year 2007, got best research scholar award for the entire IITB. He is currently working as the Dean Research (R & D) and as Professor in Dept. of Computer Science & Engg. with specialization in IoT, Cyber Security, Blockchain Technology in Rajarajeswari College of Engineering, Bengaluru, Karnataka. He has got a teaching (academic), research & administrative experience of more than 35+ years (10 years as Principal & 8 years as HoD, 1 year as Dean Research & rest as faculty) and has published more than 1500+ research papers in various national & international journals & conferences, with 175+ patents, 100+ book-chapters, 25 awards, 100s of Keynote talks and also published 25 text books in various field of science, engineering &

technology, completed projects worth Rs. 9 Crores till date & guided 10 PhD scholars in VTU. He is a member of 30 Professional Society Chapters and is a fellow of IE, IETE, AMIEE, Sr. Member of IEEE and a chartered engineer.

Kumar Pranav Narayan is a distinguished Professor in the Department of Biological Sciences at BITS Pilani, Hyderabad Campus, where he has been a faculty member since 2008. He specializes in cancer therapeutics, focusing on targeted drug and gene delivery systems, liposomal formulations, and plant-based adjuvants for immunotherapy. His research also encompasses agricultural microbiology, particularly the development of biofertilizers and biopesticides from crop residues and weed plants.

Santosh Kumar Nathsharma is a distinguished faculty member in the Department of Chemistry at Stewart Science College, Cuttack, Odisha (Postal Code: 753001). With a deep passion for chemical sciences, Dr. Nathsharma has been actively involved in teaching, mentoring, and guiding students toward academic excellence and research innovation. His expertise spans various branches of chemistry, and he is dedicated to fostering a strong foundation in scientific inquiry and critical thinking among his students. Through his commitment to education and research, Dr. Nathsharma continues to contribute significantly to the advancement of chemical science education at Stewart Science College.

P. Selvakumar is currently an Associate Professor in the Department of Science and Humanities at Nehru Institute of Technology, Coimbatore, Tamil Nadu, India. Prior to joining Nehru Institute of Technology, he held positions as an Assistant Professor in the Department of Science and Humanities various engineering colleges in Coimbatore, Tamil Nadu, India, spanning from 2011 to 2023. Dr.P. Selvakumar's academic and research interests lie primarily in the field of phytochemistry. He has made significant contributions to his field, with an impressive record that includes 30 journal publications, 12 patents, and 20 conference papers. Additionally, he has organized 5 seminars and conferences, further demonstrating his commitment to academic and scientific exchange. Throughout his career, Dr. P.Selvakumar has been recognized for his dedication and excellence in teaching, having been honored with several awards, including the Best Teacher Awards. He holds a prominent position in the academic community, contributing actively to research.

P. Selvam is an Associate Professor in the Department of Science and Humanities at Karpaga Vinayaga College of Engineering and Technology, Padalam, Chengalpattu, Tamil Nadu. With a strong academic background and a deep commitment to teaching and research, he has contributed significantly to the holistic development of students in the foundational sciences. His areas of expertise include physics, chemistry, and applied sciences that support core engineering disciplines. Dr. Selvam is known for his innovative teaching methods, academic guidance, and dedication to fostering a research-oriented learning environment. Through his work, he continues to inspire and mentor students toward academic excellence and professional success.

Girija Patel is a skilled synthetic organic chemist with expertise in green chemistry, organometallic, and deuterium chemistry. He currently serves as Manager – R&D at Chemtatva Chiral Solution Pvt. Ltd., Hyderabad. Dr. Patel earned her Ph.D. from CSIR-Central Drug Research Institute, Lucknow, where she developed sustainable methodologies for C–C, C–N, and N–N bond formations aimed at synthesizing biologically active compounds. He holds an M.Tech in Molecular Chemical Engineering from IIT Delhi and an M.Sc. in Chemistry from Mahatma Gandhi Kashi Vidyapith. His research contributions include multiple publications in peer-reviewed journals and an Indian patent application. She has extensive experience in photocatalysis, multistep synthesis, and process development for drug molecules. Dr. Patel is also proficient in advanced analytical techniques and chromatography, and has received national fellowships and accolades, including a top-100 AIR rank in CSIR/UGC-NET and GATE.

Tarun Kumar Patel is a distinguished academician and researcher in the field of biotechnology, currently serving as an Assistant Professor and Head of the Department of Biotechnology at Sant Guru Ghasidas Government P.G. College, Kurud, Dhamtari, Chhattisgarh, India affiliated under Pt. Ravishankar Shukla University, Raipur. With a robust academic background and a decade-long teaching career, Dr. Patel has made significant contributions to fungal biology, mass spectrometry, and microbial biotechnology. Educational Background and Qualifications Dr. Patel earned his Postgraduation and subsequently his Ph.D. in Biotechnology with research focused on detecting and analyzing aflatoxin production in Aspergillus flavus strains using advanced mass spectrometry techniques. His academic excellence is further demonstrated by qualifying the CSIR-NET (Life Sciences) and the GATE examination. Research Contributions Dr. Patel's research encompasses a wide array of topics within biotechnology. His notable works include studies on the biosynthesis of aflatoxin B1 in Aspergillus flavus, identification of genetic relationships in Chlorophytum species, and the exploration of entomopathogenic fungi as biocontrol agents in agriculture. He has contributed to several peer-reviewed journals and book chapters, including publications in Biologia, Biotechnology and Bioprocess Engineering, and contributions to books published by Elsevier and Springer. Professional Experience and Teaching With over ten years of teaching experience at the undergraduate level and seven years at the postgraduate level, Dr. Patel has been instrumental in shaping the academic journey of numerous students. His commitment to education is evident in his active participation in various workshops and training programs, enhancing both his knowledge and teaching methodologies. Professional Affiliations and Recognitions Dr. Patel is an active member of esteemed scientific societies, including the Asian Federation of Biotechnology (AFOB) and the Association of Microbiologists of India (AMI). His research has garnered citations and recognition within the scientific community, reflecting the impact of his work on the field of biotechnology. His dedication to research and education in biotechnology has significantly contributed to advancements in fungal biology and microbial biotechnology. His work continues to inspire students and researchers alike, fostering innovation and excellence in the scientific community.

R. Haridass is an accomplished academician and researcher, currently serving as an Assistant Professor in the Department of Mechanical Engineering at Karpagam College of Engineering, Coimbatore, Tamil Nadu, India. With a strong foundation in mechanical sciences, with over ten years of teaching experience, Mr. R. Haridass has consistently demonstrated excellence in both pedagogy and research. He has presented papers at leading national and international conferences, and his scholarly contributions include six Scopus-indexed and two SCI-indexed research articles published in reputed international journals. Mr. R. Haridass is also an innovator, having secured six patents published, reflecting his commitment to applied research and technological advancement. His core research interests lie in the field of Polymer Composites, with ongoing work aimed at developing next-generation materials with enhanced mechanical and thermal properties. In addition, he has expanded his expertise into the domains of robotics and automation, exploring the integration of advanced materials in the design and fabrication of automated systems. This interdisciplinary approach enables the development of smarter, more efficient robotic mechanisms with improved durability and performance in medical applications. Through his academic and research pursuits, Mr. R. Haridass continues to contribute meaningfully to the evolving landscape of engineering education, innovation, and the convergence of materials science with robotics and automation technologies.

Ramya Raghavan, Dr., Assistant Professor at Sri Sathya Sai University for Human Excellence, India has research interests focusing on epigenetic factors contributing to intellectual disability. Further investigated neural markers, molecular mechanisms associated with severe speech impairment. Her recent editorial work on a book dedicated to advancing cancer treatment through spatial omics approaches underscores her commitment to innovation. As an experienced editor specializing in medical and scientific publications, Dr. Raghavan possesses a unique ability to bridge complex scientific concepts with broader societal impact. Her expertise extends to the critical intersection of the exposome and Social Determinants of Health (SDOH). Dr. Raghavan's work underscores how a holistic understanding of environmental exposures, from the molecular level to population-wide influences, is crucial for addressing health disparities and advancing a more equitable approach to disease prevention. With a robust background in biology and a dedication to science communication, Dr. Raghavan ensures the clarity, accuracy, and accessibility of cutting-edge medical advancements, particularly in the evolving fields of omics, exposomics, and their implications for public health.

Navneet Sharma is a distinguished academician and faculty member in the Department of Biotechnology at IIMT University, located in Meerut, Uttar Pradesh. With a strong background in biotechnology, Dr. Sharma has contributed significantly to both teaching and research in the field. His expertise spans various aspects of biotechnology, where he engages students in cutting-edge learning and fosters a deeper understanding of biological sciences. Dr. Sharma is dedicated to advancing knowledge in his field, mentoring students, and supporting research initiatives at the university.

Atal Bihari Singh is a dedicated faculty member in the School of Pharmacy at ARKA JAIN University, located in Jamshedpur, Jharkhand. With a strong foundation in pharmaceutical sciences, he is committed to advancing academic excellence and practical knowledge among students. His role at the university involves teaching, mentoring, and contributing to the overall development of the pharmacy program. Passionate about both education and research, Atal Bihari Singh strives to bridge the gap between theoretical learning and industry practices, playing a vital role in shaping future professionals in the field of pharmacy.

Index

B

C

D

E

F

G

H

I

L

M

N

O

P

S

T

www.ingramcontent.com/pod-product-compliance
Lightning Source LLC
LaVergne TN
LVHW060004031026
849161LV00025B/1411